GLOBAL
CONNECTIONS

GLOBAL CONNECTIONS

INTERNATIONAL
TELECOMMUNICATIONS
INFRASTRUCTURE AND POLICY

Heather E. Hudson

VAN NOSTRAND REINHOLD
I(T)P® A Division of International Thomson Publishing Inc.

New York • Albany • Bonn • Boston • Detroit • London • Madrid • Melbourne
Mexico City • Paris • San Francisco • Singapore • Tokyo • Toronto

I(T)P® Van Nostrand Reinhold is an International Thomson Publishing Company. ITP logo is a trademark under license.

The ideas presented in this book are generic and strategic. Their specific application to a particular company must be the responsibility of the management of that company, based on management's understanding of their company's procedures, culture, resources, and competitive situation.

Printed in the United States of America

http://www.vnr.com   Visit us on the Web!

For more information contact:

Van Nostrand Reinhold
115 Fifth Avenue
New York, NY 10003

Chapman & Hall GmbH
Pappalallee 3
69469 Weinham
Germany

Chapman & Hall
2-6 Boundary Row
London SEI 8HN
United Kingdom

International Thomson Publishing Asia
60 Albert Street #15-01
Albert Complex
Singapore 189969

Thomas Nelson Australia
102 Dodds Street
South Melbourne 3205
Victoria, Australia

International Thomson Publishing Japan
Hirakawa-cho Kyowa Building, 3F
2-2-1 Hirakawa-cho, Chiyoda-ku
Tokyo 102 Japan

Nelson Canada
1120 Birchmount Road
Scarborough, Ontario
M1K 5G4, Canada

International Thomson Editores
Seneca, 53
Colonia Polanco
11560 Mexico D.F. Mexico

1 2 3 4 5 6 7 8 9 10 QEBFF 02 01 00 99 98 97

Library of Congress Cataloging-in-Publication Data available upon request

Hudson, Heather E.
  Global connections : international telecommunications infrastructure and policy / Heather Hudson.
      p.   cm.
  Includes index.
  ISBN 0-442-02362-6
  1. Telecommunication—Technological innovations.   2. Telecommunication—Social aspects.   3. Information technology.   4. Information networks.   I. Title.
HE7631.H83   1997
384—DC20   96-41396
                                                                      CIP

For Rob

Composed of interconnected local, national, and regional networks, the GII has the potential to provide people around the world with the means and opportunities to participate fully in the information age. We firmly believe that improved access to information can and will facilitate improvements in the human condition, regardless of geographic location, income, or level of education.

—ALBERT GORE, JR.
U.S. Vice President
The Global Information Infrastructure: Agenda for Cooperation, 1995

CONTENTS

ACKNOWLEDGMENTS

Many individuals and organizations have contributed in some way to this book. Only a few can be mentioned here.

I owe a great deal to the International Telecommunication Union for both inspiration and information. The ITU's recent *World Telecommunication Development Reports* have provided a valuable source of current data and insightful analysis. I have also been privileged to work with two Secretary Generals of the ITU: Richard Butler, who drew global attention to the connection between telecommunications and development, and envisioned a major role for the ITU in fostering telecommunications growth in the developing world; and Pekka Tarjanne, who has guided the ITU into the era of competition and convergence.

The award of a Fulbright Distinguished Lectureship for the Asia-Pacific Region enabled me to present some of the material in this book to Asian telecommunications planners and practitioners, and also to gather information on current trends in telecommunications policy and the growth of new services in Asia. I also benefited from periods spent as a Senior Research Fellow at the East-West Center in Honolulu at the invitation of Meheroo Jussawalla, and at CIRCIT (the Centre for International Research on Communication and Information Technologies), in Melbourne, then under the direction of William Melody.

The World Bank funded a review of the research on the role of telecommunications in rural development that allowed me to update my knowledge, and also to rethink earlier findings and assumptions on telecommunications in developing countries. Many thanks to

Rogati Kayani, Bjorn Wellenius, and Richard Stern for their support and helpful suggestions.

The Ford Foundation and the Aspen Institute funded two studies I directed on rural telecommunications in the United States that were the origin of much of the analysis in this book on telecommunications policies for rural areas. Special credit is due to Edwin Parker, my co-author for both of the books that were based on those studies, for innovative recommendations to U.S. federal and state agencies that I believe are also relevant for policy makers in other countries.

The Telecommunications Management and Policy Program in the McLaren School of Business at the University of San Francisco served as the incubator for this book. Earlier drafts were used in graduate courses in international telecommunications. Students in these classes also gathered information on business applications and corporate strategies. Several members of the Telecommunications Program's Advisory Board provided valuable contacts and insights concerning the telecommunications industry and international business applications.

PREFACE

In a speech commemorating World Telecommunications Year in 1983, Arthur C. Clarke stated: "We have now reached the stage when virtually anything we want to do in the field of communications is possible. The constraints are no longer technical, but economic, legal, or political."

The 1990s have brought visions of information superhighways and global information infrastructure. The communication satellites that Arthur Clarke envisioned in the 1940s now link nearly 200 countries, and satellite dishes are perched on rooftops from Sweden to Sri Lanka and Brazil to Borneo. Fiber-optic cables with immense capacity span continents and cross ocean floors. Wireless networks link people on the move to their colleagues and their computers. Already, these technologies have created a global "corporate village," as multinational companies hire researchers and programmers in Russia and India, assemble products from parts manufactured on several continents, and monitor shipments and inventories for thousands of retail outlets worldwide. Entrepreneurs in developing regions are beginning to use these tools to find niche markets in other countries, while "footloose" businesses in North America and Europe are locating in rural areas, and relying on electronic links to their suppliers and customers. These technologies, and others yet to be introduced, can also enable people almost anywhere to keep in touch with family and friends, find and share information on the Internet, and watch news and entertainment from their own country or across the globe.

The operative word here is "can." As Arthur Clarke pointed out, technology is the easy part. Whether these and other new services will be used by people in remote areas, developing regions, and inner cities depends more on institutional incentives and constraints than on technology. This book examines the applications of these new technologies and their impacts on business, education, health care, and rural development. It also reviews the current structure of the telecommunications sector in major industrialized and developing regions, analyzes models for restructuring the sector, and proposes strategies to create incentives needed to introduce further reforms and to maximize benefits from investment in information infrastructure.

The inspiration for *Global Connections* comes from the pioneers of this new era, particularly the users of communications technologies who are finding new ways to harness and share the power of information, and policy makers who are struggling to ensure that this new information infrastructure benefits everyone, including the rural, disabled, and disadvantaged in their societies. It is my hope that this book will provide some helpful insights for these pioneers, and will encourage others to join them.

Heather E. Hudson
San Francisco
September 1996

SECTION *1*

Setting the Context

C H A P T E R

1

Introduction

> *"Telecommunications played as much of a role as pickaxes and shovels in bringing down the Berlin Wall and the barbed wire of the Iron Curtain."*
>
> —BILL McGOWAN, founder of MCI [1]

THE TELECOMMUNICATIONS REVOLUTION

We are in the midst of a global information revolution driven by the convergence and proliferation of telecommunications and information technologies. Via satellite, we have witnessed the fall of the Berlin Wall, war in the Middle East, famine and starvation in Somalia and other parts of Africa. International news channels open a window on the world that shows us events as they are happening, while video entertainment is also becoming global through satellite distribution of music, movie, and sports channels received via backyard antennas or fed into cable systems.

Yet this is not a one-way revolution. Phone calls, facsimile messages, and electronic mail are linking individuals, organizations, and businesses around the world. A salesperson in the field can send an order from a portable computer over a wireless data link. A researcher can communicate with colleagues and access libraries and databases around the world via the Internet. And a doctor in Africa can seek information on treating critically ill patients with a satellite-based electronic mail network.

Four major technological trends are driving these changes:

- **Capacity:** New technologies such as optical fiber have enormous capacity to carry information. They can be used for anything from entertainment and distance education to transmission of highly detailed images for remote medical diagnosis. Satellites also offer a tremendous amount of bandwidth.

- **Digitization:** Telecommunications networks are becoming totally digital, so that any type of information, including voice and video, may be sent as a stream of bits in compressed form and reconstructed for use at the receiving end. Compressed digital video can be used to transmit motion video over as few as two telephone lines (128 kbps) for teleconferencing and to deliver hundreds of digitized television channels via satellite.

- **Ubiquity:** Advances in wireless technology such as cellular radio, Personal Communications Services (PCS), and low earth-orbiting (LEO) satellites will provide mobile and personal communications virtually anywhere. These technologies also make it possible to serve rural communities without laying cable or stringing copper wire.

- **Convergence:** The convergence of telecommunications, data processing, and imaging technologies is ushering in the era of multimedia, in which voice, data, and images may be combined according to the needs of users, and distinctions between traditional sectors of telecommunications, information processing, and broadcasting are increasingly arbitrary and perhaps irrelevant.

This global electronic web is contributing to profound changes that we are just beginning to understand. Telecommunications is changing the way business does business. Organizations are being restructured into global businesses. We are witnessing the growth of transnational corporations as global corporations are transformed into a transnational model—essentially stateless and drawing their collective identities through webs of communication networks and information systems linking widely separated divisions. Global business optimizes the comparative advantage of each operation within a national setting; this specialization requires coordination—and therefore the need for exchange of information.

Businesses are organized for fast response, often combining a telecommunications network that allows monitoring of global activities with a decentralized model that enables field units to respond rapidly to local conditions. As Bill McGowan, the founder of MCI, noted: "We are seeing the emergence of a new corporate reality, free of geographical impediments to doing business. Telecommunications and information technologies are helping these global companies enter new markets, create new products and services, improve the speed and quality of decision making, boost productivity, and improve customer service."[2]

During the past decade we have also witnessed the dramatic impact of international telecommunications on human events. Technologies ranging from satellites and cellular phones to facsimile machines and computer modems have rendered national boundaries invisible. When asked how the reforms in Eastern Europe came about, Polish President Lech Walesa responded: "That's the result of civilization—of computers and satellite TV and other innovations which present alternative solutions. Is it possible for a new Stalin to appear today who could murder people? It's impossible."[3] Atrocities in Rwanda and the former Yugoslavia displayed on our television screens may lead us to challenge Walesa's optimism. Yet there is compelling evidence that global telecommunications is not only influencing human events by delivering them live to our living rooms, but helping people in countries where information access has been tightly controlled to communicate with each other and with the outside world. As noted above, access to information from beyond the Iron Curtain helped to bring down the Berlin Wall. Telephone calls and faxes helped Chinese dissidents to tell the world what was happening at Tiananmen Square and to feel the support of people around the world for democratic change in China.

THE GROWTH OF THE
INFORMATION SECTOR

It is no coincidence that the rise of international telecommunications parallels the emergence of a global economy. Modern telecommunications and information technologies are being used to get information where it is needed, when it is needed, and in its most useful form. Former President Gorbachev offered this explanation about why the Soviet Union was lagging behind the West: "We were one of the last to realize that in the age of information science, the most valuable asset is knowledge."[4] In 1987, the Soviet Union had only one telephone line per 300 people.

In the 1960s, researchers began to identify and study the role of knowledge in economy activity. The classical division of the economy into resource-based activities such as agriculture, mining and fishing, manufacturing, and services masked the growing role of information in all of these activities. That early research led to the designation of the "information sector" as that part of the economy that is concerned with the production and distribution of information services and of information technologies. The information sector is divided into two subsectors: primary and secondary. The primary information sector includes activities directly involved with information as an output— including education, research and development, finance (which is concerned with information about money), insurance (concerned with information about risk), as well as print and electronic media, publishing, and the production of telecommunications and information equipment. The secondary information sector includes activities in which information is an intermediate good, but not the final product. For example, information is required in many steps of producing an automobile: from research and development to just-in-time delivery of parts, to computer-assisted manufacturing, to shipping of vehicles, marketing, and dealer support.

In the late 1970s, it was estimated that about 50 percent of the economic activity in the United States was information related. Studies replicated by the Organization for Economic Cooperation and Development (OECD) found similar percentages in other industrialized countries. In the past twenty years, we have witnessed growth in both the primary and secondary information sectors. Information-based services from databases to electronic banking to telemarketing have proliferated, as have producers and marketers of telecommunications equipment, computers, and home and industrial electronic equipment. But perhaps more significantly, virtually every economic activity now involves information—from agriculture to manufacturing, mining to transportation. Large agribusinesses use telecommunications to buy and sell commodities and to manage their holdings. Individual farmers use modems to check market prices and to get advice from agricultural extension services. Oil companies use data from remote sensing satellites to aid exploration, and satellite terminals for communications from remote camps and offshore oil rigs. Using Electronic Data Interchange (EDI), manufacturers and parts suppliers can communicate directly for ordering, shipping, and billing. Freight companies monitor their fleet via radio and satellite, while airlines use computer networks for scheduling, reservations, and rewarding frequent flyers.

The telecommunications industry has both contributed to and benefited from this growth of the information sector. The growing market for telecommunications equipment and services, coupled with the growth of competition, has resulted in tremendous growth in the market for telecommunications equipment and services, which in turn has stimulated growth of the information sector. In addition, the expansion of telecommunications networks means that information-based activities can operate virtually anywhere. For industries specializing in the provision of information goods and services, reliable telecommunications infrastructure can make location and distance irrelevant. Singapore, for example, has become a major financial and trading center. Taiwan and Korea are now major suppliers of electronic equipment to world markets. Brazil has specialized expertise in satellite communications and aviation technology. Some developing countries have attracted "back office" information industries. American Airlines's ticket data is keypunched in Barbados; legal data for the Lexis database is entered in South Korea. Indian engineers transmit software code to Texas Instruments in Dallas from Bangalore. Entrepreneurs in these countries need instant access to global information to monitor market trends and to keep up with the most recent technological innovations.

GROWTH IN INTERNATIONAL TELECOMMUNICATIONS TRAFFIC

The ability to access and share information is key to any business activity. As telecommunications and information technologies converge, and as businesses seek global markets, telecommunications is playing an increasingly significant role in business activities from research and development to manufacturing, marketing, shipping, and inventory control. Telecommunications is also becoming increasingly important in education, enabling students far from universities to take courses, and employees to keep up-to-date through continuing education delivered to their workplace. Telecommunications can also contribute to health care delivery by enabling health workers in the field to consult with experts, and medical researchers around the world to share information.

International telephone traffic has been growing at a rate of 17 percent per year over the past decade, fueled by growth in international trade, immigration, travel, and the expanding global economy.[5] As international trade increases, telecommunications traffic reflects communities of interest and may serve as an indicator of economic development; for example, more than 75 percent of traffic originates in the industrialized countries of North America and Western Europe. However, traffic from the Republic of Korea, Hong Kong, Taiwan, and Singapore now exceeds the traffic of industrialized countries in the region (Japan, Australia, and New Zealand).[6] In contrast, although outgoing international traffic from Russia increased almost 20 times between 1984 and 1994, Russia still generated only about 2 percent of outgoing traffic generated by the United Kingdom and the United States in 1994.[7]

Migration to industrialized countries, for example, from Latin America and Asia to the United States, and from Eastern Europe, the Middle East, and North Africa to Western Europe, has created new communities of interest and demand for electronic communication. Among the ten fastest growing U.S. international routes in the 1980s were Guatemala, El Salvador, India, Taiwan, and the Philippines. There were more calls from Germany to Turkey and Yugoslavia than from Germany to Italy and Belgium.[8] Many studies have indicated that there is a high positive correlation between economic development and telecommunications investment. Yet correlations do not indicate causality. While countries are likely to invest more in telecommunications as their economies grow, there is evidence that investment in telecommunications can in itself contribute to economic development (see chapter 8).

TELECOMMUNICATIONS: STILL A MISSING LINK IN THE DEVELOPING WORLD

Telecommunications plays a vital role in the social, political, and economic development of every country. Yet, despite rapid technological change and growth in telecommunications in the past decade, there are still vast regions of the world that have no access to telecommunications, and others where access is extremely limited (see Table 1.1). The vast majority of telecommunications facilities are located in the industrialized world. Three-quarters of the world's telephones are in eight countries. Africa has only 1.8 percent of the world's main lines, although it accounts for more than 12 percent of the world's population. China and India have a combined population of more than 2 billion, or almost 40 percent of the world's population, but less than two lines per 100 people. As telephones tend to be concentrated in cities, access in rural areas is much more limited, and nonexistent in many developing regions.

As discussed in section 3 of this book, telecommunications is a vital component in the development process, as a complement to

TABLE 1.1
WORLD PERCENTAGE SHARES OF POPULATION VS. TELEPHONE LINES

Region	Percent of Population	Percent of Main Telephone Lines	Average Teledensity (lines per 100 population)
Africa	12.5	1.8	1.7
Asia	59.3	33.0	4.8
Europe	14.1	39.0	32.0
Latin America	8.5	6.2	8.4
North America	5.2	26.8	59.9
Oceania	0.5	1.7	38.7

Derived from ITU, *World Telecommunication Development Report*, 1995.

other infrastructure investments such as transportation and electrification. Reliable telecommunications networks can improve the productivity and efficiency of agriculture, industry, and social services, and enhance the quality of life in developing regions. Farmers in Brazil can check coffee futures prices on the world market; flower growers in Kenya can fill orders faxed from Europe. Programmers in India transmit computer code via satellite to companies in the United States and Europe. Also via satellite, university students studying by correspondence in the South Pacific can participate in real-time tutorials, and African researchers can access medical journals in the United States.

Until recently, telecommunications was considered a luxury by many development planners. This view was supported by evidence that extending telecommunications networks to rural and remote areas, where most of the developing country population lives, was prohibitively expensive. However, innovations in satellites and wireless telephony, coupled with solid state components for digital switching and end-user equipment, have dramatically lowered the costs of providing telecommunications facilities to virtually any location, from an inner-city neighborhood to a rural village or an isolated island. New approaches to financing telecommunications in developing regions are also creating incentives for investment which should help to close the information gaps between industrialized and developing regions. The number of main lines in the world is expected to increase from 647 million in 1994 to more than 1 billion by the year 2000, requiring an investment of more than $567 billion, mostly targeted at the developing world.

WHAT IS DRIVING THE REVOLUTION?

During the last twenty years, technological innovation, new regulatory frameworks, market structures, and the proliferation of telecommunications services have caused profound changes in the telecommunications sector. Until the early 1970s, the worldwide telecommunications market consisted of national markets whose needs were primarily met by local companies, and networks were operated by regulated monopolies or government departments. Competition for equipment sales took place primarily in developing countries or markets where there was no existing local industry. Users had very little influence on the development of the infrastructure and services or equipment. "In this way the telecommunications world was the opposite of the computer world, where users and manufacturers together dictate the development of technology and applications."[9] Electronics and software, two very competitive industries, became integrated in telecommunications. "The telecommunications world, the more it went digital, using computers in switches, and the more it started transmitting data, had to adapt to this competitive milieu; in the marriage between computer and telecommunications the young bride mobilized the older groom."[10]

Now "telecommunications has passed from an environment oriented by technology and driven by supply to an environment oriented towards applications and driven by the users."[11] The markets have become increasingly global, with numerous mergers, acquisitions, and joint ventures, and a rapid increase in expenditure for research and development. The benefits for users should be lower prices, wider choice, and better service. On the industry side, new players will enter, and the telephone corporate giants will invade each other's turf.

The changes in technology and the shift to a more competitive environment are resulting in phenomenal growth in the telecommunications sector worldwide. Global telephone lines are increasing by 6 percent per year. International telecommunications traffic increased 442 percent in the decade from 1984 to 1994 to 53.5 billion minutes per year.[12] The total revenues of the 50 largest telecommunications operators exceeded $478 billion in 1994.[13] World expenditures for telecommunications equipment and products exceeded $240 billion in 1995, about two and one half times the size of the market a decade earlier.

The U.S. telecommunications industry has become highly competitive, and many other countries are now introducing competition into their telecommunications sectors. Bill McGowan, a tireless proponent of competition, noted that: "Competition has spurred technical advances and expanded network capacity while lowering prices and fostering the widespread use of new business tools built around the distribution and effective use of information." Less than a decade after the 1984 divestiture by AT&T of the Bell Operating Companies, the U.S. telecommunications industry had grown into a $185 billion business: an astounding growth rate of 60 percent. Long distance telephone business had expanded into a $55 billion per year industry since divestiture, and long distance prices had dropped approximately 45 percent.[14]

As a result, carriers are shifting from a technology-oriented to a needs-oriented approach to service development. However, this new competitive environment is raising new challenges for providers and users alike. Coupled with changes in technology, the new competitive environment is eliminating the predictability of telecommunications. Competition has spawned new service operators with different ideas and ambitions; and upset the comfortable relationship between users, operators, and manufacturers; and raised questions about the viability of existing regulatory frameworks. New technologies have introduced a host of new equipment and services, and accelerated trends toward shorter and shorter product lives.

With competition has come reduced regulation, as the marketplace takes on the functions of controlling price and service options through customer demand. However, the policy-making role of regulators and legislators is perhaps more important today. As Geoffrey Vincent points out: "Regulators are in many ways the kingpin of the emerging market. Their role is important as it has never been before. In a monopoly situation, any weakness in the regulatory framework can be overcome by negotiation. In a competitive environment, any changes to regulations could have a dramatic effect on the business plans of several players, and might be subject to legal or political challenges. Regulators will have very limited opportunity to adjust a framework that is not working. The challenge will be to get it right the first time."[15] However, what works in one country can't necessarily be applied to another. "Each country has a different starting point. How the PTT [Post Telephone and Telegraph] is structured, whether there is an indigenous telecommunications industry, geography and population spread, the state of existing networks, the demands of industry and consumers, government policy and the culture and attitudes of business and individuals—all these contribute to the mix and must be taken into account, as well as worldwide developments in technology and service provision. The trend toward a global telecommunications industry is inexorable, but there will be very distinct local and regional flavors."[16]

THE PURPOSE OF THIS BOOK

This book is designed to provide an overview of the new international telecommunications environment and trends in telecommunications policies, an assessment of the impacts of these changes on the industry and users, and an analysis of the socioeconomic impact of these changes in both industrialized and developing societies. In short, the book attempts to address the *What?, Why?,* and *How?* of international telecommunications:

- *What* are the major new developments in international telecommunications?

- *Why* are these new developments taking place?
- *What* trends are shaping the future of the telecommunications industry around the world?
- *What* are the implications of these trends for telecommunications users? For planners and policy makers?
- *Why* are some parts of the world still without access to telecommunications?
- *Why* are political and economic factors more important than technology in shaping the telecommunications environment?
- *What* is the current status of the telecommunications sector in major industrialized countries?
- *How* is the telecommunications sector changing in the emerging economies and developing countries of Central and Eastern Europe, Asia, Latin America, and Africa?
- *How* do we create a global information infrastucture?

We first review some of the major technological innovations that are resulting in new products and services and the major players in the international telecommunications equipment and services industries. We then look at some of the major applications of these new technologies and services for business, education, social services, and economic development.

The book then examines the interwoven trends of restructuring of the telecommunications sector, changing policies including deregulation and standards setting, and globalization of the industry. The first section covers deregulation and the introduction of competition in the United States, and examines these trends in the United Kingdom and Japan, and other major industrialized countries. We then address the role of the European Union in setting common goals and standards for its member states.

The focus then shifts to the developing world, beginning with an overview of the role of telecommunications in the development process. We then examine various strategies that may be used to reform the telecommunications sector in developing regions, to spur investment and encourage applications to support development. This section then reviews changes in the telecommunications environment and remaining policy challenges in the emerging and industrializing economies of Asia and Latin America, and the special problems facing the least developed countries.

The final section of the book examines the changing international telecommunications environment, including international satellite and submarine cable networks, the growth of international competition, and the proliferation of regional carriers. It also addresses the changing policy environment including the reorganization of the International Telecommunication Union, the emergence of regional standards bodies, and issues of trade in telecommunications services and tariff reform. Finally, we examine technological and policy trends that are likely to shape the international telecommunications environment of the early 21st century.

2

New Technologies and Applications

"A revolution in communications technology is taking place today, a revolution as profound as the invention of printing."
—ITHIEL DE SOLA POOL, MIT professor of Political Science[1]

THE ERA OF CONVERGENCE

The telecommunications era began in 1844 when Samuel Morse transmitted his historic telegram: "What hath God wrought!" The following decades witnessed the growth of telecommunications as both wire and wireless technologies were developed—first using copper wire and high-frequency radio, and then adding coaxial cable, microwave, satellites, and optical fiber. The result has been an enormous increase in the capacity of communication networks. A satellite transponder can now transmit the equivalent of more than 2,500 voice channels, while a single optical fiber can transmit more than 10,000 voice channels or 50 television channels.

The revolution announced by Morse and later pioneers Marconi and Bell followed a limited vision of simply moving information farther and faster. Other information functions such as manipulation, storage, and duplication were considered to be totally separate, and remained so until the past quarter century. After all, mechanical adding machines, typewriters, paper files, and printing presses seemed to have nothing to do with telecommunications. But with the introduction of the computer, which could process vast amounts of information, and with the development of magnetic and electronic media for storage including magnetic tape, floppy disks, CD-ROMs, and optical disks, came the potential to merge the transmission function of telecommunications with functions of processing, storing, and reproducing information—the era of convergence.

Coupled with the increase in transmission capacity is the dramatic increase in processing speed and miniaturization of information technologies. The transition from analog to digital communications applies not only to data, but to voice and video as well. Information in digital form can now be transmitted through telecommunications networks and stored and retrieved at will. The processing power of integrated circuits is now merged with telecommunications through the convergence of telecommunications and information technologies. For example, digital switches are really specialized computers that can store and process data to be used for specialized services such as call forwarding, automatic number identification, automatic call return, selected number blocking, individual ring patterns for specified numbers, and voice message processing, commonly known as voice mail. Other now commonplace examples of converging technologies include facsimile transmission; electronic mail; voice mail; teleconferencing; automated attendant systems accessible using a touchtone phone; "smart cards" that store personal, medical, or financial information; information services accessing centralized databases; paging services; and numerous other "value added" services that combine information access, storage, and transmission. On the horizon are widespread applications of speech synthesis, voice recognition, and text-to-speech.

The buzzwords of this new digital world are "multimedia" and "interactivity." With high-capacity conduits such as optical fiber, digitized video as well as voice and data can be transmitted to the home or office. Magnetic and optical storage media allow the information to be stored and accessed at will. And the two-way capabilities of telecommunications networks let users interact—whether to order a movie or an advertised product, to pay their bills, or to participate in a conference or collaborate on a project.

Although these applications require many times the bandwidth of a telephone call, digitization offers the ability to compress the signal in order to reduce bandwidth requirements, thereby increasing even more the capacity of telecommunications networks to transmit information. The result is that video signals may use only a fraction of a broadband channel, and voice signals may also be compressed. For example, video conferences may now be held using two dial-up 64 kbps channels, whereas a few years ago, at least ten times the bandwidth was required. And toll quality digitized voice is now possible using 8 kbps, with near toll quality at 4.5 kbps, so that it is possible to obtain at least eight voice channels from a single 64 kbps circuit that previously carried a single voice channel. Digital compression is also expanding options for entertainment video transmission. Movies, television shows, and video games will be compressed for transmission to home receivers via satellite and optical fiber. Another technique called ADSL (asymmetrical digital subscriber line) allows several channels of motion video to be transmitted over a single pair of copper wires.

Facsimile use exploded in the past decade, while transmission speeds increased and the price of facsimile machines dropped. There are more than 30 million fax machines worldwide. Facsimile use has spread from large organizations to hotels, small businesses, offices in the home, and offices on the move—with the ability to transmit from cars over cellular networks and from aircraft via satellite. In the United Kingdom, about one business in three has a fax machine.[2] The growth has accelerated as a result of plain paper machines, the inclusion of fax capability in computer communications software, and special features offered by carriers such as AT&T's Enhanced Fax, which allows customers to broadcast to as many as 10,000 fax machines via nodes in the United States, United Kingdom, Canada, Japan, and Hong Kong. This service came to prominence in 1990, when more than 1 million messages were sent to troops in the Middle East through Operation Desert Fax.

Many businesses and trade organizations use faxes to update their clients. For example, the U.S. Chamber of Commerce sends alerts on legislative developments to its members. *USA Today* transmits its sports wire overnight to sports fans. Business and financial information services offer special fax packages to subscribers.

Video teleconferencing was introduced in the past decade for meetings involving several sites and for dissemination of information and training, typically to a large number of sites. However, high transmission costs kept the market small. Now costs are declining, as compression technology is making more video available on narrowband, with basic video transmission requiring as few as two digital telephone lines. This dial-up video service is available over ISDN in some European and Asian countries, and over switched 56 kbps service, and increasingly over ISDN; in the United States, higher bandwidth transmission can be via optical fiber or satellite, the latter being used primarily for point-to-multipoint video distribution. Whereas analog video was limited to one video signal per transponder, digital video can have six to eight video channels per transponder with high quality audio and video.

Technological convergence is leading to industry convergence, as telephone companies, cable television companies, wireless communication companies, and software providers are forming alliances. Their goal is to tap the potential of these various technologies to deliver enormous amounts of information and entertainment, and to introduce interactivity, so that users not only select programs, but can customize information services, shop from home, and interact both with the content they receive and with other users.

Technological convergence also gives more control to business users, who can operate their own virtual private networks over public network facilities. They can also choose whether to use their own PBXs or obtain internal switching services from the telephone company (this service is known as Centrex in the United States); to install their own voice messaging systems or obtain them from the telephone company or third party vendors; or to bypass all or part of the public switched network. The extent to which these services are available varies from country to country, depending on the policies of the telecommunications operators and of national regulators. These issues are discussed in the following chapters.

WIRELESS COMMUNICATIONS: TOWARD MOBILITY AND UBIQUITY

In addition to dramatic increases in transmission capacity, primarily through optical fiber and digital compression, we are witnessing the "rebirth" of radio technologies, first demonstrated by Marconi, which hold the promise of alternatives to wire and cable to reach the user, and of ubiquitous portable communications.

The Growth of Cellular Networks

Although radio has long been used for mobile communications, for example, in taxis, ambulances, and police cars, and on ships and airplanes, at one time the constraint of available frequencies appeared to be a severe limitation. Cellular networks are designed to overcome this limitation by dividing a region into a honeycomb of cells, each with specific frequencies that can be reused in other, nonadjacent cells. The traffic is "handed off" from one cell to another as the caller moves, and the cellular network is interconnected with other terrestrial networks to link fixed and mobile callers.

The first cellular system began operation in Chicago in 1983. Mobile communications was one of the success stories of the 1980s. The cellular markets boomed first in Scandinavia, with rapid growth in the United States and United Kingdom. In the 1990s, cellular phones have proliferated in Singapore and Hong Kong (see Table 2.1.). Growth rates are declining in industrialized countries, but accelerating in Southeast Asia and Latin America, where subscribers are often using cellular as an alternative to the fixed network. For example, in Cambodia, about two-thirds of all telephone subscribers are cellular users because the wireline network is so limited (see Table

TABLE 2.1
COUNTRIES/TERRITORIES WITH
HIGHEST DENSITY OF CELLULAR SUBSCRIBERS 1994

Country/Territory	Cellular Subscribers per 100 Population
Sweden	15.8
Norway	13.6
Finland	12.8
Denmark	9.7
United States	9.3
Singapore	8.4
Iceland	8.2
Bermuda	7.7
Hong Kong	7.4
Australia	7.0
Canada	6.5
New Zealand	6.5
United Kingdom	6.5

Derived from ITU, *World Telecommunication Development Report,* 1995.

2.2.). Another region where growth will be high is Eastern Europe, which, as noted above, is similar to many developing regions in its lack of telecommunications facilities. The first Eastern European systems went into service in Hungary in 1990 and Estonia in 1991. Cellular networks also operate in Poland, the Czech Republic, and Russia.

As with other forms of telecommunications, mobile communications are going digital. The system being installed in Europe is known as GSM (Global System for Mobile Communications, originally Groupe Special Mobile, after the committee that conceived the system in 1982), a standard that is being adopted by many other countries. In Europe, travelers will carry a plastic "smart card" that allows them to use advanced cellular phones in more than fifteen countries.

TABLE 2.2
COUNTRIES/TERRITORIES WITH
CELLULAR SUBSCRIBERS AS HIGHEST PERCENTAGE OF TOTAL SUBSCRIBERS

Country/Territory	Cellular Subscribers as Percentage of Total Subscribers
Cambodia	66.2%
Brunei Darussalam	20.2
Norway	19.8
Thailand	18.9
Finland	18.8
Sweden	18.8
Malaysia	16.6
Philippines	15.3
Macau	14.5
Sri Lanka	14.2

Derived from ITU, *World Telecommunication Development Report,* 1995.

Paging and Mobile Data Services

Radio paging began as a local service, and is now nationwide in the United States, regionwide in Europe, and globally accessible via satellite. Radio paging is more than a beeper: paging can deliver alphanumeric messages to users, and act as a transport service for mobile value-added services such as in-car guidance and traffic information. As microchips get more powerful, terminals become both cheaper and more intelligent: alphanumeric pagers cost about the same to produce as a simple beeper did ten years ago. In some regions, paging is threatened by cellular; there are now only half as many paging as cellular subscribers in Western Europe. However, pagers are very popular in parts of Asia; for example, in Singapore, one person in three has a pager; in Hong Kong, almost one in four carries one.[3] In recent years, there has also been growth due to national satellite paging and international roaming. For example, satellite paging networks in the United States and Canada use the same frequencies, and there is a joint venture between the Skytel paging operator in the United States and Televisa of Mexico; there are also agreements with several countries for international roaming. Europe plans to introduce its own system, known as ERMES (European Radio Message System).

Mobile data services will give the same mobility to data communications as cellular gave to voice: communication from notebook and palmtop computers, and from handheld devices known as PDAs (personal digital assistants). In the United States, mobile digital packet networks now offer specialized services for mobile computing and communications. Sales representatives can transmit orders while on the road, couriers can transmit the status of parcel pickup and delivery from their trucks, and itinerant health care workers can update patient records from the patient's bedside.

Personal Communications Services

New systems known as PCNs (Personal Communications Networks) or Personal Communication Systems (PCS) are being introduced, based on "microcells" that may be as small as a city block. The United Kingdom licensed three consortia to operate PCNs (one group has changed hands, and the other two have agreed to set up a common network). Both the United States and Canada began with experimental licenses. PCS frequencies were auctioned in the United States in 1995; networks are being built by local telephone companies, cable television companies, and cellular operators. In many countries, PCN start-up will be delayed until the late 1990s, when digital cellular networks may start to fill up, and operators look for new technologies to meet demands for portable communications.

PCS technology may also open up the local loop, the last remaining "natural monopoly" in most jurisdictions. Wireless services offer an alternative to wire-based public networks as a way to reach subscribers. AT&T's acquisition of McCaw Cellular could enable it to re-enter the local telephone business. Joint ventures between wireless and cable television networks could compete with the telephone companies in providing two-way switched services.

Lessons from Mobile Communications

The experience with mobile communications may offer lessons for the future viability and penetration of new mobile services. The mobile environment is uneven: penetration in some industrialized countries ranges from 2 to 5 per 100 population; in others with similar or better economies, the percentages are lower. In Europe, Sweden has 15 users per 100 population, while Germany has three, and France has less than two.[4] Penetration of cellular may be determined by the way the service is tariffed and priced relative to national income (GDP). In some countries, the provider has positioned cellular as a limited, high-priced service targeted at a small elite. In other countries, tariffs are set to encourage growth; competition may spur lower pricing. Availability of spectrum and pricing of interconnection to the public network can also affect mobile pricing and growth.[5]

Third generation systems may usher in the era of the "universal mobile telecommunications system," offering low cost personal communications which can be used at home, at work, or in public places, and which combine the functions of present-day cellular, cordless, and radio paging services.[6] However, tariffing may shape demand in new services. For example, in several countries, operators plan to price digital cellular above analog service to reflect a "premium service" that will attract subscribers either because of its high quality or its snob appeal. But in the United States, some operators are pricing digital cheaper than analog, or at least offering special discounts to large users, in order to build the market and encourage subscribers to switch to digital from congested analog networks.

SATELLITE COMMUNICATIONS

International Satellite Communications

Satellites are another radio-based technology that continue to play an important—and evolving—role in international telecommunications.

There are more than 100 satellites with more than 2,000 transponders and 72 million hertz of bandwidth worldwide. The current value of investment in satellite communications is approximately $15 billion; the value of business dependent on satellites exceeds ten times that amount.[7] While fiber-optic undersea cables are now carrying the majority of global voice and data traffic, the broad coverage area of satellites makes them ideal for other forms of communication, for example, point-to-multipoint data and video. Thus, satellites deliver television to viewers in many parts of the world through direct reception or transmission to local cable and rebroadcast systems. Satellites also distribute wire service and financial data through networks operated by Reuters, Dow Jones, and China's Xinhua News Agency, among others.

The ability to transmit from virtually any location simply by setting up a small antenna has made satellites extremely useful for multipoint interactive networks and for network video transmission. For example, a VSAT (very small aperture terminal) network with nearly 600 nodes transmits data for India's Ministry of Informatics. VSAT networks transmit voice and data from offshore oil operations in Mexico and Indonesia. And VSAT networks with thousands of nodes link the retail locations of companies such as Wal-Mart, Target, Safeway, and Chevron in the United States. Portable satellite stations also bring news of world events to our homes, with live coverage of natural disasters such as earthquakes in Japan and California, volcanic eruptions in the Philippines, and floods in the United States and Bangladesh, as well as starvation in Africa, civil war in the former Yugoslavia, and protests in China's Tiananmen Square.

Intelsat, the International Telecommunications Satellite Organization, was established in 1964. Intelsat now has 139 national members, and more than 60 other users. It remains a major international communications link for voice and data, especially for many developing countries without reliable terrestrial links to the rest of the world. The U.S.S.R. established the Intersputnik system, which now links Russia and other members of the Commonwealth of Independent States (CIS) with Eastern European countries, Algeria, Cuba, Vietnam, and other countries that were within its sphere of influence. Inmarsat was established in 1979 to provide global maritime communications: it now also provides aeronautical communications, and plans to offer personal communications with a new generation of satellites. Regional satellite systems were established following the Intelsat model. Eutelsat was set up in 1977 to link European nations and now has 45 members. Arabsat covers the Middle East and northern Africa.

The past decade saw the introduction of private international satellite systems, the first of which was PanAmSat, a U.S.-based company that began by providing voice, data, and video services for Latin America, with links to the United States and Europe, and now operates a global network also covering Asia and Africa. The regional AsiaSat system, based in Hong Kong, offers video services throughout Southeast and South Asia and domestic communications to several countries in the region. Astra, a private satellite system based in Luxembourg, distributes video programming throughout Europe.

Developing countries with domestic satellite systems include Argentina, Brazil, China, India, Indonesia, Malaysia, Mexico, South Korea, and Thailand. Industrialized countries with satellites include the United States (with several systems), Australia, Canada, France, Germany, Japan, Russia, and Sweden. These systems, and the growth of competition in satellite communications, are discussed in chapters 14 and 15.

VSAT Networks

There are approximately 100,000 VSATs (very small aperture terminals) installed worldwide, with about 75 percent of them located in the United States.[8] The major vendors in the United States are Hughes Network Systems, AT&T Tridom, GTE Spacenet, and Scientific Atlanta. VSAT networks are used for banking, point-of-sale transactions, reservation systems, remote monitoring, and inventory control.

While most VSAT networks are domestic, there are a growing number of international networks. A transatlantic VSAT service built by Hughes and operated by MCI links the United States with Europe via Intelsat. Holiday Inn Worldwide was the first customer for its international reservations service. AT&T provides a global VSAT service that operates through shared hubs, serving multinational firms with headquarters in the United States and operations worldwide. One advantage is that these shared hubs can replace leased lines and virtual private networks requiring extensive regulatory approvals by PTTs.[9] Scientific Atlanta has built hubs in Toronto for Canada and outside Paris for pan-European networks.

These VSAT systems typically bypass terrestrial networks in order to provide cheaper or more reliable service than is available over the public network. Brazilian banks operate VSAT networks using Brasilsat; India's National Stock Exchange links its members with VSATs via the Insat satellite. In rural and developing regions, VSATs are also used for communication where no terrestrial facilities are available. However, VSAT networks may operate only where governments authorize competition with established carriers. The issues of competition and bypass are discussed in the following chapters.

Applications for Remote Areas and Developing Countries

A disadvantage of using satellites for voice communications is the delay resulting from the transmission of the signal more than 44,600 miles (72,000 kilometers) to and from the satellite, and twice that distance in double-hop systems (where two satellite links are required, either using two different satellites or a hop from a VSAT to a master station, and another hop to a second small terminal). Therefore, in the industrialized world, voice telephone service has migrated to terrestrial technologies except in remote areas such as Alaska, northern Canada, and Siberia, where satellites remain the only viable solution for telephone service, as well as for monitoring pipelines, power grids, oil rigs, and forests.

In developing countries, satellites will most likely continue to play an important role in the provision of telephone and data services to remote areas as well as in television distribution. The advantages of satellite systems in terms of reliability, flexibility of capacity, and most importantly, the ability to install communication facilities wherever they are needed without the need for extension of terrestrial networks, will continue to make them valuable for developing countries.

More satellite terminals will be solar powered as costs per watt of photovoltaics are reduced. Improvements in radio technologies such as digital microwave, cellular radio, and rural radio systems (known in the United States as BETRS: Basic Exchange Telephone Radio Service) may enable rural satellite stations to function as hubs with the spokes or "last miles" to surrounding communities connected via terrestrial wireless networks.

Satellites for Mobile Communications

The market for mobile communications is growing, with nearly 55 million cellular telephone subscribers worldwide by the mid 1990s.[10] However, what share of mobile services will be offered by satellite versus terrestrial technologies remains uncertain. In the United States, more than 15,000 truckers are linked with two-way mobile data services, and both the United States and Canada have dedicated mobile satellite systems. In Europe, Euteltracs is a satellite-based, two-way data communications and position-reporting service for mobile services.

Alternatives to terrestrial wireless networks for personal mobile communications are systems of low earth-orbiting (LEO) satellites that will make it feasible to communicate worldwide using handheld transceivers. Several systems have been proposed, the most publicized of which is Motorola's planned Iridium system of 66 satellites, which will require intersatellite links to move traffic around the world. Others will use satellites only for originating links, and hand the traffic off to existing terrestrial networks. However, global demand may depend on price as well as convenience. Motorola's estimated charge for Iridium usage is more than three dollars per minute, which would appear to sharply limit global demand. Competitor GlobalStar plans much lower prices, and handsets that can also be used as cellular phones.

At the other end of the continuum are very cheap store-and-forward satellite systems. One such system, Healthsat, uses a single LEO to provide communications to health care providers in Africa and other developing regions. The satellite passes over at least once per day; messages can be transmitted in bursts while the satellite is within view before disappearing over the horizon. The system is used for electronic mail and transmission of text, such as articles from medical journals.

Global positioning systems (GPS) triangulate satellite data to calculate position. Originally developed for the military, slightly degraded (in terms of accuracy) systems are now commercially available and widely used on airplanes and boats, even by hikers carrying hand-held GPS locators in their backpacks. Some automobile companies are building GPS terminals into dashboard navigation systems, so that drivers can instantly find their location.

Direct Broadcasting Satellites

Direct broadcasting satellites (DBS) are designed to transmit television signals directly to small rooftop antennas. European DBS systems now offer specialized channels for news, sports, music, and movies. Pizza-pan-sized antennas have proliferated largely because of the limited choices of television channels available in most European countries, and the lack of cable networks in many areas.

Although DBS technology was originally proposed in the United States in the 1970s, it has been slow to take off, primarily because of competing technologies for distributing television signals. For example, in the United States, there are more than 60 million cable subscribers, and more than 2 million rural residents have installed backyard antennas to receive signals directly from existing satellites. The first U.S. direct broadcasting satellite, built by Hughes, began operations in 1994. It uses digital video compression to deliver up to 175 television and movie channels. Hughes is also introducing direct-to-home (DTH) satellite services in Latin America. Rupert Murdoch plans to deliver television directly to Japanese households via satellite.

Some analysts think the future of DBS is tied to high definition television (HDTV), which requires enormous bandwidth to deliver the digital information necessary to display a film-quality television picture. At present, there is considerable debate about the future of this technology, generically known as Advanced Television Technology (ATV). Proponents see a major role in ATV as vital to U.S. interests in the next phase of the electronics revolution: "There is no doubt that ATV will offer viewers pictures of exceptional, perhaps mesmerizing caliber. However, underlying the development of ATV is a technological revolution whose consequences for the entire electronics industry, and those industries dependent upon electronics, are of staggering proportions. . . . The market for the myriad interrelated industries—products, services, processes—could be worth tens of billions of dollars. Hundreds of thousands of jobs are at stake."[11]

Japan has already introduced analog HDTV, but market penetration is very limited. The United States has opted for a digital system, developed by a consortium of industry and academic researchers, after competition among several different proposals. The U.S. approach builds more on advances in computer technology than on upgrading analog broadcasting systems. Yet the transition from analog to digital television will likely be phased in over an extended period because of the tremendous investment in facilities required to transmit the content, the additional frequencies required, and the concern of regulators that programming should be accessible to all viewers, regardless of whether they have analog or digital equipment, just as color programs could also be viewed on black and white TV sets.

INFORMATION SUPERHIGHWAYS

Industrialized countries are now advocating the construction of information superhighways, broadband digital networks that would bring vast amounts of information within the reach of every resident. In the United States, government, industry, educators, and public interest groups are debating how to build a national information infrastructure (NII) that would make the information available on the superhighways universally accessible and affordable. While the eventual goal appears to be linking every home to interactive broadband networks, interim solutions include providing access to schools, libraries, hospitals, and community centers. Optical fiber, which now carries traffic across the country and links digital telephone switches, is gradually being extended into the local loop. However, other means of reaching end users such as coaxial cable, wireless technologies, and existing copper wire may also be used to provide some additional capacity and interactivity. Deregulation of the telecommunications industry will hasten the convergence of telecommunications and broadcasting, and telephone companies, other carriers, broadcasters, cable television companies, satellite companies, and the film and computer software industries will vie for a role in the digital world.

A global information infrastructure (GII) has been proposed to eventually link people everywhere to digital superhighways. But simply providing access to basic telecommunications remains an enormous challenge. The ITU estimates that $466 billion will be required between 1993 and the year 2000 to meet current projections of telecommunications growth. However, that investment, while resulting in more than 310 million new lines, will only raise global teledensity from 10.5 to approximately 14 telephone lines per 100 population, with projected density of only 2.6 in Africa and 6.2 in Asia.[12]

In addition to providing global access to information, extension of Internet access to all countries would be a step toward the proposed GII. The Internet is a network of networks originally designed to link research universities in the United States with support from the Defense Advanced Research Projects Agency (DARPA). Today, the Internet has more than 27 million users, with an estimated growth rate of 113.1 percent compounded annually. More than 5 million host computers are located in at least 60 countries, connecting more than 50,000 networks. A further 68 countries are interconnected via electronic mail or other networks.[13] The fastest growing application is the World Wide Web, which offers graphics, color, and hyperlinks connecting sites with information on related topics. With the introduction of the Mosaic Web browser in 1992, Web usage began to take off. As commercial Web browsers became available, Web use exploded, growing 1,400 percent in 1994 alone.

Although the Internet links users around the world, access remains limited in many developing countries. Those with limited electronic mail connections cannot access the World Wide Web. Some countries, primarily in Africa, still have no Internet access. In many developing countries that are connected, only a few sites, such as universities in major cities, are linked to the Internet.[14] Of course, users also need access to a computer or terminal to get on the Net. Table 2.3 shows that there is a strong relationship between availability of personal computers and number of Internet users. High-income countries have an enormous advantage, with almost seven times as many personal computers per 100 as in upper-middle-income countries, and 130 times as many as in low-income countries. But the gap in Internet users is even greater, with high-income countries having 35 times as many Internet users per 10,000 population as upper-middle-income countries and an astounding 15,000 times as many as the poorest countries.

Cheaper and more powerful personal computers are likely to spread through developing countries, but true global access will require an enormous investment in extending and upgrading telecommunications networks throughout the developing world. Price is also a criti-

TABLE 2.3
ACCESS TO INFORMATION TECHNOLOGY

Country Classification	Estimated PCs per 100 pop. (millions)	Estimated Internet Users per 10,000 pop.	Estimated Internet Hosts (thousands)
High-Income	18.26	309.97	4,800
Upper-Middle-Income	2.68	8.76	81
Lower-Middle-Income	0.72	1.83	37
Low-Income	0.14	0.02	1

Derived from ITU, *World Telecommunication Development Report,* 1995.

cal factor in accessibility, even in high-income countries. High telecommunications costs imposed by national monopolies are cited as the the chief cause of low Internet penetration into European corporate offices. German Internet providers pay $75,000 per year to lease a 64 kbps circuit, compared to approximately $15,000 per year for a T1 line (1.544 Mbps) in the United States.[15] Countries with high-quality networks and competitive service providers will likely see the fastest growth in Internet use.

CHAPTER 3

The Major Players: The Telecommunications Industry and Users

". . . telecommunications has passed from an environment oriented by technology and driven by supply to an environment oriented towards applications and driven by the users."
—GERARD SANTUCCI, *XIII Magazine*[1]

THE INTERNATIONAL TELECOMMUNICATIONS INDUSTRY

In 1994, the inhabitants of the globe generated more than 40 billion minutes of international telephone traffic, an average of 85 minutes per subscriber. The high-income countries generated ten times as much traffic as upper-middle-income countries. Geographically, Europe originated the most international traffic, followed by the Americas and Asia (see Table 3.1). The compound annual growth rate in international traffic between 1984 and 1994 was 15.2 percent, with Asia's traffic growing the fastest, at an annual rate of 21.9 percent.

This growth in international traffic, plus the growth in domestic traffic, has fueled demand for telecommunications equipment and services. The three major groups in the telecommunications industry are the carriers or public network operators, which own and operate the transmission networks, the equipment suppliers or vendors, and the value-added service providers.

The major global competitors in international telecommunications include the carriers, the equipment vendors, and the value-added services providers.

The Carriers

The telecommunications carriers, also known as operators, are the most visible players in the industry. Most carry both domestic and international traffic. A few, such as Teleglobe of Canada and KDD of Japan, are strictly international carriers.

Table 3.2 shows the major international carriers based on revenue from international traffic. As international competition heats up, several of these carriers have entered into strategic alliances to capture new business and retain current customers. Each aims to provide "one-stop shopping" for multinational business customers by offering a single source of international telecommunications capabilities. The major alliances are:

- **Unisource:** a consortium of European telecommunications operators including PTT Telecom (Netherlands), Telia of Sweden, the Swiss PTT, and Telefonica of Spain.
- **WorldPartners:** established in 1993 by AT&T, KDD of Japan, and Singapore Telecom to provide a comprehensive package called the WorldSource family of customized business services. WorldPartners has added European members through equity participation by Unisource. WorldPartners has recruited nonequity partners including Unitel of Canada, NTT of Japan, Telecom New Zealand, Telstra of Australia, Korea Telecom, PLDT of the Philippines, Hong Kong Telecom, and Bezeq of Israel.
- **Concert:** a joint venture between BT of the United Kingdom and MCI, created as part of BT's 20 percent investment in MCI

TABLE 3.1
INTERNATIONAL TELEPHONE TRAFFIC 1994

Country Classification	Traffic (million minutes)
High-Income	40,615.4
Upper-Middle-Income	6,844.4
Lower-Middle-Income	3,800.7
Low-Income	2,290.8
Region	
Europe	23,811.4
Americas	18,809.7
Asia	8,809.6
Africa	1,118.9
Oceania	1,001.6

Derived from ITU, *World Telecommunication Development Report*, 1995.

in 1994.* Concert combines the second largest U.S. international carrier with BT, which is attempting to move aggressively into international markets. The billion dollars' worth of assets contributed by the two companies include BT's Syncordia, a single-source telecommunications outsourcing unit.

- **Global One:** formed by major operators that were left out of the above alliances. Global One represents an effort to catch up with the other global alliances by Sprint, the third largest U.S. international carrier, France Telecom, and Deutsche Telekom. Announced in 1995, Global One stalled for nearly a year while the U.S. Federal Communications Commission (FCC) and European Union reviewed the proposed venture. Sprint gained not only a global presence but also an infusion of $4.2 billion from its partners, to be invested in U.S. domestic ventures (see Table 3.2).

* In November 1996 BT and MCI announced plans to merge. The merged company will be called Concert.

TABLE 3.2
TOP PUBLIC TELECOMMUNICATIONS OPERATORS
BY OUTGOING INTERNATIONAL TRAFFIC 1994

Operator	Country	Minutes of Traffic * (billions)	International Revenue (billions)
AT&T	U.S.	7.95	$5.11
Deutsche Telekom	Germany	4.96	4.66
MCI	U.S.	3.52	1.83
France Telecom	France	2.50	3.50
BT	U.K.	2.37	2.96
Telecom Italia	Italy	1.76	1.48
Swiss Telecom PTT	Switzerland	1.65	1.71
Hong Kong Telecom	Hong Kong	1.58	2.11
Sprint	U.S.	1.47	0.71
PTT Telecom BV	Netherlands	1.35	1.35
DGT	China	1.17	1.40
Belgacom	Belgium	1.05	0.60
KDD	Japan	0.95	1.72
Telefonica	Spain	0.95	1.05
Teleglobe	Canada	0.86	0.30
Austrian PTT	Austria	0.81	0.80
Mercury	U.K.	0.77	0.74
Telia	Sweden	0.70	0.70
Telstra	Australia	0.69	1.02
Telmex	Mexico	0.66	1.76

* Minutes of international traffic on public telephone networks.
Source: ITU, *World Telecommunication Development Report,* 1995.

The partners in these strategic alliances include most of the largest international carriers such as:

- **AT&T:** AT&T, the largest U.S. long distance carrier and the world's largest international carrier, derives about 15 percent of its international revenue of $37 billion from overseas, and employs 22,000 outside the United States.
- **BT (formerly British Telecommunications PLC):** BT is the United Kingdom's largest carrier, and is a partner with MCI in Concert. BT also owns Tymnet, an international data network,

and has established Syncordia to manage global corporate networks.

- **Deutsche Telekom (formerly Deutsche Bundespost):** Europe's largest operator, Deutsche Telekom has absorbed the telecommunications networks of the former East Germany, which it is upgrading and extending, and is now expanding into international markets. The German carrier is positioned to provide multimedia services on the "information highway," with an extensive fiber-optic network, cable television distribution to 15 million homes, and wireless networks.
- **France Telecom:** France Telecom, the world's fourth largest international operator, is known for operating a very modern network in France, and for introducing Minitel, a videotex system with 6 million subscribers. FT has offices and subsidiaries in more than 30 major cities in Europe, North America, Latin America, and the Asia-Pacific region.
- **KDD (Kokusai Denshin Denwa Co.):** Established in 1953, KDD is Japan's largest international carrier.
- **MCI:** MCI is the second largest U.S. carrier, and provides direct dialing from the United States to 180 countries. It formed an international unit after buying Western Union International in 1983; in 1988, it bought RCA Global Communications from General Electric. MCI owns 29.5 percent of Clear, New Zealand's second carrier.[2] In 1993, British Telecom bought a 20 percent stake in MCI.
- **Sprint:** The third largest U.S. long distance carrier, Sprint owns SprintNet, a public data network with links to 108 countries. It also owns 50 percent of PTAT, the Private Transatlantic Telecommunications fiber cable with Cable & Wireless.
- **Telefonica:** The Spanish carrier is a major shareholder of privatized operators in Argentina, Chile, Puerto Rico, and Venezuela, and is expanding into other Latin American countries.

Other major operators include:

- **Cable & Wireless PLC:** Cable & Wireless, established in the 19th century as the exclusive provider of British international telecommunications, now owns Mercury, the second largest carrier in the United Kingdom, and is the largest shareholder in Hong Kong Telecom. It is also a partner in Optus, Australia's second carrier, and owns a major stake in AsiaSat. C&W also has majority interests in operators in British Commonwealth countries in the Caribbean and the South Pacific.

- **Hutchison Telecommunications Ltd.:** Hutchison Telecommunications is part of the telecommunications group of Hutchison Whampoa Ltd. of Hong Kong, which also has interests in property, retailing, port projects, and utilities. Hutchison installs and operates mobile telephone and paging networks. Hutchison claims 55 percent of Hong Kong's cellular phone market, and has formed strategic alliances in Australia, the United Kingdom, Malaysia, Thailand, South Korea, Taiwan, and Bangladesh.[3]
- **Satellite systems:** International and regional satellite systems including Intelsat, Inmarsat, PanAmSat, and AsiaSat and the new LEO systems (see chapters 14 and 15).

However, among the world's largest telecommunications carriers are companies that provide domestic services primarily or exclusively. The largest is NTT, Japan's domestic carrier, with the greatest number of domestic main lines and highest revenue of any carrier. The other large domestic players are U.S. companies including GTE, which operates local and mobile services in the United States, and the "Baby Bells," the companies spun off in the divestiture of AT&T (see Table 3.3).

Restricted from entering new markets at home after the breakup of AT&T, the Bell Operating Companies (BOCs) sought new opportunities abroad. Among the international ventures of the BOCs are:

- **Ameritech:** mobile communications in Poland; directory publishing in Germany;
- **Bell Atlantic:** mobile communications in the Czech Republic and Russia;
- **Ameritech and Bell Atlantic:** in 1990 acquired Telecom New Zealand for $2.46 billion; divested to 49.9 percent in 1993;

TABLE 3.3
TOP PUBLIC TELECOMMUNICATIONS OPERATORS 1994

| By Revenues | | | By Main Lines | | |
Operator	Country	Revenue (billions)	Operator	Company	Main Lines (millions)
NTT	Japan	$60.1	NTT	Japan	59.87
AT&T	U.S.	43.4	Deutsche Telekom	Germany	39.20
Deutsche Telekom	Germany	37.7	France Telecom	France	31.60
France Telecom	France	23.3	DGT	China	27.23
BT	U.K.	21.3	BT Telecom	U.K.	27.07
Telecom Italia	Italy	18.0	India	India	24.52
GTE	U.S.	17.4	MPT	Russia	24.10
BellSouth	U.S.	16.8	BellSouth	U.S.	20.22
Bell Atlantic	U.S.	13.8	Bell Atlantic	U.S.	19.20
MCI	U.S.	13.3	Ameritech	U.S.	18.24

Derived from ITU, *World Telecommunication Development Report,* 1995.

- **Bell Atlantic and US WEST:** in 1990 signed partnership agreement with the Czech PTT to build cellular network;
- **BellSouth:** major shareholder in Optus, Australia's second carrier; mobile communications in Denmark; paging services in Switzerland and the United Kingdom;
- **NYNEX:** provides project management services to PLDT (Philippine Long Distance Telephone Company); owns 50 percent of Gibraltar telephone company; has cable TV interests in the United Kingdom; sells software in several countries; is a major partner in FLAG (Fiber Optic Link Around the Globe);
- **Pacific Telesis:** has joint ventures in mobile communications in Germany, Denmark, Portugal; investment in paging services in Thailand; interests in Tokyo Digital Phone and Kansai Digital Phone, Japan; partner in several U.K. cable consortia;[4]
- **SBC (formerly Southwestern Bell):** acquired a 5 percent equity share in Telefonos de Mexico (Telmex) for $485 million; has

cable television interests in the United Kingdom; sells telephone equipment in the United Kingdom;

- **US WEST:** partner in mobile communications in the Czech Republic, Hungary, Russia; in joint venture with United International Holdings Ltd. to construct cable TV network on Malta.

Telecommunications Equipment Vendors

Telecommunications equipment consists of two main product categories:

- network equipment (transmission and switching such as optical fiber, microwave equipment, satellites, wireless networks, switches, routers, etc.);
- terminal equipment, ranging from telephone handsets to facsimile machines to networked computers to satellite earth stations.

As noted in chapter 1, the telecommunications equipment market is growing dramatically, as a result of both technological innovation

TABLE 3.4
TOP GLOBAL TELECOMMUNICATIONS EQUIPMENT VENDORS 1994

Company	Country of Origin	Telecom Equipment Sales (billions)	Foreign Sales (percentage of total sales)
Alcatel	France	$20.4	72.4%
Motorola	U.S.	14.4	43.9
AT&T*	U.S.	14.3	9.8
Siemens	Germany	12.8	57.7
Ericsson	Sweden	10.7	90.0
NEC	Japan	9.5	15.5
Nortel	Canada	8.2	87.1
Fujitsu	Japan	4.8	29.5
Bosch	Germany	3.4	54.0
Nokia	Finland	2.5	40.2

* Now Lucent Technologies.
Derived from ITU, *World Telecommunication Development Report*, 1995.

and convergence, and of privatization and the introduction of competition in communications in many countries. The major markets for telecommunications equipment are Europe, which accounts for 37 percent of the market, the United States, which accounts for 32 percent, and Japan, which accounts for 16 percent.

Among the major equipment suppliers are (see Table 3.4):

- **Alcatel:** Alcatel is France's largest equipment maker, a subsidiary of Alcatel Alsthom, with equipment sales worldwide. Alcatel led the industry with sales of $20.4 million in 1994. In 1987, it purchased the international equipment operations of ITT Corporation. Alcatel has pursued a strategy of external growth that has taken it to first place in the world in public switching equipment, optical clusters and cables, and second place in line transmission systems.
- **Motorola:** Motorola is the world's largest maker of radio communications equipment, and sells cellular network equipment and cellular telephones worldwide. It also has developed wireless LANs (local area network), and has plans for a low earth-orbiting (LEO) satellite system known as Iridium.
- **AT&T:** AT&T's equipment arm was split off as a separate company called Lucent Technologies in 1995. It accounts for more than 12 percent of the global telecommunications equipment market.
- **Siemens AG:** Siemens of Germany supplies equipment to more than 100 telecommunications authorities. Siemens is a major supplier in eastern Germany and is moving into Eastern Europe. In the United States, Siemens owns Stromberg-Carlson and 50 percent of Rolm, a maker of switching equipment. Siemens products are also installed in Latin America, Asia, and Africa.[5]
- **Ericsson:** Ericsson of Sweden has relied on export markets for decades, and is a major global supplier of switching equipment. Ericsson's AXE switching system is operating in more than 80 countries, with more than 50 million lines installed.[6] Ericsson is also a major global supplier of cellular communications equipment.
- **NEC (Nippon Electric Corporation):** NEC of Japan is a major

global supplier of telecommunications and computer equipment, and an early proponent of converging technologies, through its goal of uniting "C and C," computers and communications.

- **Nortel (Northern Telecom):** Nortel, a Canadian firm partially owned by Bell Canada, has become a major global vendor with markets in the United States, Europe, Eastern Europe, and Japan. In 1990, it acquired STC, one of Britain's biggest equipment manufacturers. Nortel aims to be the world's largest telecommunications equipment supplier by the year 2000.[7]

Value-Added Services

The third major group in the telecommunications industry consists of the value-added networks (VANs) or value-added services (VAS). These companies provide specialized telecommunications services such as high-speed data networks for large corporate customers, dial-up packet data networks, and Internet access (see Table 3.5).

TABLE 3.5
MAJOR VALUE-ADDED NETWORK SERVICE PROVIDERS 1994

Service Provider	Headquarters	VAN Revenue U.S. (millions)
IBM Global Network	U.S.	$1,500
Transpac	France	900
SITA	Belgium	780
SprintNet	U.S.	650
GEIS	U.S.	600
Concert	U.S.	550
Unisource	Netherlands	510
AT&T	U.S.	500
SWIFT	Belgium	330
Infonet	U.S.	290

Note: Some revenue figures are ITU estimates.
Derived from ITU, *World Telecommunication Development Report*, 1995.

The major global network for the airline industry is SITA (Société Internationale des Télécommunications Aeronautiques), based in Belgium. SITA links airline reservation networks around the world. Global funds transfers for international banking are handled through SWIFT (Society for Worldwide Interbank Financial Telecommunications), also based in Belgium. GEIS (General Electric Information Services) is owned by a nontelecommunications company, General Electric, which serves numerous corporate clients. Other major VANs are operated by individual telecommunications operators such as AT&T, Sprint (SprintNet), and France Telecom (Transpac), and telecommunications consortia such as Concert (BT and MCI), Unisource (several European operators), and Infonet (owned by ten operators).

The proliferation of personal computers and the growth of the Internet are likely to fuel the growth of the data communications industry. In addition to linking corporate networks, major operators such as AT&T and MCI are joining the host of start-up companies providing Internet access to businesses and individuals.

Other participants in the telecommunications sector include companies that make products such as:

- the building blocks of computer networks including routers and ATM (Asynchronous Transfer Mode) switches to switch high-speed data and video;
- voice processing and storage for voice messaging and voice recognition;
- video servers for video-on-demand systems;
- videoconferencing for stand-alone or desktop computer systems;
- groupware to enable teams to collaborate at a distance;
- Internet access and navigation.

Most of these products did not exist a decade ago. Many of the companies that produce them began as start-ups founded by high tech entrepreneurs. Their origins and business styles are very different from the major telecommunications firms that have dominated the industry for decades.

THE CONVERGENCE GAME

Chapter 2 presented a functional analysis of convergence including information transmission, processing, storage, and reproduction. Convergence can also be considered in industry terms. The telecommunications industry, as shown previously, is traditionally thought of as network operators and communications equipment vendors. But as voice, data, and video are converging, the industries that have developed these services are also seeking to compete and to form strategic alliances.

New Players and Strategic Alliances

Who are likely to be the major players in these strategic alliances? Table 3.6 lists the major global players in the computer and audiovisual industries, many of which are now collaborating with telecommunications companies. All of the top ten computer companies are American or Japanese. In fact, only three companies in the top 25 are European (Siemens, Olivetti, and Bull). While American and Japanese companies also dominate the audiovisual sector, European companies are better represented there than they are in computing. The top 25 include three from the United Kingdom (BBC, Thorn EMI, and Carlton), as well as companies based in Germany, the Netherlands, Italy, and Luxembourg.[8] Another major group of players in the convergence game is the cable television industry. Like the computer industry, it is dominated by American companies; 16 of the top 20 cable companies in the world, in terms of number of subscribers, are U.S. firms.[9]

Technological convergence is now forcing these companies to recognize new competitors. The cable industry can compete with telecommunications carriers to offer telephone service and Internet access. Conversely, the carriers are racing to upgrade their networks to distribute video. Computers are no longer simply data crunchers, but "multimedia workstations" that can create as well as process sound,

TABLE 3.6
CONVERGENCE: THE NEW PLAYERS

Top Computer Companies: 1994			Top Audiovisual Companies: 1994		
Company	Country	Revenue (billions)	Company	Country	Revenue (billions)
IBM	U.S.	$64.05	Time Warner	U.S.	$8.46
Hewlett Packard	U.S.	25.35	Sony	Japan	7.72
Fujitsu	Japan	20.94	ARD	Germany	5.82
NEC	Japan	17.43	Matsushita	Japan	5.74
DEC	U.S.	13.45	Capital Cities/ABC	U.S.	5.28
Hitachi	Japan	13.15	Viacom	U.S.	5.21
Compaq	U.S.	10.87	NHK	Japan	4.98
Apple	U.S.	9.19	TCI	U.S.	4.94
Canon	Japan	9.10	Disney	U.S.	4.79
Toshiba	Japan	8.60	PolyGram	Netherlands	4.73

Derived from ITU, *World Telecommunication Development Report*, 1995.

graphics, and motion video. And the Internet has evolved from being an academic network linking researchers to a global public network linking businesses and individuals as well as universities, schools, and libraries.

Survivors will see this new environment as an opportunity rather than a threat. In an effort to take advantage of that opportunity, companies are diversifying by developing new information products and services, as well as participating in mergers, acquisitions, and joint ventures—seeking partners with complementary technology and expertise. For example:

Telephony and Cable Television:
- US WEST, which provides telephone service in the western United States, bought out Continental Cablevision and invested in a 25 percent stake in Time Warner Entertainment to gain access to 13 million cable subscribers.

- Sprint, the third largest long distance company in the United States, has joint ventures with three of the largest U.S. cable companies, TCI (TeleCommunications Inc.), Comcast, and Cox Communications, giving it access to 39 million homes.[10] Together they are building a national wireless network using PCS technology.
- Cable companies in the United Kingdom have been permitted to offer local telephone service since 1987. Foreign companies that were not allowed to provide both telephony and cable services in their home markets have found the United Kingdom to be a valuable testbed. They include North American cable operators TCI of the United States and Videotron of Canada, "Baby Bells" NYNEX and US WEST, and Canada's BCE (Bell Canada Enterprises).[11]
- Telephone companies are experimenting with video-on-demand (VOD) to compete with cable. VOD projects are planned by Bell Atlantic, BT in the United Kingdom, Hong Kong Telecom, and Singapore Telecom.

Internet Access:
- MCI was the first major U.S. phone company to offer an Internet service, and is generating $100 million per year from Web surfers.[12]
- AT&T has established an Internet service called Worldnet. To gain a major market share quickly and build "brand loyalty," AT&T offered its residential customers five hours of Internet access per month free for the first 12 months.
- Sprint is targeting business customers with its Sprint Intranet Dial Service, and estimates that the corporate market for business access to the Internet could reach $1 billion per year by the end of the decade.[13]
- MFS Communications, the largest competitive access provider (CAP) of local optical fiber networks in the United States, purchased UUNet Technologies, an Internet access provider, so that it can package voice, data, video, and online services for business customers.

Telecommunications and Media:

- Rupert Murdoch's publishing giant, News Corporation, has holdings in the Fox television network, British BSkyB, and Asian STAR TV satellite television services, and the Delphi on-line and Internet service. MCI has purchased a 13.5% stake in the News Corp. media empire, gaining access to publishing and broadcasting channels and content, which it will use in building its own DBS service.
- Rogers Communications, Canada's largest cable operator, owns Cantel, a national cellular operator, is part owner of Unitel, Canada's second long distance carrier, and has interests in broadcasting stations, video rentals, and a home shopping network.
- Disney has a joint venture with Baby Bells Ameritech, BellSouth, and SBC, and with GTE, to develop and deliver new entertainment and interactive services.

Implications for Users and Regulators

New technologies and services will continue to proliferate in a more competitive environment. The result should provide more choices and better prices for users. Yet, paradoxically, the opposite may also be true. While few monopolies will survive, consolidation through mergers and joint ventures may result in new oligopolies that have few incentives to compete on price.

The positive side of convergence and consolidation may be more opportunities for "one-stop shopping." The global carrier alliances now promote this advantage to their corporate customers. Telephone companies such as AT&T, Sprint, and MCI are also bundling services so that residential customers can obtain not only telephone service but also cellular service, Internet access, and cable or satellite television from the same source.

Regulators will face new challenges in "leveling the playing field" for competition. Antitrust and antimonopoly laws are already being invoked in some countries to prevent remonopolization of the telecommunications industry. Regulators will also have to rethink the traditional classifications of broadcasting and telecommunications. In the United States, cable television franchises are awarded by cities and municipalities, while local telephone service is under state jurisdiction. In Singapore, cable is regulated by the Singapore Broadcasting Authority, while VOD, a proposed service of Singapore Telecom, comes under the Telecommunications Authority of Singapore. But are interactive cable television and video-on-demand services provided by telephone companies really two separate industries? From the users' perspective, they are simply two ways of delivering multiple channels of video to the home.

APPLICATIONS OF TELECOMMUNICATIONS

While voice still accounts for the majority of international telecommunications traffic, the fastest growing components are data and image, as corporate and institutional users increasingly rely on facsimile, data transmission, electronic mail, and teleconferencing to manage their activities and reach their customers. The following subsections illustrate how users are applying telecommunications around the world.

Finance and Banking

Time is money! Electronic funds transfer using telecommunications moves money instantly—for credit card transactions, for withdrawals and deposits using ATMs (automatic teller machines), for transfers within banks from branch to branch, and for global transfers among banks, brokerages, and investors. Investing has become a 24-hour activity, as analysts and brokers monitor the exchanges in Europe, North America, and Asia. The SWIFT global financial network moves the equivalent of $2.3 trillion around the world every day.[14] SWIFT, which is jointly owned by a cooperative of more than 200 U.S., Canadian, and European banks, operates packet-switched network services that support interbank communications applications such as funds transfers for some 3,000 financial institutions world-wide.

In 1987, investors learned about the downside of instantaneous information about global markets, as the plunge on Wall Street triggered massive selling in London, Tokyo, and Hong Kong. Investors have also learned to monitor satellite television for breaking news that could affect the markets. In 1990, financial institutions in Singapore lost millions of dollars in the first few minutes of the Gulf War because they didn't have CNN (Cable News Network). They then persuaded the Singapore government to make an exception for financial institutions to its ban on satellite antennas.

Bank of America, one of the world's ten largest banks, is an innovator in telecommunications applications for both commercial and individual customers. Bank of America's Money Tracking Service (BAMTRAC) provides global money transfer services and cash management reporting services for its customers. BAMTRAC cuts the information float for customers so that they have up-to-date information on all transactions and accounts, and cuts costs associated with clerks, mailing, and paper.[15] Bank of America and other banks expect to offer online banking via the Internet when they determine that encryption technology affords sufficient security to protect financial information.

The Hong Kong and Shanghai Banking Corporation (the HSBC Group) operates a global network from its headquarters in Hong Kong. HSBC recognizes that the bank of tomorrow will be in the information business to produce, sell, and distribute information to customers anywhere in the world. The branches of tomorrow's global banks "will be the information devices, electronic gateways, and information networks through which that information is delivered, and by which the customer relationship with the bank will be defined."[16]

In Brazil, banking became very time- and information-dependent when the volatile inflation rate required that economic indicators such as salaries, loans, and bank accounts be indexed daily. Brazilians therefore habitually deposit their cash into bank accounts and draw out only what they need each day, resulting in an enormous volume of daily transactions that must be collected and processed as soon as possible. Also, Brazilians pay many of their bills at banks, adding to the need for transaction processing.

Banco Bradesco, Brazil's largest bank, with 1,702 branches and more than 15 million deposit accounts, has implemented a VSAT network to link its sites. The bank negotiated with the Brazilian government to implement legal and regulatory changes, enabling it to establish its own private network, which uses Brazil's domestic satellite, Brasilsat, to transmit data from the branches to the central data center in São Paolo. A local company, Victori Communicacoes, upgraded the network, and provides local support and maintenance. The network allows the bank to link remote branches without telephone service, whereas previously transaction reports had to be picked up by boat. In other areas with telephone service, the bank installed VSATs instead of waiting at least three years to have terrestrial lines installed.[17]

Global Companies

Pepsi-Cola International (PCI) is present in 165 countries and maintains offices with reporting responsibilities in 80 countries. The telecommunications facilities available vary from ISDN to telex. Pepsi must maintain contact throughout the former Soviet Union with bottlers, many of which lack access even to a fax machine. Since 1989, the company has installed digital PBXs to integrate voice, data, fax, and telex. Today, as more PCI offices in Eastern Europe are becoming autonomous country headquarters, PCI is upgrading its facilities. In addition to the ISDN functionality, PCI has installed a voice mail server that can leave messages in several languages, as part of a "first-aid" strategy to PCI locations in Eastern Europe. The company plans to introduce a communications platform that can meet all needs from basic telephony to videoconferencing.[17]

Levi Strauss and Co., famous around the world for its blue jeans, is using an information-intensive strategy to improve customer service, forge stronger relationships with its customers and suppliers, and reduce the time needed to develop products and fill customer orders. In a single private network, Levi's connects its headquarters in San Francisco with more than 110 sites in Europe, Asia-Pacific, and Latin America.

Retailing

In North America, point-of-service applications are reaching down to the level of the corner gas station and convenience store. Grocery chains use applications ranging from inventory monitoring to cash machines. Where there is low-volume traffic, VSATs are often an appropriate solution, even where telephone service is available.

- **Wal-Mart,** the discount retailing chain, operates more than 2,400 stores in the United States, Canada, and Mexico. Via VSATs and terrestrial frame relay networks Wal-Mart links its distribution centers, more than 4,000 vendors, and every cash register in its stores to get up-to-the-minute data on stocks and purchases, make immediate pricing adjustments, and obtain daily information for central buyers, product line managers, and business unit executives. Wal-Mart also uses its VSAT network for video distribution of management and market information and training.
- **Chevron** has installed VSATs at 5,000 service stations in the United States that enable drivers to "pay at the pump": when credit cards are inserted, they are automatically verified, and sales are tracked at each gas station.
- **General Motors** has become the largest VSAT user in the world, with a 10,000 node network called Pulsat linking GM operations and support with dealers. (General Motors owns Hughes, the largest manufacturer of satellites and satellite equipment, and a satellite network operator.)

Telecommunications has become critical to managing costs and increasing sales in the cutthroat world of discount retailing. As discount retailers adopt just-in-time inventory systems, vendors need to share inventory responsibilities with retailers to cope with more frequent orders that must be on time and error free. Wal-Mart, Kmart, and other high-performance retailers are using information sharing with vendors to implement "continuous replenishment." Key vendors are allowed to connect to the retailers' computer networks to obtain point-of-sales data via Electronic Data Interchange (EDI). Vendors are responsible for replenishing inventory as it is depleted, rather than waiting for purchase orders.[18]

Credit Card Verification

Credit card verification is a major deterrent to fraud and abuse of credit cards. In the past, clerks hunted through pages of "bad" credit card numbers before authorizing credit card purchases. Now, in more than 90 countries, merchants check cards electronically, using point-of-sale terminals connected to telephone lines or satellite terminals. Small retailers and restaurants use stand-alone terminals, while chain stores, gas stations, and supermarkets may use systems integrated with cash registers, pump controllers, or scanners. Kmart's VSAT network has helped to cut customer waiting time by reducing the time needed for credit card and check authorizations from 45 seconds to seven seconds.[19]

Linking Suppliers via EDI

Electronic Data Interchange (EDI) is the computer-to-computer exchange of intercompany business documents in a public standard format. EDI can save time and money by replacing labor-intensive transactions and reducing human errors in ordering and payment functions. EDI also permits companies to control the timing of payments so that they can budget more accurately; however, EDI may eliminate the float benefits associated with traditional check payments. For example, Haworth Inc., a U.S.-based office furniture company, now uses EDI instead of mail and faxes to communicate with about 400 suppliers around the world. Using EDI, Haworth can send out orders to suppliers from its personal computers, saving about $1 million per year in processing time and effort. In addition, the company receives instant feedback on market trends from its suppliers, and can adjust its purchasing and production schedules accordingly.[20]

Delivery Services

The express transportation industry relies on telecommunications to maintain efficiency, keep prices competitive, and provide new customer services. Federal Express now delivers nearly 2.4 million items daily to more than 200 countries and territories. FedEx was the first courier to install computers in its vans, automate its mailrooms, and develop shipping and tracking software. In an effort to build business and maintain customer loyalty, FedEx introduced FedEx Ship and Powership, which link customers' computers directly to FedEx to send and check on shipments. FedEx claims that 60 percent of its shipping traffic is currently initiated by customers from their desktop computers.

DHL's DHLNET provides information on routing, delivery times, and system capacity for efficient sorting and distribution of goods. DHL customers can also use DHLNET to track the status of shipments and check delivery schedules and pricing. United Parcel Service's UPSnet links distribution sites in 46 countries to track more than 821,000 packages per day. UPS also enables customers to ship and track UPS parcels using MaxiShip and MaxiTrac desktop software. Its TotalTrac system links 55,000 delivery people to the network through an electronic mobile clipboard and vehicle-mounted transmitter. These three global shippers have also added Internet access so that customers can use the Internet to place orders, track shipments, and view billing information.

Telework and Telecommuting

The ability to communicate "anytime, anywhere" is changing both where and how people work. Self-employed individuals may now choose to live in rural areas and communicate with their clients via telephone, fax, and e-mail. Employees may choose to "telecommute," to work from their homes or a nearby satellite office rather than commuting to the workplace. And companies themselves are taking advantage of telecommunications to work around the clock by sharing tasks among teams around the world. Others are equipping their sales and maintenance staff with notebook computers and wireless modems so that they don't need permanent office space.

Digital Equipment Corporation, a major manufacturer of networked computer systems and associated peripheral equipment and software, now does business in 95 countries outside the United States. Digital has one of the largest distributed networks in the world, with over 85,000 DECnet nodes and 24,000 TCP/IP hosts. The network is used to provide support to distributed work teams responsible for all elements of a project or service. Digital uses its enterprise network to support geographically distributed work teams with electronic mail, electronic bulletin boards, and computer, audio, and videoconferencing as required. The creative use of electronic mail, computer conferencing, and data exchange between computers enabled a workgroup to develop a successful new product without face-to-face interaction. Digital is now using workgroup computing, a set of tools and services that can foster better communication among groups, to improve creativity, productivity, and profitability.[21]

India has combined computer expertise with telecommunications to develop a software export industry. Indian software exports are growing by 25 percent per year, accounting for more than 40 percent of technology exports.[22] Using satellite links, highly educated engineers can work for foreign companies without leaving India. Indian software programmers transmit computer code to companies overseas from software technology parks in Bangalore and four other Indian centers. An integrated network service called SoftNET provides electronic mail, file transfer, and videoconferencing used by software developers.

"Telecommuting" refers to the ability to work at a distance, rather than travel to an employer's premises. Telecommuters may work from home, or from specially equipped centers set up close to where they live. In Japan, several companies have set up telework centers to enable some of their employees to work close to home, rather than commuting for two hours or more into Tokyo. Pilot projects in North America and Europe have also provided facilities for people to work from their homes or from regional centers rather than commuting to corporate offices.

Yet despite the availability of modems and fax machines, telecommuting in industrialized countries has been slow to take off, partially because employers are reluctant to give more autonomy to employees and to supervise them at a distance, and also because some employees feel that not being on-site at the workplace harms their chances of career advancement, and leaves them isolated. However, as traffic jams worsen and countries invoke stronger pollution controls, companies may have greater incentives to offer telecommuting options to their employees. Also, the growing number of small companies and individual entrepreneurs who work out of home offices (often starting new careers as a result of corporate layoffs or downsizing) is contributing to a greater popularity of telecommuting.

Rural Applications

Entrepreneurs in information-intensive businesses may now live virtually anywhere with good telecommunications. Architects, software programmers, writers, and consultants are moving to rural areas in the United States, Canada, Scotland, and Scandinavia. In Montana, entrepreneurs who have moved to the high country are called "modem cowboys." In Colorado, they are called "lone eagles."

In Scandinavia, "telecottages" have been established in rural communities to enable the residents to gain computer skills and do information-related work. The telecottage is typically equipped with personal computers, modems, telephones, and facsimile machines. A company that got its start through a telecottage now does telemarketing research from a small town in northern Sweden. Other telecottage activities may include bookkeeping, data entry, and word processing that can be done for distant clients. The telecottage model is spreading to other countries, including developing regions that have access to reliable telecommunications.

Access to information is becoming increasingly important to agriculture. In Washington State, farmers log into a computer bulletin board to find a market for their soybean crop. By monitoring global weather reports and crop data online, farmers can also determine what to plant and where and when to sell. The use of telecommunications in agriculture is not limited to industrialized countries. In Brazil, a rural cooperative used its newly installed telephone line to contact the commodity futures market in Chicago, in order to decide whether to hold or sell its coffee crop. In China, rural cooperatives use village telephones to contact suppliers and to take orders for their produce.

Reporting from the Field

Reporters are now able to cover events from the field by filing their stories directly by satellite. For example, a reporter covering the war in the former Yugoslavia in 1992 transmitted his story from a shoe-box-sized terminal via the Inmarsat satellite. Other reporters had to trek 12 hours to the nearest telephone. This service, called C-link by Comsat, provides store-and-forward messaging including text transmission, facsimile, and electronic mail. Suitcase-sized Inmarsat A terminals offer voice as well as high-speed data, facsimile, and telex. They are used at remote construction sites and oil rigs in regions of the world with poor or nonexistent communications.[23]

Portable satellite uplinks enable viewers around the world to witness natural disasters and calamities such as earthquakes, floods, and volcanic eruptions, famine in Africa, wars in the Middle East, and uprisings in Russia, China, and Eastern Europe. Telecommunications links have also shown the world the signing of peace accords for the

Middle East, the pledge to end apartheid in South Africa, and the beginnings of democracy in Russia and Eastern Europe. And coverage of international sporting events such as the World Cup and the Olympic Games has attracted global audiences now measured in billions.

Distance Education

Telecommunications technologies now make it possible for students to interact with teachers and other remote students from their home or workplace. Applications range from televised courses carried over broadcast or cable systems with students interacting via telephone or computer modem, to instruction and tutorials using computer conferencing, to fully interactive motion video linking several classrooms.

In the United States, five basic models have been developed to use telecommunications in education. The *curriculum-sharing* model links schools so that courses available at one school can be taught to students at another location. This approach typically connects classrooms in a local area or county using microwave and, now more commonly, fiber-optic links between the schools.

The second approach, the *outside expert* model, involves identifying course content that is not available in many rural schools, developing specialized instructional programming, and delivering the programs to the schools. These projects are typically regional or national in scope; many use satellites to transmit the courses to the schools and phone lines for interaction with students. For example, the TI-IN satellite network was founded to provide specialized high school courses to rural Texas schools when the state government mandated new curriculum requirements in mathematics, science, and foreign languages. TI-IN now offers courses via satellite to schools in 46 states.

The *consortium* model has been applied in higher education, so that several universities join together to deliver courses to remote students. The major example of this approach is the National Technological University (NTU), which delivers graduate technical courses via satellite to engineers at their workplaces throughout the country.

Employers requested that universities collaborate to establish the network so that their engineers could update and extend their knowledge without having to leave their jobs to attend courses on campus.

A fourth model is the *educational broker,* which delivers seminars and courses via satellite from a wide variety of sources. An example is NUTN, the National University Teleconferencing Network, based in Oklahoma. NUTN offers a wide range of adult and continuing education programs from many sources.

A fifth model is the *virtual classroom,* in which students and instructors communicate via the Internet. Instructors post assignments on the network; students turn in papers and receive feedback via electronic mail. Some courses are entirely electronic; others require students to attend some face-to-face classes. These virtual classes are designed for adults who can use computers at home or work or while traveling.

Universities in several countries use satellites to reach distant students. The University of the South Pacific, based in Fiji, offers tutorials via satellite to students studying correspondence courses in ten island nations and territories. The University of the West Indies also offers instruction to students at extension centers throughout the Caribbean using a combination of satellite and terrestrial audio links.[24] In China, employees can take courses at their workplace via satellite. And in Mexico, the Monterrey Institute of Technology and Higher Studies links 26 campuses using the Morelos domestic satellite.

Health Care

In Africa, often telecommunications facilities simply do not exist in rural areas, and, where available, are often both unreliable and expensive. A simple satellite network called Healthnet, which provides basic communications data, is operated by the nonprofit SatelLife organization. SatelLife was founded by the International Physicians for the Prevention of Nuclear War to reflect their belief that the greatest threat to our common humanity is the gap that exists between health conditions in the developing world and those in industrialized countries. The SatelLife satellite provides simple, inexpensive store-and-forward communications, using a single low earth-orbiting satellite that passes overhead once per day. The satellite is used by medical schools and researchers in Africa to send data from the field and access remote databases such as the National Library of Medicine in Bethesda, Maryland, and the electronic version of *The New England Journal of Medicine.*

Why did physicians feel compelled to raise funds for their own satellite? Because, despite modern technology, telecommunications facilities in the poorest regions were either unavailable or unaffordable. "The need in Africa for electronic mail not dependent on traditional communications infrastructure is desperate: In Zambia, international calls are billed at US$6 per minute. In Kenya, a fax costs $7.70 per page outgoing. In Tanzania . . . the minimal cost of a telex [is] a little more than US$25." [25] African researchers are able to use the satellite for free for the first three years, enabling them to gain access to medical libraries and other sources of expertise. [26]

In industrialized countries, the growing field of telemedicine is introducing transmission of x-rays and other medical images from rural clinics to medical specialists, electronic monitoring of patients from their homes, and computerized patient record systems that can be accessed by all health care and insurance providers. The results could be both reduction in costs, as information can be moved rather than patients, and improvement in care, especially in rural and impoverished areas which suffer a shortage of physicians.

OVERCOMING DISTANCE:
THE PARADOXES OF VIRTUALITY

Connection and Isolation

The telecommunications networks spanning the globe are overcoming the barriers of distance—between relatives in other countries, employee teams working in different time zones, researchers tackling the same problem on different campuses, people sharing common interests in political and social issues, hobbies, and recreation. These virtual communities are overcoming social as well as physical isolation. Yet, dependence on the electronic links to virtual communities may erode the contacts that form the basis of our personal communities in the neighborhood and the workplace. Telecommuters have to adjust to an environment without chats at coffee breaks and lunchtime, and without the visible presence and personal interactions that are often necessary to advance in a career.

Centralization and Decentralization

"Road warriors" equipped with notebook computers and cellular phones can work longer in the field, perhaps without needing any permanent office. Information workers, from software programmers to data entry clerks, can work across continents from their employers. And yet these technologies also allow more centralized control. Gone are the days when distance meant autonomy. Field staff can no longer escape monitoring by head office managers, and supervisors can monitor the keystrokes of distant data clerks.

Productivity and "Working Smarter"

As shown in many of the above examples, telecommunications and information technologies can help people to "work smarter," using information to increase output and improve quality. Yet overall productivity in industrialized countries where these tools are used most does not seem to have increased in the past decade. There have been productivity gains in the secondary information sector, where information is an intermediate input, but not the final output. For example, the automobile industry has increased productivity by using computerized design and manufacturing. However, the service sector and its information-intensive industries seem less successful. Part of the problem may be that we don't know how to measure the productivity of many kinds of information work. We can measure keystrokes entered or memos typed, but tangible results of these activities are harder to identify than cars assembled or packages delivered. Or perhaps we need to be patient. Economist Paul David points out that a quarter century elapsed before the electric motor transformed the workplace.[27] We may need to wait a little longer to witness the full potential of converging telecommunications and information technologies.

4

Structural Models of the Telecommunications Industry

"Infrastructure provides, if not the engine, then certainly the wheels of economic growth."

—PEKKA TARJANNE, Secretary General of the ITU[1]

THE NATURAL MONOPOLY MODEL

The early days of telecommunications were marked by amazing discoveries and inventions—of ways to send pulses and then words over wires, and then through the air—followed by disputes over patents and a scramble by entrepreneurs to set up communications companies and sign up customers. The results were often chaotic. Pictures of New York City at the turn of the century show dozens of wires strung on poles by competing telephone companies. Out of this apparent chaos emerged the sentiment that telecommunications should be considered a natural monopoly—that it made no economic sense to have more than one set of wires running down a street, or more than one provider of communication services. The analogy was to other public utilities such as water and electricity, where duplication of water mains and power lines seemed wasteful and inefficient.

The Public Monopoly Model

Both Europe and the United States adopted this model of natural monopoly, although their monopolies took very different forms. The
United Kingdom and continental Europe adopted the public monopoly model, with telecommunications run by the post office, as an extension of its mail services. The new departments became known as
PTTs, for Post, Telephone, and Telegraph. Putting telecommunications under a government department meant that there was no profit
incentive; indeed, what profits were generated were used to subsidize
the postal service. The relatively high rates and limited accessibility
perpetuated the view that telecommunications was a luxury rather
than a convenience and virtually a necessity.

Another consequence was the lack of independent government regulation. The implicit assumption of the PTT model appears to be:
"What is good for the PTT is good for the country." Stated another
way, since the PTT is part of the government, and the government exists to serve the needs of the public, the PTT must therefore be serving the public interest. An implied corollary is that there is no need
for participation by users, and besides, they wouldn't have the expertise to make any informed contribution.

Most developing countries have adopted the PTT model, usually
because they were once colonies of European powers that introduced
the PTT structure in their colonial governments. Communist countries such as the U.S.S.R., China, and their satellites also chose to
operate telecommunications through government ministries. International telecommunications revenues provide an important source of
foreign exchange for many developing countries. In many of these
countries, telecommunications profits not only subsidize the postal
service, but also contribute to the national treasury.

The Private Monopoly Model

In the United States, telecommunications services were provided by
private (i.e., nongovernment) monopolies, the largest being American

Telephone and Telegraph (AT&T), which grew out of the early Bell Telephone company and parts of Western Union. Smaller monopolies, which became known as "independent companies" (independent from the Bell System), provided services in their own territories. In the 1940s, when the Rural Electrification Administration began to offer low-cost loans to extend telephone service, rural cooperatives were established to provide telephone service in areas that were still without telephones. However, the overwhelming majority of telephone subscribers in the United States were served by private monopolies.

The profit motive of the private monopolies drove them to expand their networks, because more customers meant more revenue. More subscribers also increased the value of the network for everyone connected to it. Theodore Vail, president of AT&T in the early 20th century, recognized that what was good for AT&T could also be good for America. It was he, and not government policy makers, who first propounded the concept of "universal service," i.e., the goal of providing service to all via the Bell System. Yet the same profit motive unchecked led AT&T to drive other companies out of business so that it could take over their customers, and left the rural and other apparently unprofitable areas unserved.

In a competitive environment, the marketplace serves as referee—customers who are not satisfied with the price or quality of goods and services can choose another supplier. Policy makers in the United States decided that regulation was needed to replace the discipline of the marketplace. State public utilities commissions were established to regulate power and water utilities and local telephone companies. At the federal level, the Interstate Commerce Commission was the first to regulate interstate communications. The United States also turned to the courts to control the telephone monopoly. Antitrust laws passed early in the century to curb the abuses of the railroads and the petroleum industry have been repeatedly applied to telecommunications.

However, the complex and technical issues raised by the growing telecommunications industry needed more specialized and dedicated regulation. Chaos in the wireless industry, where radio stations were using any frequencies they liked to reach audiences and drown out competitors, added to the problems created by unbridled telephone monopolies. The Communications Act of 1934 established the Federal Communications Commission, which took over regulatory responsibility for interstate and international telecommunications (both of which are federal responsibilities under the Commerce Clause of the United States Constitution). This legislation set the ground rules for the telecommunications industry for more than 60 years; it was superseded by the Telecommunications Act of 1996 (see chapter 5). Public participation is an integral part of the American regulatory process; comments are solicited on proposed rule changes, and hearings provide opportunities for the public to participate through user and consumer organizations and individual testimony.

STEPS TO PRIVATIZATION

Changes in telecommunications technology and the growth in demand for services have driven many governments to take steps to privatize their telecommunications monopolies. The goals are typically twofold: to increase efficiency and to raise capital for new investment in telecommunications or sometimes for the national treasury. The first step toward creating incentives to improve efficiency and innovation in the telecommunications sector is usually to remove telecommunications from a government department and to create an autonomous government organization operated on business principles. The end goal in many cases may be privatization as part of a national strategy to turn government-operated enterprises over to the private sector, or simply as a means of freeing the sector to raise its own capital and to introduce management policies that are rarely tolerated in a public enterprise.

In countries with a PTT structure, there may be several steps on the path to privatization. An intermediate phase of autonomous public enterprise is preferred where national governments or powerful constituencies such as unions are opposed to selling off the telecommunications operator. National airlines and government-owned industries are often similarly structured. This model allows the telecommunications sector to be run on business principles, able to reinvest its own profits, and at arm's length from the government bureaucracy, without giving up government ownership. However, a government-owned corporation may still be subject to many political and bureaucratic constraints such as personnel and tariff policies. Some PTTs are required to hire a yearly quota of employees set by the government, and do not have the authority to set pay scales or performance incentives. They may also be required to maintain rate structures that include major subsidies.

In Europe, the United Kingdom was the first to adopt this restructuring strategy: BT was split off from the Post Office to become a Crown corporation before being privatized. Several developing countries have taken this intermediate step of corporatization. Others have gone a step further by inviting private investment in joint ventures with government-owned operators. European and American carriers are investing in telecommunications operations in Central and Eastern Europe, Asia, and Latin America.

At the far end of the continuum are countries that have fully privatized their public network carriers. Both the United Kingdom and New Zealand, for example, split telecommunications off from the post office, and then privatized the carriers, in BT's case by issuing shares, and in New Zealand by first allowing foreign investment by Ameritech and Bell Atlantic in Telecom New Zealand (TCNZ). In the developing world, several countries, including Chile, Mexico, Argentina, Venezuela, and Singapore, have all at least partially privatized their telecommunications systems.

INTRODUCING COMPETITION

Beginning in the late 1950s, technological change began to undermine the natural monopoly model, with key developments in customer premises equipment (CPE) (also known as terminal equipment), in microwave and satellite communications, and in computer communications. In the United States, despite the protests of

AT&T, other parts of the telecommunications industry and users argued that none of these new technologies nor the services they could deliver fit the natural monopoly model. Why should telephone sets attached to AT&T lines be manufactured only by its subsidiary, Western Electric? Couldn't the dangers of harmful interference that AT&T feared from what it called "foreign attachments" be overcome by setting standards that all manufacturers would have to meet? Eventually, the users and the new competitors prevailed; the result was the birth of the competitive terminal equipment industry, which now includes everything from answering machines and fax machines to modems and PBXs.

The computer industry had never been considered a natural monopoly that needed special regulation. Why, then, should computer communications over telephone lines be regulated? The Federal Communication Commission's decision to distinguish between basic services, which were to be regulated, and enhanced services involving some form of information processing, which were to be unregulated, stimulated the growth of the competitive value-added network (VAN) industry.

In the early 1960s, a start-up company called Microwave Communications Inc., now known simply as MCI, proposed to link business customers in Chicago and St. Louis by building its own microwave network and interconnecting to the local Bell network. Prolonged disputes between MCI and AT&T were resolved by regulatory and court decisions that led to the authorization of specialized common carriers and eventually to competition in public long distance and other services.

In Europe, these pressures built more slowly, but Europeans eventually had to confront the challenges of technological change. The European Union has been a driving force for liberalization of the telecommunications industry, recognizing that its economies need innovative and responsive telecommunications services to compete in global markets. Western European countries have introduced competition in various facilities and services, typically starting with terminal equipment and value-added services. Many European governments have tended to protect their telecommunications monopolies from competition, responding to the concerns of their telecommunications administrations rather than to the complaints of consumers and corporate users. By January 1998, members of the EU are to abolish all public network monopolies (with additional transition periods for countries with small or less developed networks).

Developing countries have also been hesitant to open the door to competition, fearing that newcomers would "cream skim" the most lucrative customers, leaving the government operator with the burden of serving the less profitable rural and low-income customers. However, even the most cautious now allow competition in terminal equipment and value-added services, and several have licensed multiple providers of paging and cellular services.

The United Kingdom and New Zealand are the most liberalized markets, with all of their telecommunications services open to competition. The United States has removed most barriers to local competition under the 1996 Telecommunications Act. Hong Kong is also among the most liberal jurisdictions; although Hong Kong Telecom has a monopoly over international telecommunications until the year 2007, Hong Kong has authorized competition in virtually everything else, including mobile, paging, and personal communications services, and local telephone service.

How Much Competition?

Table 4.1 shows liberalization trends around the world in the mid 1990s. The majority of countries in all regions had introduced competition in telephone sets. The Americas and Europe also had considerable competition in cellular service. International competition was not available in most countries, including all of the African countries that responded to the survey.

Table 4.2 summarizes the status of telecommunications competition in industrialized countries in 1994. All countries had introduced competition in customer premises equipment (CPE), and most had also liberalized value-added services such as X.25 data networks and

TABLE 4.1
LIBERALIZATION TRENDS*

Region	International Voice Competition			Cellular Competition			Telephone Set Competition		
	Yes	No	Partial	Yes	No	Partial	Yes	No	Partial
Africa	0%	100%	0%	19%	74%	7%	48%	33%	19%
Americas	20	80	0	60	36	4	84	12	4
Asia	10	80	10	35	60	5	65	25	10
Europe	4	88	8	40	40	20	92	8	0
Oceania	25	75	0	25	75	0	50	50	0
Former U.S.S.R.	0	75	25	50	50	0	50	25	25

* Based on responses from 105 countries.
Derived from ITU, *World Telecommunication Development Report*, 1994.

leased lines, and some mobile communication services, particularly paging. As noted above, competition was most open in the United Kingdom, New Zealand, and the United States (the United States moved toward opening competition in the local loop in 1996). Sweden and Japan are also shown as being very open to competition. However, in Japan, NTT is still the dominant domestic carrier, and smaller competitors find the regulations do not facilitate open entry (see chapter 5). In Sweden, and some other countries that have opened their markets to competition, there may actually be few new entrants.

Many countries, including the United States, have established analog cellular service as a duopoly. Australia has also licensed duopolies in local and toll services. However, customers generally find that duopolies offer little difference in quality or price after an initial competitive blitz to gain market share. Many countries are opening their digital cellular networks to competition, but face problems of allocating frequencies among carriers and finding sites for multiple cellular antennas.

The extent to which markets are actually open depends on the rules of interconnection and access. As noted above, interconnection with dominant carriers must not only be required, but technical issues, pricing of access, and unbundling of network components are critical to establishing truly open competition. Equal access for customers is

TABLE 4.2
STATUS OF FACILITIES COMPETITION IN OECD COUNTRIES 1994

Country	Public Switched Network			Data and Leased Lines		Mobile Communications			Terminal Equipment (CPE)
	Local	Trunk	Int'l	X.25	Leased Lines	Analog	Digital	Paging	
Australia	D	D	D	D	D	D	C	C	C
Austria	M	M	M	M	M	M	M	M	C
Belgium	M	M	1998	C	M	M	M	M	C
Canada	M	C	M	C	C	RD	D	C	C
Denmark	M	M	M	C	M	D	C	M	C
Finland	C	C	C	C	C	D	D	D	C
France	M	M	1998	C	M	D	D	D	C
Germany	M	M	1998	C	M	M	D	C	C
Greece	M	M	2003	1997	M	—	D	M	C
Iceland	M	M	M	M	M	M	M	M	C
Ireland	M	M	2003	C	M	M	M	M	C
Italy	M	M	2003	C	M	M	D	M	C
Japan	C	C	C	C	C	RD	C	C	C
Luxembourg	M	M	1998	C	M	M	M	M	C
Netherlands	M	M	1998	C	M	M	C	C	C
New Zealand	C	C	C	C	C	C	C	C	C
Norway	M	M	M	C	M	M	D	C	C
Portugal	M	M	1998	1997	M	M	D	C	C
Spain	M	M	2003	1997	M	M	M	C	C
Sweden	C	C	C	C	C	C	C	C	C
Switzerland	M	M	M	M	M	M	M	C	C
Turkey	M	M	M	M	M	M	M	M	C
U.K.	C	C	C	C	C	D	C	C	C
U.S.	PC	C	C	C	C	RD	C	C	C

C = Competitive

D = Duopoly

M = Monopoly

PC = Partially Competitive

RD = Regional Duopoly

199X, 200X = year competition is to be introduced

Derived from OECD, *Communications Outlook 1995,* and ITU, *World Telecommunication Development Report,* 1994.

also important. In many countries, customers have to dial several digits to make toll calls if they select a nondominant carrier.

BYPASS:
COMPETITIVE OPPORTUNITY OR THREAT?

What if telecommunications administrations are not responsive to customers or innovative in offering new services? Monopolies can create technical or financial bottlenecks that impede access to facilities and services. The monopoly utility model that was adopted to protect the public interest may now act as a constraint to retard growth of the telecommunications sector, and as a result, the economy as a whole.

If users become frustrated with trying to go through the bottleneck, they may decide to go around it instead. As noted in chapter 2, banks, retailers, and oil companies are among those using VSAT networks to link their operations rather than going through the public network. In industrialized countries, the choice is made primarily on financial grounds: the VSAT network is cheaper. In countries with less developed infrastructure, including those in Eastern Europe, reliability may be the key factor. In most cases, users would rather not have to develop expertise in telecommunications and set up their own networks; generally, they would rather deal with the carriers and leave the technical details to them. They are simply looking for affordable and reliable service. But out of frustration, and sometimes desperation, users have turned to setting up their own networks.

Some countries ban bypass, typically arguing that it is harmful to the sector as a whole because it syphons off the most lucrative business (often called "cream skimming"), forcing the carrier to raise its rates to cover the costs of serving other subscribers. Carriers may disparage bypass as at best unfair, and at worst a threat to their survival. Rather than seeing bypass as the users' revenge, carriers can help to make it unnecessary by offering services users want at prices that are both fair and affordable. The following are pricing options available from carriers in some countries that can make bypass less attractive:

- **Discounts:** Carriers may offer reduced rates for high volume or off-peak use. For example, many domestic carriers now offer discounted rates for calls on evenings and weekends when traffic volumes are relatively low. International carriers also discount calls during nonbusiness hours.
- **Reduced Rates for Spare Capacity:** Some carriers offer reduced rates to lease capacity that would otherwise be unused. For example, some telephone companies offer "dark fiber" rates for access to optical fiber capacity that would otherwise be unused ("dark"). Intelsat offers reduced rates for use of transponders on its backup satellites; this service is described as "preemptible," although it has never actually had to be preempted.
- **"Virtual" Networks/Volume Discounts:** "Virtual" networks use software to allow users to customize capacity on the public network for use as needed. The user controls the parameters as if it were a private dedicated network, but in fact shares public facilities with other customers.
- **Fractional Tariffs:** Some carriers in North America offer "fractional T1" tariffs for channel capacities up to T1 rates (1.544 megabits per second). Thus, a customer who needs only one quarter of a T1 (for example, for videoconferencing) would pay a discounted rate, rather than the full retail price for six 64 Kbps channels.
- **Distance Insensitivity:** Packet data networks typically charge by the volume of usage (number of packets) rather than the distance transmitted. VSAT networks also typically use distance-insensitive tariffs. Internet providers charge by hours of usage. Public networks could also adopt distance-insensitive tariffs rather than traditional distance-based pricing.

These options are not yet universally available even in industrialized countries. However, they show that carriers can be innovative in pricing as well as services in response to regulatory or competitive pressures.

THE FALLACY OF DEREGULATION

The changes in the telecommunications policy environment are often referred to as "deregulation." In fact, in virtually every country except the United States and Canada, the result of changing the telecommunications structure has been the need to introduce regulation where none existed. Under the PTT model, there is no separate regulatory function: decisions on frequency allocation, standards, and prices are typically made by the PTT or a related arm of the government.

In general, private corporations offer the greatest commercial benefits in terms of efficiency, competitiveness, and capital acquisition. They also are leaders in technology development and innovation in equipment and services. But, as the United States found early in the century, a monopoly operator needs some form of regulation to ensure that the single provider does not abuse monopoly power in setting rates or discriminating between customers. Where competition is adopted as a goal, liberalization of itself cannot assure competition absent structural and behavioral restraints on former, or "transitional" monopolists.[2] However, even a competitive environment is likely to require some oversight to ensure that service providers are competing fairly, and that their collective activity is serving the national interest.

Setting the Rules of the Game

Introducing competition in telecommunications requires much more preparation than simply opening the doors to anyone who wants to get into the business. Competition will be very limited if new entrants are forced to build entire end-to-end networks before getting into business. Instead, they will need to interconnect with existing networks, as MCI did when it built its microwave links between Chicago and St. Louis. Companies that start by building local networks such as wireless overlay networks in countries where local service is inadequate, or rural links where service was not previously offered, will also need to interconnect to the national networks to complete toll calls and reach existing subscribers.

Monopoly carriers may attempt to protect their monopoly by refusing to interconnect or by pricing access beyond the reach of new carriers. Interconnection rules are therefore critical to introducing competition successfully; these rules must specify both the technical requirements for interconnection and pricing guidelines. Otherwise, new entrants will be forced to offer inferior services, or will be unable to afford connection to the existing network. Interconnection rules are also necessary to ensure technical interoperability of networks, so that users do not have to worry about variations in standard quality of service depending on which technology is being used (for example, wire, wireless, coaxial cable, or optical fiber) or which carrier is providing the service.

Governments should also set policies that create incentives to use existing capacity rather than encouraging carriers to duplicate facilities. Such policies may be especially important in developing countries where there are many requirements for capital-intensive development projects, and telecommunications equipment imports may contribute to foreign exchange shortages. An effective strategy to encourage use of existing capacity is authorization of resale, so that new operators can lease excess capacity from existing carriers. For example, rather than duplicating a fiber backbone, they can lease capacity at wholesale rates from a carrier that has built the fiber links. If resale at discounted rates is not authorized, the new entrant is more likely to invest in building a duplicate network so that it can own and depreciate its assets.

Resale also spurs competition in industrialized countries by enabling new service providers to lease excess capacity at wholesale rates from existing carriers, and to offer discounts or special service packages to targeted customers. Customers will benefit from lower prices and customized services. The dominant carrier may lose some revenue initially, but it will then have an incentive to compete in both price and service quality to retain or win back subscribers. Resale also provides an economic incentive to use excess capacity on existing networks, such as optical fiber trunks, that otherwise would go unutilized. In the United States, most long distance companies, are actually resellers of capacity owned by the largest companies such as AT&T, MCI, and Sprint. Local competition in the United States is also beginning with resale. Rather than duplicating local networks, new local carriers can offer discounted or customized services by leasing capacity from the local exchange carriers.

Regulators can make the game fairer for new competitors by requiring number portability (the ability to keep the same telephone number when changing carriers). For example, since businesses list their toll-free numbers in advertisements and promotional materials, they are unlikely to switch carriers even if the price is lower, if it means changing their toll-free numbers. In the United States, toll-free (800 number) services must provide number portability, so that customers can change carriers without changing their 800 numbers. To foster local competition, U.S. regulators now require local carriers to provide number portability. These rules are now possible because of intelligence in the network; a computerized switching system (known as Signaling System 7) separates address data from the message of the call, and routes it to databases to look up the service providers to use for routing and billing the call.

Equal access is also important if markets are to be truly open. Equal access means that customers can access any carrier they choose without dialing extra digits. For example, in the United States, the FCC requires "1+" access for interstate long distance, so that callers can access the carrier they have preselected by dialing 1 plus the telephone number. A caller who has to dial five or more digits to reach an alternative carrier is unlikely to select that carrier.

Other Regulatory Functions

Other functions that a regulator will need to perform include:

- **Pricing:** Where competition exists, the marketplace may ensure that prices are reasonable. However, where one carrier dominates the market, or where demand far exceeds capacity, regulatory intervention may be necessary. The problem for users is that prices will be set too high, either to obtain excess profits or to ration scarce capacity. High prices may also conflict with national goals of providing affordable access to services.

However, oversight may also be needed to ensure that prices are not set too low! Pressure to keep rates low may leave a carrier without adequate resources to maintain and upgrade facilities. For example, some U.S. rural cooperatives face severe pressure from their members to maintain very low rates. They have found that it is important to have arm's length oversight to ensure that rates are not set too low.

- **Explicit Subsidies:** Internal cross-subsidies are often used to provide affordable services in less profitable areas. With the introduction of competition, it is important to make such subsidies explicit where they remain necessary, so that a dominant carrier cannot cross-subsidize competitive services with profits from monopoly services to drive out competition.

- **Quality of Service:** With privatization comes the potential for degradation of service in less profitable areas, as carriers strive to expand and upgrade the most profitable services. Rural Americans have expressed concern about the deterioration of rural telecommunications, as have consumers in Canada, the United Kingdom, and Australia. Monitoring of service quality is likely to be particularly important in developing countries, where one of the main objectives of restructuring the sector is to improve quality of service. It may also be possible to tie service quality standards to other incentives such as pricing flexibility.

- **Criteria for Access by Multiple Providers:** Where service competition is allowed, oversight is necessary to ensure that all have equal access to the network. Moreover, a provider of facilities must not gain an unfair advantage if it also offers services by restricting access to its network or charging inflated prices to other service providers who want to use it.

- **Standards:** Uniform standards are needed to ensure that equipment is compatible and of acceptable quality. An impartial standards agency can also ensure that a dominant carrier or supplier does not introduce standards that unreasonably discriminate against other vendors.

UNIVERSAL SERVICE

When countries introduce competition, they must grapple with the question of how to ensure that telecommunications services are accessible to the entire population. Typically, a monopoly, whether private or government-owned, uses internal cross-subsidies to keep the prices of some services lower than cost. Some governments, including several U.S. states, have mandated that local service should be very inexpensive, in order to make telephone service affordable for almost everyone. With competition, cross-subsidies must be eliminated so that former monopolists cannot price below cost to drive out new entrants.

One solution is to designate one operator as the "carrier of last resort" that is to provide service if no one else will do so. A "universal service fund" can be established to cover excessive costs of serving these subscribers. But the carrier of last resort may have no incentive to keep these costs as low as possible. However, an obligation without any cost-sharing mechanism may squeeze the former monopolist excessively. For example, NTT in Japan argues that the burden of universal service must be shared with the NCCs which have taken 40 percent of its long distance market in some areas.

REDEFINING UNIVERSAL SERVICE

To set telecommunications policies that will foster socioeconomic development, policy makers must recognize that information—access, sharing, and dissemination—will contribute to achieving their country's development goals. These general development goals must be translated into specific telecommunications goals.

In industrialized countries, the goal of universal service to telecommunications has come to mean provision of basic telephone service to every household. However, as more advanced services become widely available, we need to broaden the definition to encompass access to services that may not be provided in every household but are widely accessible. Thus, there could be a multileveled definition of access, identifying requirements for households, for social service institutions, and for communities. For example:

- **Level One:** household access;
- **Level Two:** institutional access (for example, schools, hospitals, and clinics);
- **Level Three:** community access (for example, libraries, post offices, community centers).

In industrialized countries, connections to the Internet could be required for public libraries to serve the needs of community residents, while Internet access and other services such as videoconferencing could be available for schools and health centers. For example, the U.S. Telecommunications Act of 1996 states that schools, health care providers, and libraries should have access to advanced telecommunications services.[3]

Developing country planners would need to modify these goals, but may also find that the community or the institution is the appropriate unit of analysis for providing universal access to telecommunications services. Targets might include:

- **Universal access to basic communications:** Access may be defined using a variety of criteria such as:
  - population: e.g., a telephone for every permanent settlement with a certain minimum population;
  - distance: e.g., a telephone within a certain number of kilometers of all rural residents;
  - time: e.g., a telephone within an hour's walk or bicycle ride of all rural residents.
- **Reliability:** Standards for reliable operation and availability; quality sufficient for voice, facsimile, and data communications.
- **Emergency Services:** A simple way to summon help immediately, so that anyone, including children and illiterate adults, would be able to get assistance in an emergency such as an accident or illness, fire or natural disaster.
- **Pricing:** Pricing based on communities of interest, that is, where people have the greatest need to communicate; for exam-

ple, to regional centers where stores and government offices are located; to other places where most relatives are located (surrounding villages, regional towns, etc.).

The advances in communications technologies discussed above now make it possible to provide affordable access to telecommunications and information services in rural and remote areas comparable to what is available in urban areas. In fact, the U.S. Telecommunications Act of 1996 states that consumers in all regions of the U.S. "including low-income consumers and those living in rural, insular, and high cost areas, should have access to telecommunications and information services . . . that are *reasonably comparable* to those services provided in urban areas and that are available at rates that are *reasonably comparable* to rates charged for similar services in urban areas."[4] [Italics added.]

While planners may want to modify this goal for lower-income countries, there is no longer a compelling technological or financial reason to limit rural services. The same technologies that are used to transmit voice can also transmit facsimile and data, and, through digital compression, video as well. Access criteria in rural areas may actually be comparable to access criteria in high-density urban areas, where the goal is not to provide a line for every dwelling, but access for everyone through public phones in kiosks, shops, and common areas.

It is important to note that these goals are in effect "moving targets": they do not specify a particular technology, but assume that as facilities and services become widely available in urban areas, they should also be extended to rural areas. Indeed, the technologies used to deliver the services in rural areas may differ from those installed in urban areas; for example, satellite links and radio networks may be less costly for rural communications than optical fiber or even copper wire.

Four fundamental criteria are critical in implementing this goal of universality:

- Accessibility: The widest range of telecommunications facilities and services should be available throughout the country, and everyone should have access to basic services.

- Equity: There must not be major disparities in availability and price of telecommunications technologies and services. That is, in addition to maintaining universal access to basic services (however they are to be defined), policies are needed to ensure that people are not penalized because of where they live or what companies offer services to them. For example, information services need to be available in rural as well as metropolitan areas, in inner cities as well as suburbs. And rates for access to these services should not vary significantly throughout the country even if they are provided by different companies or using different technologies.
- Connectivity: In an era of new technologies and competing providers, there must be universal connectivity, so that people can communicate with each other and with information sources regardless of who provides their services or what technology links them to networks.
- Flexibility: Policies must recognize that changing technologies and the introduction of new services mean that we will have to be flexible in setting targets and adjusting to change.

Monitoring Progress

It is likely that the marketplace will be the best mechanism for bringing innovative and affordable services to most rural regions in industrialized countries and to many areas of developing regions. However, policy makers in all countries will need to monitor progress to determine whether there are disparities in access, quality of services, or pricing that need to be addressed. The following are proposed indicators for a "National Universal Service Report Card." The intent is to develop some simple, ongoing way to measure progress, especially when staff and funding resources for extensive data collection are limited.

Proposed indicators to monitor for the report card include:

Availability of Service

- national teledensity: While imperfect, these data are routinely available from telecommunications administrations. To provide a better estimate of urban/rural access without thrashing out a universal definition of rural (rural India and China are very different from rural Papua New Guinea or Mongolia), these data could be disaggregated to show:
  - teledensity in cities over 500,000 population (or largest city where there is only one major urban center in the country);
  - teledensity in the rest of the country (while this figure will overestimate rural teledensity because it will include major towns, its advantage is that data are likely to be available to make this calculation).[5]

Quality of Service

- average length of time to obtain service: urban and nonurban;
- average time to repair service: urban and nonurban;
- percentage of lines connected to digital switches;
- percentage of lines with direct dial service (subscriber trunk dialing): national and international;
- percentage of multiparty lines: urban and nonurban.

Price

Here, the relevant variable is not absolute price for access but price relative to the income of the users. For example, if a line costs more to install than a family's annual income, telecommunications cannot really be considered accessible.

- price of installation: as percentage of annual average per capita income;

- monthly connection charge: as percentage of monthly average per capita income;
- price of 3-minute 100-km domestic daytime call: as percentage of monthly average per capita income;
- price of 3-minute 500-km domestic daytime call: as percentage of monthly average per capita income.

Internet Access

- number of Internet gateways: per million population;
- percentage of universities with Internet connection;
- percentage of secondary schools with Internet connection.

Mobile Communications

- percentage of land area covered by mobile services;
- percentage of population in areas covered by mobile services.

National governments would set targets for each of these benchmarks, perhaps following guidelines adopted by the ITU or other organizations for countries at various stages of development. These data could be collected and published annually by the ITU and/or by regional organizations. The data could be useful for numerous purposes, such as:

- for the countries themselves, to determine how they rate compared to other countries in their region or with countries having similar economic profiles;
- for use in setting national goals for telecommunications access;
- for use in monitoring the effects of restructuring of their telecommunications sectors, for example: what changes have there been in availability of service? pricing of service? availability of new services?;

- for investors such as financial institutions, entrepreneurs, and telecommunications companies to determine where there are opportunities to invest in provision of facilities or services;
- for international development organizations such as the World Bank, regional development banks, and bilateral aid agencies, to prioritize needs for financial and technical assistance.

The Telecommunications Sector in Industrialized Countries

CHAPTER *5*

Restructuring the Telecommunications Sector: the United States, the United Kingdom, and Japan

"Never before has the interaction between technological forces and societal systems been so critical to the shape of the communications infrastructure."
—ROBERT LUCKY, Corporate Vice President, Bellcore [1]

INTRODUCTION

In this chapter, we examine the structure of the telecommunications sector in three major industrialized countries: the United States, the United Kingdom, and Japan. These countries are chosen not only because they are major economic powers, but because they have taken different paths in restructuring their telecommunications sectors. In the following chapters in this section, we examine the evolving telecommunications policies of the European Union and of selected industrialized countries including France, Germany, Canada, Australia, and New Zealand.

THE UNITED STATES

With a population of more than 260 million and a telephone density of 56.5 access lines per 100 people, the United States is the world's largest single market for telecommunications equipment and services. Its history of technological and regulatory innovation, and its evolution from domination by a private monopoly to a pluralistic and increasingly competitive environment, make the United States telecommunications sector both a compelling and enigmatic model.

Policy and Regulation

As noted in chapter 4, the United States introduced government regulation to ensure that private monopoly carriers served the public interest by providing reliable services at affordable prices. Interstate and international communications are constitutionally federal responsibilities, and are under the jurisdiction of the Federal Communications Commission (FCC), established by the Communications Act of 1934. The FCC is responsible for licensing of international carriers and national communications systems, for allocation of frequencies, and may also set national standards where necessary.

Another federal agency with responsibility for telecommunications is the National Telecommunications and Information Administration (NTIA) in the Department of Commerce. NTIA advises on national and international telecommunications policy, and is responsible for the federal government's civilian telecommunications requirements. NTIA has taken an active role in promoting a national information infrastructure (NII), with the goal of linking all Americans through high-capacity "electronic superhighways."

Congress also plays a central role in setting telecommunications policies through enacting legislation, which is implemented by the FCC and enforced by the courts. In 1996, Congress passed the new Telecommunications Act, the first major overhaul of telecommunications legislation since the Communications Act of 1934. In late

1992, Congress passed the Cable Reregulation Act, which reintroduced regulation of the cable television industry, after evidence of widespread significant increases in rates since the industry had been deregulated in 1984. (President Bush's veto of the bill was overridden by Congress; it was the only presidential veto that was not upheld during his administration.)

The courts can also play a major role in U.S. telecommunications policy. The antitrust laws have been invoked several times to prevent alleged abuse of monopoly power by AT&T, finally resulting in the Modified Final Judgment of 1982 that broke up AT&T. The federal courts also review decisions of the FCC under appeal. In the United States, the federal courts have played a major role in upholding or overturning many of the key decisions that led to the introduction of competition. In many other countries, telecommunications regulations are subject to political review by appeal to the minister of communications or the cabinet rather than to the courts.

Intrastate commerce is the responsibility of the states, which typically regulate telecommunications through state public utilities commissions (PUCs), agencies that are generally also responsible for other utilities such as water, gas, and electricity, and transportation in some cases. This dual federal and state authority makes it difficult for the United States to establish a cohesive national telecommunications policy. The FCC has preempted several state regulations to allow more competition in specific telecommunications markets, although in 1986, the Supreme Court ruled that the FCC could not preempt state regulations merely because they conflict with federal telecommunications policy. However, the Telecommunications Act of 1996 allows the FCC to preempt enforcement of state or local policies that restrict competition.

Technological Challenges: The Origins of Competition

As discussed in chapter 4, the U.S. telecommunications industry evolved differently from the sector in most other countries. From its inception, U.S. telecommunications was privately owned, whereas

European governments operated telecommunications, usually in association with the post office. The U.S. system evolved into private monopolies, the largest of which was AT&T, also known as the Bell System.

However, the monopolies began to crack in the 1950s, with a series of regulatory decisions and court cases that ushered in the era of competition. The first case, in 1956, concerned Hush-A-Phone, a device that fit over the telephone receiver for user privacy. AT&T's tariff did not permit any "foreign attachments" to its network, including the Hush-A-Phone. The appeals court ruled that AT&T's tariff restriction was "an unwarranted interference with the telephone subscriber's right reasonably to use his telephone in ways that are privately beneficial without being publicly detrimental." [2]

In the late 1960s, a company called Carterfone developed an acoustically coupled device that allowed a mobile telephone user to access the public telephone network through the mobile system's base station. Again, AT&T refused to allow the device to be connected, citing potentially harmful interference to the network. Relying on the principles of Hush-A-Phone, and finding no technical harm, the FCC found AT&T's prohibition of Carterfone and other interconnecting devices unlawful. [3] Thus was born the terminal equipment or interconnect industry, i.e., customer-provided terminal equipment that could be interconnected to the public network. To protect the network from potential harm, the FCC established a registration program to certify that equipment met standards that would prevent technical damage.

The evolution of transmission technologies also brought challenges to AT&T's monopoly. In 1968, MCI won the right to offer private long distance service between Chicago and St. Louis over its own microwave network, but with interconnection to the public network at each end. [4] This was the first step to competition in long distance markets, although at the time, only private line services were authorized. As with customer premises equipment, however, the key regulatory decision was to allow interconnection with the public network.

Another new communications technology in the 1960s was the communication satellite. AT&T saw little need for satellites, as it had built a nationwide microwave network capable of carrying television signals as well as telephone traffic. However, other companies that wanted to provide transmission services argued that satellite communications should not be considered part of the AT&T monopoly. The heated debate over the potential of the technology and how it should be organized lasted until the early 1970s, although during that time the Intelsat system, initiated and largely financed by the United States, began international operations (see chapter 14). Finally, in 1973, the United States adopted the policy that became known as "Open Skies": satellite communications was to be an open and competitive industry. By the end of the decade, several domestic satellite systems were delivering cable television programming, and a domestic satellite was used for telephone service in Alaska. In the early 1980s, VSATs (very small aperture terminals) were invented, so that users could install their own private data and voice networks, effectively bypassing the entire public communications network.[5]

Yet another technological innovation that challenged the natural monopoly theory was computer communications. In the 1960s, both AT&T and IBM contemplated entering this new field. Computers had never been considered a natural monopoly, although there were few producers of mainframes, of which IBM was by far the largest. The FCC at first attempted to segregate communications services, which were to be regulated, from data processing services, which were not. In its second Computer Inquiry, in 1980, the FCC divided telecommunications services into "basic" services that were to be regulated, and "enhanced" services that were excluded from regulation.[6] This decision gave birth to the value-added services (VAS) industry, including packet data networks, database services, and other services that processed the information just passing through the network, rather than simply transporting it.

However, as computers and telecommunications continued to converge, with more computer processing built into network functions, the distinction between basic and enhanced services became increasingly blurred. For example, the storage and retrieval of messages by the telephone companies was considered an enhanced service. The FCC's third Computer Inquiry addressed ways to prevent harmful discrimination by the carriers, by preventing other enhanced services providers from accessing their networks, or by cross-subsidizing their own enhanced services in order to drive competitors out of business. The FCC required that the carriers separate their enhanced from their basic services business (known as structural separations), and that they introduce plans for Open Network Architecture (ONA), so that information on configuration of the telephone companies' networks would be available to unaffiliated companies that could choose to interconnect at any level in the networks' hierarchy.* A parallel requirement was unbundling of services, so that service providers could pay only for the parts of the network they wanted to use.[7]

Divestiture: The Breakup of AT&T

In addition to technological change, another theme influencing the U.S. telecommunications industry is the application of the antitrust laws. Several times in this century, AT&T was accused of abusing its monopoly power; principally by refusing to interconnect with other carriers (with small local companies in the early decades of the century, and with MCI in the 1960s) or by predatory pricing, that is, pricing below cost to drive others out of the market. AT&T used this strategy to bankrupt and then buy out smaller telephone companies before 1912, and to undercut MCI's attempts to establish competitive, private line long distance services in the 1960s.

In 1911, 1956, and 1982, AT&T was the subject of antitrust challenges. In 1911, under the Kingsbury Commitment, AT&T agreed to refrain from abusive practices such as predatory pricing and refusal to interconnect with other companies in return for subjecting itself to regulation, thus protecting its monopoly status. In 1956, AT&T agreed to stay out of the computer business (where it could have challenged IBM), again in return for retaining its telecommunications monopoly. However, in 1982, AT&T agreed to divest itself of its local

---

* A similar policy in Europe is called Open Network Provisioning (ONP). See chapter 6.

operating companies in return for the opportunity to enter competitive businesses.

This agreement became the most significant single event in U.S. telecommunications history: the breakup of American Telephone and Telegraph (AT&T). Under the Modified Final Judgment (MFJ) implemented in January 1984 (legally modifying the 1956 Consent Decree), AT&T divested itself of its Bell Operating Companies (BOCs), which were grouped into seven Regional Bell Operating Companies (RBOCs) or holding companies, namely: Ameritech, Bell Atlantic, BellSouth, NYNEX, Pacific Telesis, Southwestern Bell Corp. (now SBC), and US WEST (see Figure 5.1).*

Restrictions placed upon the BOCs were intended "to prevent them from engaging in any nonmonopoly business so as to eliminate the possibility that they might use their control over competitors in such businesses." Thus, the BOCs were not permitted to:

- manufacture telecommunications products and customer premises equipment;
- provide interexchange services, which were to be provided by interexchange carriers (IXCs) such as AT&T, MCI, and Sprint;
- provide information services;
- provide any other product or service that is not a "natural monopoly regulated by tariff."

In addition, they were to provide services to interexchange carriers equal in type, quality, and price to the services provided to AT&T and its affiliates.

The RBOCs have diversified into real estate, publishing, office automation, and software development. They have also extended their activities overseas, primarily through acquisitions and joint ventures. The ban on the provision of information services was lifted in July 1991. In 1993, the RBOCs were granted a generic waiver of the MFJ line of business restrictions in order to aid their efforts to compete in

---

* In 1996, Bell Atlantic and NYNEX announced plans to merge, and SBC acquired Pacific Telesis.

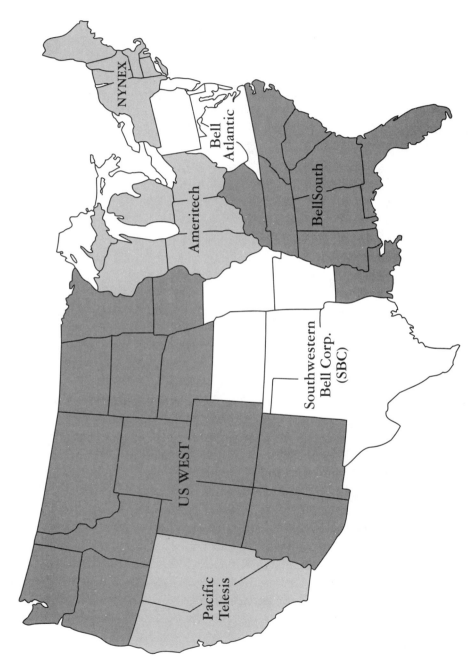

FIGURE 5.1 The territories of the Bell Operating Companies after the breakup of AT&T.

foreign markets. They were authorized to provide or own the foreign half of telecommunications services between the United States and a foreign country, and allowed to own through a foreign telecommunications entity, up to 10 percent of submarine cables and international satellite systems that are used to carry traffic between the United States and the foreign country. Thus, the RBOCs would not have to worry about the uncertainty of MFJ restrictions associated with entering foreign deals.[8]

There appear to be several reasons why the RBOCs entered into international ventures. First, they were limited to providing local services in the United States, where the market is virtually saturated and local services are still regulated in most states. Second, they were able to gain international experience in services that they would like to offer in the United States, if restrictions were lifted, such as cable television. In other cases, their strategy has been to compete internationally in ventures that build on their domestic expertise, such as management of telephone companies, network operations, and building and operating cellular networks.

Legislative Reform: The Telecommunications Act of 1996

The Telecommunications Act of 1996 was the first major overhaul of the federal communications legislation in more than 60 years. It does not completely replace the 1934 Act, but it does supersede the Modification of Final Judgment that broke up the Bell System and placed restrictions on the BOCs, as well as other major court decrees that kept GTE's operating companies from providing long distance services and set the terms for the acquisition of McCaw Cellular by AT&T.

The key elements of the 1996 act include:

- Opening local service to competition: the act seeks to eliminate legal impediments that have kept long distance carriers, cable television operators, and others from offering local telephone service;

- Removing MFJ restrictions: the Bell companies may now provide long distance services once they have met FCC conditions designed to ensure that their local markets have become competitive; they may manufacture equipment after entering the long distance market; they may engage in electronic publishing under a separate affiliate;
- Letting local phone companies distribute video: local exchange carriers (LECs) may operate "open video systems" that do not discriminate among video programmers or require discriminatory rates or conditions;
- Redefining universal service: universal service is considered to be an evolving rather than static concept that is to be periodically redefined; it includes not only households, but access to advanced service for for schools, libraries, and health care facilities.

The 1996 act thus opens the door to local competition and to competition in converging services: telephone companies will be able to distribute video, and cable companies will be able to offer local telephone service. However, while the act removes many restrictions on the industry, it does not eliminate regulation. In fact, perhaps its most significant requirement is that the FCC formulate the rules of the new game, such as a competitive checklist for opening local markets, standards for interconnection, and definitions of universal service.

Mobile Communications

In the 1980s, the FCC adopted a duopoly model for cellular communications, and divided the United States and its possessions into 734 cellular markets. The available spectrum was divided into two frequency blocks in each market, one available for local wireline carrier companies, and the other for nonwireline companies, such as radio common carriers (RCCs). The FCC has licensed approximately 1,500 cellular systems, which are held by about 150 entities.

Initially, the FCC used comparative hearings to award licenses in the top 30 markets, and lotteries to process applications where there was more than one applicant for other markets. In 1993, Congress required the FCC to modify the mechanism used to grant licenses for commercial mobile radio services (CMRS) from lotteries to auctions. In 1994, the FCC adopted a comprehensive framework for the regulation of all mobile services, including PCS. Frequencies and service areas have been established for both broadband and narrowband PCS. Frequencies for these services were auctioned. To foster rapid deployment of PCS services, the FCC required that the licensees offer services to at least one-third of the population in each market area within five years of being licensed, two-thirds within seven years, and 90 percent within ten years.

The FCC has also awarded licenses for low earth-orbiting satellites (LEOs). So-called "Little LEOs" are to offer nonvoice services such as vehicle tracking, environmental monitoring, and two-way data communications. "Big LEOs" will provide cellularlike mobile services to any location, and must offer continuous voice services throughout the United States, and may offer global services, subject to regulatory requirements of countries where they intend to operate. Five "Big LEO" systems have been licensed (these are discussed further in chapter 15.)

National Information Infrastructure

The Clinton/Gore administration has put forth a vision for building a 21st century infrastructure to meet the needs of all Americans. This national information infrastructure (NII) is to include a broadband interactive network that is to be private-sector owned and operated, and is to link communities, schools, and health care facilities, as well as businesses and homes.

The NII initiative has elements of industrial policy: its primary goal is "creation of a competitive, vibrant marketplace which will boost economic growth and job creation." It also proposes to provide access to information services to all Americans in order to improve education and medical services, and the general quality of life. The inititative consists of two major components: a technology and pilot demonstration program, and a telecommunications policy reform program, much of which was realized in the Telecommunications Act of 1996. Information infrastructure issues are discussed further in chapter 18.

THE UNITED KINGDOM

Privatization and Liberalization

The United Kingdom, an initiator of the government monopoly model of telecommunications, now has one of the most liberalized telecommunications industries in the world; it is currently the only country in the European Union with full competition in the provision of basic network as well as other services. With unlimited competition in value-added services, the United Kingdom accounts for 70 percent of the European VANs market and 50 percent of its EDI market.

The British telecommunications system was originally operated by the post office. In 1981, the United Kingdom began the move toward privatization and liberalization by separating British Telecom from the post office. To introduce competition, in 1982, the United Kingdom established a duopoly by licensing one competitor, Mercury Communications, a subsidiary of Cable & Wireless, to operate fixed link networks. In 1984, the British government passed legislation authorizing the privatization of BT, and slightly more than 50 percent of the company was sold through a public stock offering. The privatization of BT was the first initiative in the Thatcher government's plan to privatize government-run industries.

In 1983, the British government announced that it would review its duopoly policy, granting licenses to BT and Mercury for seven years, and would not issue any other licenses for public domestic and international services during that time. Mercury was to be given time to establish itself without threat of further entry, in return for which it undertook certain commitments concerning the development of its network. Other sectors of telecommunications, including customer premises equipment (CPE) and value-added services (VAS), were opened to competition. Cable television companies were also allowed to compete in the provision of local service.

Regulation

The British Parliament established a regulatory body, the Office of Telecommunications, known as Oftel, to oversee the restructured industry. The regulatory responsibility is in the hands of the director general of Oftel, not a commission, as in some other countries. The director general can take action to enforce existing license rules, and can also initiate and conclude procedures for amending licenses. If the director general wishes to amend a license and the licensee does not agree, Oftel can proceed only if it obtains the agreement of the Monopolies and Mergers Commission. Judicial review is also available; decisions of the director general can be challenged in the courts.

The vesting of authority in one person, despite the opportunities for review, may have some inherent weaknesses. For example, this approach leaves a wide opening for gentlemen's agreements without public knowledge or consultation. Other countries with regulatory agencies, such as the United States, Canada, and Australia, have several commissioners and statutory requirements for public input. In contrast, when asked by the author whether Oftel would hold public hearings to obtain input from consumers, the first director general of Oftel responded: "If I think we need hearings, we'll have hearings."

An innovative approach to pricing initiated by Oftel is known as price caps: the goal is for the regulator to give incentives to efficient operation and to mimic the effect of the marketplace. Price control is based on financial modeling: the carrier can earn a reasonable rate of return if relatively efficient; if it earns more, it can keep the extra profit. When BT was first privatized, Oftel's rules stated that prices must not increase by more than 3 percent below the rate of inflation for direct dialed calls in the United Kingdom and for exchange line rentals. This rule was revised in 1989 to cover operator assisted calls and connection, and tightened to 4.5 percent below inflation. Regulations have since been revised to include international control, with rate ceilings 6.25 percent below inflation.[9]

Oftel attempts to monitor quality of service, but there are no formal sanctions. BT publishes quality of service indicators every six months, although there is no legal requirement to do so. BT has also instituted some incentives to respond to customer demand. For example, the company gives a firm date for commencement of service and aims to meet it within two working days. If there is a delay, BT pays a penalty only if business customers can demonstrate financial loss.

Initial Results

Mercury did capture a significant share of the business market, but penetration of the domestic sector was minimal during its first decade of operations. BT remains dominant in the residential market but has lost market share among business customers. Mercury lost many of its customers in the City of London when new competitors moved into the market after 1990. In the 1980s, its profit growth was largely due to increased traffic, rather than cost cutting or internal efficiencies.[10] BT has since reduced charges for business and long distance calls, but its local charges remain among the most expensive in Europe. The United Kingdom lags behind a number of other Western European countries in ISDN access; it has fewer than half the subscribers in France and less than one-eighth the number in Germany.

Some 130 local cable television franchises have been licensed, covering two-thirds of the population. However, not all of these systems have been constructed. Only 4.5 percent of television households subscribed to cable in 1994. Satellite broadcasting is more popular, with about 3.4 million subscribers, or one-sixth of television households.[11]

Two cellular mobile operators began operations in 1985; there are now four cellular operators sharing more than 3.7 million customers. Three mobile PCN operators were licensed, but it appears that, at most, two systems will be built. The local market remains uncertain. Some think that wireless services may be more valuable than cable for local competition. However, the construction of a PCN network is a

massive undertaking, requiring about £4 billion to set up a PCN
mass-market digital system.

One of the apparent failures has been Telepoint, a system that allows outgoing mobile calls only, and was billed as being cheaper than
cellular service. It was trumpeted as a mass-market mobile system,
but had only 10,000 subscribers by 1991. One of the possible selling
points for Telepoint in the United Kingdom was the very limited
number of pay phones, and the relatively high percentage that were
out of service. However, Telepoint has turned out to be "full of sound
and fury, signifying nothing but financial loss." [12] BT was the last of
four original licensees to suspend its Phonepoint operation. Part of
the problem may have been that the various Telepoint systems were
not compatible; also, there was apparently some marketing confusion
between Telepoint and cellular services.

The Duopoly Review

In 1990, the government issued a Consultative Document entitled
"Competition and Choice: Telecommunications Policy for the
1990s," in which it stated: "We are moving from a period of managed
competition towards one of more open competition." [13] The secretary
of industry published the conclusions of the review in a White Paper
with the same title in March 1991. The White Paper appeared to be
more of a guide than a rulebook. It addressed issues such as how much
competitors should pay BT to carry part of their calls, and how much
they should contribute to local exchange costs. Oftel retained discre-
tionary powers including the right to intervene to maintain a balance
between competition and economies of scale. [14]

The government's White Paper marked the end of a decade of transition from state monopoly to one of the most liberalized markets in
the world: "The new measures seek to strike a balance between restrictions on competition . . . encouragement of competition . . . and
avoidance of excessive help for new entrants. . . ." [15] The major result
of the review was to liberalize entry by ending the duopoly policy
and considering on its merits any application for a license to offer

telecommunications services over fixed links within the United Kingdom.

Cable companies are now able to provide telephone service in their own right, and to interconnect with one another. The government has maintained restrictions on carriage of signals until 2001, with possibility for reconsideration in 1998. Telecommunications carriers cannot carry TV signals, nor market entertainment services. "It was widely perceived at the time of the review that allowing BT or Mercury to carry entertainment signals would discourage cable licensees from building their systems and thus eliminate a . . . source of potential competition in telecommunications."[16] However, they were able to bid, starting in 1994, for local cable licenses for any part of the country not then covered by a cable license.

Unlimited self-provision of circuits by individual users has also been allowed by general license. Some 60 organizations had applied for licenses to operate fixed networks or indicated interest in doing so in the first two years. The greatest interest was in new fixed radio networks and optical fiber-based networks, making use of existing utility rights of way. One of the new competitors is British Waterways, which has linked up with Sprint to provide competing long distance service, and is laying fiber along canals. Fixed operators cannot provide mobile or Telepoint services under their main licenses, whereas mobile operators can apply to provide fixed services using their radio networks. The logic is that new entrants—cable or mobile operators—need protection to get established.

Greater flexibility in pricing was also allowed, subject to a floor limit. Alternative tariffs must be available to resellers, and low users must have access to a special scheme offering low usage discounts. Oftel concluded that BT would need to adopt a transparent transfer price approach to the provision of services to competitors, who would also be, in effect, major customers. Therefore, accounting separation was to be implemented for the different parts of BT's business at the point of interconnection, to achieve a necessary degree of transparency and ensure nondiscrimination.

"The success of this drive for increased competition through entry of many more new competitors will depend on the effective reduction of barriers to entry into a market still strongly dominated by BT.

Arrangements for interconnection between existing and new networks are the most important element in helping to ensure that new competitors are able to enter the market." [17] Concerning interconnection charges, BT argued that the cost of providing access to the network was increasing. It is allowed to increase its quarterly rental and connection charges only by a maximum of 2 percent above the rate of inflation. The government proposed that carriers interconnecting with a fixed link operator be required to make an access deficit contribution. The government reserves flexibility in enforcing the payment, and can do so through the use of waivers for operators who have only a small share of the market (such as less than 10 percent). [18]

"A paradox of the government's proposals is that liberalisation will be accompanied by increased regulation." [19] Oftel has added staff; but it will face problems not only of effectiveness, but of responding quickly in a fast changing environment. The early days involved issues affecting the duopoly; now it will have new PCN, cable, and other carriers and services to monitor. Other barriers to entry that are receiving priority attention from Oftel include numbering, particularly number portability, so that customers can change networks without changing numbers, equal access, and availability of directory information on an equal basis. As former Director General Bryan Carsberg has pointed out, the government promised liberalization, not deregulation.

JAPAN

In 1985, Japan became the third country to liberalize its telecommunications, after the United States and the United Kingdom.

The Development of Telecommunications in Japan

International telecommunications began in Japan in 1871 when the Great Northern Telegraph Company (GNTC) of Denmark connected Nagasaki to Shanghai and Vladivostok via two submarine cables.

GNTC had monopoly rights over telecommunications with the outside world; even the government had to get permission to use its facilities. In 1906, the government laid cable from Tokyo to the Ogasawara islands and connected to the American transpacific submarine cable, but Japan still relied on GNTC to reach the outside world. In 1935, Japan had to pay GNTC $1.6 million, or 20 percent of its total balance of trade surplus of $7 million.[20]

After World War I, Japan decided to allow private companies to supply the equipment needed to set up a government-operated radio-telecommunications network. This private/public approach enabled the government to establish a network and to claim scarce frequencies, without having to put up the capital to build a network. When GNTC left Japan in the late 1930s, Japan at last had the autonomous right to operate international telecommunications nearly 70 years after service began.

Until 1952, telecommunications services were run by the Ministry of Communications. In 1952, Nippon Telegraph and Telephone Public Corporation (NTT) was established for domestic communications, and in 1953, Kokusai Denshin Denwa Company Limited (KDD) was established for international communications. NTT was a public corporation, but KDD, the international carrier, was made completely private because it was feared that the overwhelming size of the domestic market would lead to second-class status for international services, and because of urgent demand by trade industries for international telecommunications.[21]

Also in 1952, the Ministry of Posts and Telecommunications (MPT) was established, with responsibility for telecommunications policy and regulatory matters, and the improvement and coordination of telecommunications. Another important ministry involved in Japanese telecommunications policy is the Ministry of International Trade and Industry (MITI), which is responsible for the promotion of utilization of computers, improvement and coordination of export, import, production, distribution, and consumption of electronics machinery and equipment, and telecommunications machinery and equipment.

The Era of Competition

In 1971, the Public Telecommunications Law was amended so that computers could be connected to the network, but no further steps were taken toward allowing terminal equipment to be attached to the network (in contrast to the United States, where CPE was being liberalized). It was not until 1985, when the sale of telecommunications terminals was liberalized from the NTT monopoly, that Japanese telecommunications entered the era of competition. Also in 1985, the Telecommunications Business Law was passed, dismantling the monopolistic structure of telecommunications services.

The changes in organization in Japan may be summarized as follows: telecommunications was operated by the state when the fortunes of the new industry were strongly tied to the state's. When the goal was to bring the number of installed telephones and technical standards up to the level of the rest of the world as quickly as possible, public and private corporate monopoly structures were used to provide more managerial flexibility than was possible with government management. When telecommunications was spread throughout the country and the demand for greater service sophistication increased, competition in the form of multiple providers was introduced. MPT was the architect of the introduction of competition and gradual privatization of NTT. Japan has no independent regulator; the ministry must approve all proposals to cut or adjust telephone rates.

Now, despite competition, the integrated end-to-end services of the 19th century are making a reappearance: Sei Kageyama calls this method of service provision the "kawa" method for the Chinese character "forever": the streams of the river are represented by three parallel lines: "For me these lines symbolize the separate parallel networks which provide exclusive end-to-end services." [22]

Type I and Type II Carriers

Under the Telecommunications Law, New Common Carriers (NCCs) were categorized into two types: Type I, which own their own transmission networks and provide services, and need a license from MPT; and Type II, which lease transmission capacity from Type I to provide value-added or other specialized services to their customers. Type I operators are facility-based firms supplying regional, long distance, satellite, international, and mobile telephone services, and dedicated circuits. Type II carriers resell specialized services including electronic mail, packet switching, and dedicated circuits.

Type II carriers are further segregated into General Type II and Special Type II. General Type II carriers do not need a license; they just need to file notification with the MPT. General Type II carriers can utilize only the international public switched network in providing service to its customers. Special Type II carriers also do not need a license; they have somewhat more flexibility in that they may offer international VAN data services by utilizing and reselling international leased circuits. Special Type II carriers providing services utilizing 500 circuits or more and providing international VAN services are required to register with MPT.

There are other differences in the regulations that apply to Type I and Type II. Type I is restricted to less than one-third foreign capital; Type II has no restrictions. For Type I, tariffs must be authorized; for General Type II, only notification is required, while Special Type II must be registered (see Table 5.1).

The growth in the number of carriers since 1985 has been phenomenal. In 1985, there were only two Type I carriers (NTT and KDD).

TABLE 5.1
SUMMARY OF TYPE I AND TYPE II REGULATIONS

	Type I	Type II	
		Special	General
Facilities:			
Own	Yes	No	No
Lease	No	Yes	Yes
License Required	Yes	No	No
Foreign Capital Restrictions	Yes	No	No
Tariff	Authorization	Registration	Notification

In 1995, there were 92. In 1995 there were a total of 1,246 Type II carriers: 39 special and 1,207 General Type II carriers[23] (See Table 5.2.).

Principal investors in the new carriers are regional electric power companies, automobile companies, and trading companies. There is no limitation on foreign holdings for Type II carriers. Foreign participants with more than 5 percent holdings in Type I carriers include British Telecom, Cable & Wireless, Hughes, Motorola, and Pacific Telesis International.[24]

There are four leading long distance carriers: Dain-i-Denden Inc. (DDI); Japan Telecom, an affiliate of Japan Railway; Teleway Japan, an affiliate of Japan Highway Public Corporation; and Tokyo Telecommunications Network (TTNet), an affiliate of Tokyo Electric Power Company Inc., which provides leased line service and switched telephone service in the greater Tokyo metropolitan area.[25]

Since 1989, there has been a choice in international communications. Japan's large trading companies, "the big six," have established large equity stakes in the two new international long distance companies competing with KDD. One competitor is International Telecom

TABLE 5.2
TYPE I AND TYPE II CARRIERS 1995

Type I Carriers:		Number
Long Distance		4
Intraregional Service		10
Satellite Service		3
Mobile Phone/Cellular		36
Radio Paging		36
International		3
	Total:	92
Type II Carriers:		
Special Type II		39
General Type II		1,207
	Total:	1,246

Japan (ITJ), whose major owners are also major international users: Mitsubishi, Mitsui, Sumitomo, and Matsushita Electric Industrial, with smaller stakes held by Marubeni and Nisshoo Iwai. Six major banks, including the Bank of Tokyo, are also minor shareholders in ITJ. ITJ began offering international, private leased-circuit services and switched services in 1989, and now provides all digital service. The other international competitor is International Digital Communications (IDC), whose major owners are C. Itoh and Co., Cable & Wireless, Toyota Motor, and Pacific Telesis International.[26]

Three satellite firms offer domestic communications: Japan Communications Satellite Company (JC-SAT), owned by C. Itoh and Co., Mitsui, and Hughes Communications; Satellite Communications Corporation (SCC), owned by Mitsubishi; and Satellite Japan Corporation, controlled by Sumitomo, Kissho Iwai, and Marubeni. KDD is part owner of Telecommunications Satellite Corporation of Japan (TSCJ). The satellite operators predict that their main source of revenues in the early years will be from video services, i.e., distribution of television to cable systems or to satellite antennas of 1 to 1.5 meters in diameter, and for satellite news gathering and video feeds.

Mobile Communications

Like the United States, Japan has adopted a duopoly policy for cellular services. Under pressure from MPT, NTT cellular has been spun off as a separate company (NTT DoCoMo) MPT to insulate mobile phone service from its monopoly on local phone service. NTT offers a national analog system, with one competitor in each major market. In Tokyo and Nagoya regions, an affiliate of Teleway Japan competes against NTT; in other regions, subsidiaries of DDI compete against NTT. A digital cellular standard was established in 1992. The MPT awarded digital licenses to the Digital Phone Group and the Tu-Ka Group in addition to NTT DoCoMo.[27]

The growth in mobile phones appears to be at least partly the result of competition. There are now more than 22 times the number of cellular terminals than before deregulation. Monthly charges have dropped from as much as $231 to $85; calling charges are also lower.[28] However, the penetration of 3.45 per 100 population is lower than in the United Kingdom (6.47 per 100) and the United States (9.26 per 100), although Japan's compound annual growth rate in cellular from 1990 to 1994 of 49.2 percent was higher than growth rates in these countries. The Japanese have also introduced PHS (Personal Handyphone Service), a lower cost digital mobile service.

Some 36 local companies compete against NTT in paging services. The number of paging subscribers has increased from 1.9 million to 8.1 million, and rates have dropped substantially.[29]

The Gradual Privatization of NTT

Until 1985, NTT had been a 100 percent government-owned public corporation. In 1985, NTT Public Corporation was privatized into a joint-stock corporation with all of the stock initially held by the government. The government has sold stock gradually to the public; it still holds 65 percent of the shares, approves board appointments, and passes judgments on all major decisions. There have been numerous proposals for an AT&T-style breakup of NTT, but the government's decision on whether to divest NTT has been postponed repeatedly, apparently because of concern about diminishing the clout and revenues of the world's largest carrier, a cornerstone of Japan's high-tech industrial policy.

Meanwhile, NTT is restructuring internally by establishing regional services companies that are 100 percent owned by the parent company. Data services and radio communications have been organized as wholly owned subsidiaries; NTT International is a subsidiary with majority ownership by NTT, but with a large number of trading companies and engineering companies as shareholders. However, the bulk of the business, including local and long distance switched and leased services, remains a part of the parent company.

NTT's service vision for the 21st century is for Visual, Intelligent, and Personal (VIP) communications services. The central theme of VIP is to offer a greater variety of services than is readily available elsewhere, and that can be selected according to personal preferences. These services would include multimedia, personal ID numbers, portable, simple-to-operate visual services with images forming their core, intelligent services that make it possible to locate and communicate with the called party wherever the person is and obtain a wide variety of information, and personal services that can specifically respond to individual customer preferences. NTT plans to provide new phone services throughout the country including portable pocket phones, textmail (communications by means of characters and still images), and video phones.[30]

The Effects of Liberalization

The benefits of liberalization include price reductions, growth in sales within the entire industry, and an increase in equipment imports. Domestic long distance rates have dropped by more than half; international long distance has been reduced by one-third. The new carriers offered rates 23 percent lower than KDD's to the United States. KDD has responded by cutting rates several times. However, this price cutting has resulted in lower profits, so the emphasis is shifting to quality and more sophisticated services for business customers. For example, KDD plans to emphasize ISDN facilities and encourage broadband services.

Liberalization has also resulted in greater efficiency. NTT achieved its goals of eliminating application backlog and offering automatic dialing throughout the country. KDD has streamlined its operations and spun off subsidiaries; it now has 34 compared to 15 in 1985.[31] However, despite these changes, NTT and KDD have lost significant market share. New competitors of NTT now have as much as 40 percent of the market in heavily populated areas, while the two new challengers to KDD have 10 percent of the international market. In the first five years of competition, NTT had lost 14.5 percent of the long distance market, 27.5 percent of the mobile/portable phone market, and 27.9 percent of the pager market.[32]

Competition has stimulated growth in both services and equipment. The market has expanded; for example, sales by domestic carriers reached $46.1 billion in 1990, a 23 percent increase from 1985.

Between 1984 and 1990, domestic spending on telecommunications equipment grew an average of 10.1 percent annually compared to a 4.7 percent annual increase in GNP.[33] The Japanese telecommunications market is expected to grow to become one of the largest in the world, with estimated investments between 1995 and 2000 of more than $17 billion.[34]

The revitalization of the telecommunications industry can be seen in the enormous growth in customer premises equipment, particularly personal terminals. For example, in Japan, there are now more than 7 million cordless phones, more than 6 million fax machines, and 4.3 million mobile phones.[35] In the last five years, there was a 1,400 percent growth rate in mobile phones and 235 percent growth in pagers. In the same five years, there was an average growth rate in fax machines of 23 percent per year; an average growth in telephone sets of 36 percent per year; and in cordless telephones of 40 percent per year.[36]

New services included free dial service (toll-free service), 900 number-type pay services, and phone cards of which approximately $300 million worth are sold annually (an average of almost three cards per person).[37] However, some problems persist. There is no equal access (1+ dialing), so customers must dial four extra digits to reach an alternative carrier. They must first subscribe, wait two weeks, and pay $15 for registration, and must resubscribe when they move. The billing is complex, with charging based on a combination of services from NTT and the NCCs. Sometimes the NCCs are more expensive for short distances. It is hard for consumers to compare rates, and difficult for carriers to introduce volume discounts and flat rates. The tariff filing process for carriers also takes time and can be inflexible.[38]

Some carriers have turned these limitations into competitive advantage. For example, because users must dial a four-digit access code to reach NCCs, DDI introduced an autodialer, with a microprocessor that selects the least-cost route, and is automatically updated. DDI also provides itemized billing, which is still not universally available from NTT.

Universal Service

Under the 1953 Public Telecommunications Law, NTT was obliged "to furnish far and wide and without discrimination, to all people, rapid and reliable public telecommunications services at reasonable charges."[39] To provide services far and wide required an immense amount of capital. To keep charges reasonable required cross-subsidies because no government subsidies were provided. Is universal service in jeopardy because of competition? Will competitors "cream skim," leaving the less profitable areas to NTT? The new common carriers (NCCs) have captured 40 percent of the long distance market between the three largest cities: Tokyo, Nagoya, and Osaka. The cumulative amount of rate reductions in the first six years of competition was $1.5 billion. Without an increase in local rates, this represents an annual rate reduction of $70 per exchange line.[40] Yet NTT retains the universal service obligation; there is no requirement for contributions to a universal service fund by other carriers.

Foreign Investment

The domestic equipment market is dominated by large Japanese manufacturing companies including NEC, Fujitsu, Hitachi, and Oki. NTT has been pressured to buy more non-Japanese equipment by Japan's major trading partners, particularly the United States. In recent years, it has bought digital transmission equipment from AT&T, digital switching equipment from Northern Telecom, pocket pagers and cellular phones from Motorola, and computers from IBM and DEC. However, Japan has by far the world's largest positive trade balance in telecommunications equipment, with a surplus of almost $15 billion in 1993.

Foreign companies are also involved in services; there are several 100 percent foreign-owned Type II carriers. Foreign companies are also involved in joint ventures: for example, Cable & Wireless and Pacific Telesis have invested in IDC; Hughes is a part owner of JC-SAT; and BT and France Telecom each have a 2 percent share in ITJ.

In addition to participating in the dramatic growth of the domestic market, Japanese firms are also playing an increasing role in world telecommunications equipment manufacturing. Japan is the world's largest exporter of telecommunications equipment, with exports of more than $17.8 billion in 1993.[41] Japanese companies are also shifting from export to overseas production, for example, in the United States, and are emphasizing new product design techniques and competitive pricing.

Telecommunications as Industrial Policy

The steps toward liberalization of services, hesitation to break up NTT, and massive investment in research and infrastructure are all manifestations of Japanese industrial policy in telecommunications. Since the 1920s, Japan has turned to the private sector to supply telecommunications equipment for what was then a government-owned network. MPT long favored protecting the NTT monopoly, while MITI advocated competition to stimulate innovation and reduce costs to users. The two agencies reflect two visions of Japanese strength: one in maintaining a coordinated and government-directed telecommunications sector, the other in fostering Japanese global competitiveness by ensuring that Japanese industry would benefit from lower telecommunications costs and new services.

Japanese industrial policy is also evident in its research and development (R&D) strategies, which use public funds to develop and test new technologies that Japanese telecommunications companies can then market internationally. For example, because the government owns a majority of NTT shares, it is entitled to a 10 percent dividend, two-thirds of which is used for promotion of fundamental research in telecommunications. Also, MPT, in conjunction with MITI, administers investment of state money in R&D companies; 21 such R&D companies in telecommunications have been established since 1985. The carriers and private companies also invest in R&D. Japan has also promoted user-oriented initiatives. When Japan celebrated the centennial of the inauguration of telephone service in 1990, it commemorated the occasion by setting up a Foundation for Promoting Sophisticated Telecommunications Usages with a $42 million trust fund supported by NTT and leading companies.

MPT initiatives include a national Support Act for implementing new enterprises in the telecommunications and broadcasting fields: for example, to encourage new enterprises such as digital mobile; interactive cable TV; and B-ISDN trials. The Telecommunications Backbone Infrastructure Enhancement Act of 1991 authorizes investment in infrastructure for the 21st century. The investments include both technical and "human" infrastructure such as fiber transmission, ATM switching, and educating 20,000 telecommunications and broadcast engineers to plan and promote regional telecommunications-oriented societies.[42] A major goal of the information infrastructure initiatives is to install fiber to the home throughout Japan by the year 2010.

The government has also looked to satellites to deliver broadband services.[43] Japan began experimentation with direct broadcasting technology with the launch of an experimental satellite, BSE, in 1978. MPT took the initiative for the project and guided its development. While the new competition in domestic telecommunications is attracting private sector investment in satellites, the government continues to play a major role. The satellites that Japan calls "operational" are still largely used for experimental rather than commercial operations. A major reason for government interest has been the development of high definition television (HDTV). In the mid 1980s, the prevailing government view was: "The key to the future success of satellite broadcasting—and an important means by which to increase its popularity—is the development of HDTV."[44] Foreseeing a global market for HDTV, the Japanese sought in the early 1990s to develop and distribute Japanese technology to their consumers, in order to preempt systems developed by United States and European manufacturers. Their broadcasting satellite enabled them to be the first country to offer HDTV delivery.

However, the HDTV strategy seems to have failed. The Japanese (and Europeans) stuck with analog technology while the world was moving to digital. The United States has opted to develop a digital system, which seems likely to become an international standard. In addition, the market for HDTV has been slow to develop in Japan, where there are now less than half as many satellite television subscribers as cable subscribers, and few of the satellite users have equipment for HDTV.

Another element in Japan's global strategy is the development of satellite construction and launching capabilities to compete in global satellite markets with the United States and Europe. Japanese companies are already major suppliers of earth stations ranging from Intelsat Standard A terminals to the smallest VSATs. With the strength of the Japanese electronics and telecommunications industries, Japanese companies will be able to offer a complete range of satellite communications products and services from the satellite itself to earth stations, teleconferencing facilities, and broadcasting, data and voice equipment.

6

The Telecommunications Sector in Industrialized Countries

The Information Highway . . . is not so much about information as it is about communication in both its narrowest and broadest senses . . . Rather than a highway, it is a personalized village square where people eliminate the barriers of time and distance and interact in a kaleidoscope of different ways.

—Canada's Information Highway Advisory Council[1]

DIVERSE APPROACHES

This chapter examines the changing structure of the telecommunications sector in other selected industrialized countries. We first examine France and Germany, then Canada, Australia, and New Zealand. Several of the themes of previous chapters are echoed here, as these countries grapple with the issues of restructuring their telecommunications sectors. Yet while their economies are similar, their approaches to telecommunications differ in many ways. France and Germany have held longest to the public monopoly model, and are only now embarking on privatizing France Telecom and Deutsche Telekom. Canada and Australia share common dilemmas—recognizing the importance of telecommunications in linking relatively small populations scattered over vast areas, and striving to protect affordable access for isolated residents, while introducing competition that may benefit urban areas most. New Zealand has moved furthest and fastest from the post office model, becoming one of the world's most open telecommunications markets.

FRANCE

Steps to Modernization: "Télématique"

In the past two decades, France has moved from having one of the most backward to one of the most modern telecommunications networks in the industrialized world. When President Valéry Giscard d'Estaing was elected in 1974, he found the national telecommunications network far behind that of most Western European countries. There were only 7 million lines for 47 million people, one of the lowest penetration rates in the industrialized world.[2] The average waiting time for a telephone was four years, and most rural areas had manual switches. In mid-1974, Giscard made the political decision to make telecommunications a national priority, following three major policy guidelines:

- at the industrial level: to implement a digital network, which led to production of digital equipment and establishment of a packet network;
- at the marketing level: to develop information services and introduce them to French consumers through the telephone;
- at the societal level: to stimulate the widespread use of telecommunications and information technologies throughout the society.

Together, these policies were to result in what French government officials Simon Nora and Alain Minc called "télématique" in their landmark study *The Informaticization of Society*.[3] The télématique concept, meaning the convergence of telecommunications and information technology, became a cornerstone of the French strategy, or as one

French analyst observes, télématique was "the locomotive pulling new services and social behaviors behind it."[4]

The showpiece of the strategy was the Minitel, a small, simple computer terminal for public use that was connected to the telephone network. France Telecom (FT) promoted use of the small Minitel videotex terminals by creating an online national telephone directory and providing the terminals for free. Despite initial resistance to a mandatory electronic directory (FT had to continue to print the hard copy versions), usage grew dramatically during the 1980s.

Today, France has the world's nearest approximation to a mass-market interactive service in the Minitel. In 1994, there were 7 million terminals accessing more than 26,000 services. These service providers now make more than 6 billion francs per year.[5] Each month, users place more than 130 million calls to the Teletel network and log between 8 and 9 million hours of connect time. France Telecom also generates revenue from the billing services it provides to information providers. Business use now represents about 50 percent of Minitel traffic. The most popular services are directory inquiries and train timetables.[6] The national electronic telephone directory is the world's largest online database, and handles nearly 2 million calls per day.[7] Other popular consumer services are finance and banking, transportation reservations, and teleshopping. Teletel has also earned considerable notoriety for its "services roses," online explicit and pornographic services.

The Minitel network uses the Transpac data network, which allows France Telecom to offer videotex services at a single tariff nationwide and to provide uniform service access codes for users in all parts of France. Among the other features that contributed to the Minitel's success were the provision of terminals that were small, free, and easy to use; a strategy that emphasized the Minitel as an extrapolation of the telephone, not the less familiar computer; billing included on the telephone bill; and anonymity for service providers with recordkeeping done by France Telecom.

Restructuring the Telecommunications Sector

Laws passed in 1990 modified the structure of the telecommunications sector following the guidelines of the European Community's 1987 Green Paper (see chapter 7). The law split telecommunications and postal services; both report to the Ministries for Industry, Post and Telecommunications, and Foreign Trade. The new laws assigned specific roles to each of the players:

- the government is responsible for regulation through the Direction Générale des Postes et Télécommunications (DGPT);
- France Telecom as national operator is the sole provider of basic voice telephony and telex services to the general public, but also operates in areas open to competition;
- private carriers may operate in areas that are not the exclusive domain of France Telecom, either under regulated competition for data transmission and wireless communications, or in open competition for all other services.[8]

Competition is possible for terminal equipment, value-added services, and networks for closed user groups. Cellular is a duopoly where the competitor is SFR (Société Française du Radiotéléphone). Simple data transport services (called bearer services), which are open to the general public, must be licensed by the Ministry, ostensibly to ensure network integrity and prevent cream skimming of the most profitable specialized services. Regulatory oversight for audiovisual services is the responsibility of the Audiovisual Council (CSA), while the DGPT retains regulatory authority for cable- and satellite-based telecommunications services.

Despite announcements in the early 1990s about privatization of key government-owned industries, France Telecom appeared destined to remain government-owned for the immediate future. However, in mid-1993, the French government released a report recommending that it should float a minority stake in France Telecom on the stock market.[9] Today, the Chirac government appears committed to privatizing France Telecom as part of its overall privatization strategy for government-run industries and services.

France Telecom Today

France Telecom is now an autonomous, state-owned entity, whereas previously it operated as an arm of the French government. Its budget is no longer subject to parliamentary approval. The management of France Telecom is now the responsibility of the chairman, the CEO, the executive directors, and a newly created board of directors.[10]

France Telecom now operates one of the world's most modern networks, with more than 28 million lines providing service to 97 percent of French households. By 1994, the local French network was composed entirely of electronic stored-program controlled switches, 89 percent of them digital.[11] Its Transpac service is Europe's largest packet-switching network, and provides the basis for its value-added services, including Teletel, electronic messaging, and EDI. It also provides nationwide ISDN service, known as Numeris. France Telecom has deployed the digital cellular GSM network, stressing its advantages of high transmission quality and international mobility. The investment in modernization came at a heavy price: FT borrowed $22 billion and must live with price caps as it tries to reduce the debt.

Since 1985, France Telecom has operated the Telecom 1 national satellite system, which provides data broadcasting services at speeds up to 2 Mbps, television and radio transmission, and support for telecommunications between France and its overseas territories in the Caribbean and Atlantic (St. Pierre and Miquelon). Telecom 2 was launched in 1992.

By the year 2000, France Telecom plans to generate 20 percent of its revenues from international activities such as expansion of overseas investments and its range of business products for use by multinational companies.[12] Its activities abroad include major shareholdings in telecommunications carriers in Argentina and Mexico. In 1990, France Telecom and the Italian operator STET led a consortium that included J.P. Morgan and the Argentine group Perez Companc, which successfully bid for the northern half of Entel, Argentina's national telephone network. Also in 1990, France Telecom, along with Southwestern Bell Corp. (now SBC) and Mexican investor Carlos Slim, paid $1.76 billion for a 51 percent voting stake in Telmex, the Mexican telecommunications operator. That transaction provides the group with a 20.4 percent interest in the company.[13] FT is also trying to market the Minitel overseas, although joint ventures in the United States have not been successful. In addition, FT is participating in a paging network in the Czech Republic and a cellular network in Poland.[14] In 1994, France Telecom and Deutsche Telekom announced joint ventures in value-added and international services; they are now both partners in a consortium with Sprint known as Global One (see chapter 3).

GERMANY

Restructuring the Telecommunications Sector

Like France, Germany adopted the public monopoly model, with telecommunications the responsibility of the post office, the Deutsche Bundespost (DBP). Change began with the 1989 law on Posts and Telecommunications in the Federal Republic of Germany (FRG), which introduced organizational reform and market reform. The business side of the DBP was divided into three enterprises: Telekom, Post Services, and Postbank. The law separated the political and regulatory functions of the DBP from its entrepreneurial and operational functions. The former are performed by the federal minister of Posts and Telecommunications, while the latter are the responsibility of Deutsche Bundespost Telekom (known as Telekom).

The 1989 liberalization opened up all nonvoice service to the private sector. Terminals were also to be competitive. However, the Telekom monopoly on all voice services was temporarily suspended to allow the private sector to offer company-to-company satellite links between eastern and western Germany to alleviate bottlenecks. This step may be difficult to reverse. The only remaining monopolies are the establishment of telecommunications networks and the operation of voice telephone service. Users are able to lease lines and private operators are admitted as competitors for mobile services and planned for PCN. There are two competing digital mobile phone systems: D1 (Telekom) and D2 (Mannesman—a joint venture with Pacific Telesis). Anticipating the opening of basic communications under European Union directives in 1998, Vebacom, owned by Veba and Cable & Wireless, plans to invest $2.2 billion over five years, and aims to be Germany's second major phone company.[15] Other consortia are also planning to compete.

Under the German constitution, Posts and Telecommunications must be a public administration, directly assigned to the federal government. It was not possible to amend the constitutional provision without a two-thirds majority in both houses. In June 1993, the opposition agreed to a government plan that would turn Telekom into a joint stock company, but questions on timing and technicalities remained. Helmut Ricke, head of DBT, urged German legislators to fix a firm timetable for selling shares in the telecommunications enterprise.[16] Ricke's wish was finally granted; Telekom was privatized in November 1996.

As a result of the constitutional restrictions, Germany adopted an interim model that may be characterized as liberalization without privatization: the German postal reform of 1989 has initially resulted in liberalizing telecommunications markets without privatization of a previous government monopoly. The German approach has thus shown that opening up markets to competition is possible without privatizing a state-owned enterprise. (Other countries that have followed this path, at least for the interim, are France and Australia, where public and union sentiment against privatization have been strong.)

Regulation

The Ministry for Posts and Telecommunications is responsible for regulation. The Ministry is also legal supervisor of Telekom, and licenses competitors and approves tariffs for leased lines. The second level of regulatory tasks is controlling the conduct of Telekom: preventing discrimination, predatory pricing, and inefficiency in its monopoly business, which, at present, is 80 percent of its operations. Third, the regulator sets the "rules of the game" in areas of limited competition, such as in mobile communications. Fourth, it ensures obligatory services, at present, it is allowing cross-subsidies.[17] Frequency management is also now under the jurisdiction of the Federal Ministry of Posts and Telecommunications, whereas previously, it was under the Deutsche Bundespost.

There may, however, be a need for additional regulatory reform. The German Monopolies Commission has pushed for more reform, such as hiring out the right to build and run networks to private operators in eastern Germany. There have also been disputes over the pricing of the lines that Mannesman must lease from Telekom.[18]

Deutsche Telekom Today

Deutsche Telekom is Europe's largest carrier. Globally, it is third in revenues after NTT and AT&T, and second in main lines behind NTT (see chapter 3). Despite remaining government-owned in the 1990s, Telekom's corporate culture has changed significantly, as it now operates as a commercial enterprise in most respects. Its organizational structure now consists of 21 main business divisions. Telekom is a public administration, but is run by a board of directors who are not civil servants. The second level is also open to managers from the private sector. However, unlike most private enterprises, it is wholly state-owned and obliged to contribute fixed shares of revenues to the federal government.

To finance German reunification, Telekom was called upon to make a special surrender of profits, "revealing reluctance on the part of the State to relinquish the source of revenue provided by telecommunications. . . ." in spite of recent postal reforms.[19] Until 1996, Telekom was required to balance the deficits of the post office and postal bank, and to surrender 10 percent of its turnover to the minister of finance. From 1996, it will be treated like a private company for tax purposes and will also be subject to profit distribution regulations like a private company. However, once privatized, it will be free to decide its own business policy on issues such as recruitment, performance-related pay, and cost management.

Telekom faces competition in all services except networks and voice switching for third parties. Thus, it faces a mandatory dual role of being a market-led enterprise and provider of basic public services to all. Its major challenges are to strengthen its competitive position nationally and internationally; to internationalize services in response to globalization of customers; and to modernize the infrastructure in the five new states in eastern Germany.[20]

Upgrading Infrastructure in Eastern Germany

Telekom is using income from its telecommunications services to improve infrastructure, especially in the "new Laender" (eastern Germany). The Telekom 2000 program plans to build and modernize infrastructure in the five new federal states of eastern Germany by 1997, requiring an investment of 55 billion DM and creating 100,000 jobs. Telekom is installing 7 million new telephone lines and 400,000 fax lines, and providing 50,000 Datex-P lines. An overlay network of more than 30,000 lines between east and west is being constructed. Telekom will also link 2.2 million households to cable TV and establish cellular mobile networks offering full geographic coverage. Telekom management believes that this experience in "managing change" will also be useful in participating in the upgrading of the infrastructure in Eastern Europe; it is seeking partnerships with other companies to gradually build a homogeneous telecommunications infrastructure in Eastern Europe.[21] (For more detailed analysis of this infrastructure investment, see chapter 11.)

The enormous capital requirements in eastern Germany have contributed to the government's reluctance to relinquish financial control over the Telekom "cash cow." Also, the government's decision to relax restrictions on internal voice and data satellite transmission, allowing private competitors with the public switched network to speed up modernization, may be difficult to reverse, and may signal wider competition in voice and bypass services.

THE TELECOMMUNICATIONS SECTOR IN OTHER WESTERN EUROPEAN COUNTRIES

All European Union member countries must meet the EU guidelines of opening their networks to competition by January 1, 1998 (see chapter 7). An overview of the competitive status of OECD countries, which includes EU members, is given in chapter 4. The three largest members of the European Union, France, Germany, and the United Kingdom, have been discussed in earlier chapters. The following is a brief synopsis of the telecommunications environment in some other Western European countries.

Belgium is a major regional telecommunications hub: it is the European headquarters for many multinational corporations; the European Commission is also located in Brussels. Belgacom, the former telecommunications arm of the Belgian PTT, was corporatized in 1993. In 1996, three international carriers, Ameritech, TeleDanmark, and Singapore Telecom, bought 49.9 percent of Belgacom. There are two mobile licenses, Proximus GSM, a joint venture of Belgacom and AirTouch, and Mobistar, owned by France Telecom. The regulator, IBPT (Institution of Postal Services and Telecommunications), is defining the new framework for the opening of the voice and infrastructure markets in 1998, as required by the EU. A second fixed line carrier, Telenet Flanders, has been proposed by the government of Flanders, and would link all of the country's cable television networks by 1998.[22]

Denmark sold off 49 percent of the state-owned operator, TeleDanmark, in 1994. TeleDanmark has enjoyed a monopoly in public network services, including not only local, national, and international networks, but transmission of radio and television channels. There are two cellular networks, one owned by TeleDanmark, and the other, Dansk Mobil Telefon, owned by a consortium of Scandinavian companies and BellSouth. The regulator, the National Telecommunications Agency, is opening the sector to competition, for example, by requiring interconnection between mobile operators and between private broadband networks.[23] The monopoly on infrastructure provision, long distance and international voice telephony will remain in place until 1998.

Finland's telecommunications sector differs from others in Europe in that it has numerous small local companies. Telecom Finland became a joint stock company in 1994. There are also 49 local telephone companies (both private companies and cooperatives) and 13 other telecommunications operators.[24] Like other Scandinavian countries, Finland has both a high teledensity (55 lines per 100 inhabitants) and high cellular penetration (12.8 subscribers per 100 population).[25] Finland-based Nokia has become one of the world's largest cellular equipment companies. Finland has a duopoly structure for cellular communications, with Telecom Finland as one of the operators; all other services are open to competition.

Italy has been slow in opening its telecommunications market. STET, the holding company that controls Telecom Italia, has been partially privatized. Telecom Italia retains a monopoly in analog cellular service. It holds one of two GSM cellular licenses; the other was awarded to Omnitel-Pronto Italia, a consortium including Olivetti, Bell Atlantic, and AirTouch. Telecom Italia is investing in fiber in the local loop, so that it will be able to carry video programming. Foreign operators including Bell Atlantic, BT, and US WEST have formed alliances with Italian companies to operate competing telecommunications and cable television networks when the sector is liberalized.[26]

The **Netherlands'** national telecommunications operator, KPN, has been partially privatized; the state still retains 70 percent ownership. KPN has had a monopoly on local, national, and international services. A second national operator is to be up and running before 1998. Virtually all Dutch households are connected to cable television networks to receive programs from neighboring countries. These cable networks may become the platform for new cable telephony services. There are two GSM operators, including KPN and a Vodafone joint venture.[27]

Spain has moved faster to privatize its operator than to open up its market. The government now retains only 32 percent of the national carrier, Telefonica, and plans to reduce its stake to a token share. Telefonica operates two analog cellular networks, and competes in GSM services with Airtel, a consortium of AirTouch, BT, and local companies.[28] Spain is eligible for a five-year exemption from the EU competition rules, and may therefore defer liberalizing its public networks until 2003. Meanwhile, Telefonica has invested in telecommunications ventures in several Latin American countries including Argentina, Colombia, Ecuador, and Peru.

Sweden has the world's highest teledensity, with 68.3 lines per 100 inhabitants (including telephones in the summer cottages that dot the countryside and the Stockholm archipelago).[29] Sweden has corporatized its state-owned operator, now known as Telia (formerly Televerket), but Telia has not been privatized. Sweden has never had a legalized monopoly in telecommunications, although Televerket enjoyed a de facto monopoly. The major long distance competitor is Tele2, which was established in 1992 by Kinnevik and Cable & Wireless. France Telecom's Transpac and BT are major competitors in data communications. Sweden also has the world's highest penetration of cellular phones, with 15.8 per 100 inhabitants, and 30 per 100 in Stockholm. Major cellular operators include Telia Mobitel, Comvik (a Kinnevik subsidiary), and NordicTel, in which AirTouch and Vodafone are key investors.[30]

CANADA

Overview

Canada is the world's second largest country in land mass, but has a population of only 29 million, or slightly more than one-tenth of the population of the United States. Canada's population density averages 2.62 per square kilometer; however, most of the population lives within 100 miles of the U.S. border, and much of the country is very sparsely populated, with residents scattered on large farms and ranches, and in isolated villages. Canada's population distribution is similar to Australia's, in that the majority of the population lives in a thin belt on the edge of the country, and there are vast areas with very few inhabitants.

Canada has one of the highest telecommunications penetration rates in the world, with 98.2 percent of households having telephone service, or 57.5 access lines per 100 population. The Canadian long distance market is valued at $7 billion.[31] Most Canadians view toll calls as a necessity rather than a luxury; in 1994, they averaged 140 long distance calls per person, almost double the number in 1984.[32]

The Canadian telecommunications industry generated approximately C$18 billion (about US$13 billion) from Canadian equipment and services in 1992, and exported more than US$2 billion in equipment in 1993. Northern Telecom (Nortel) controls 60 percent of the Canadian customer premises equipment market, and is also a major global supplier of telecommunications equipment. Founded as Northern Electric in 1895, the company was the equipment supplier for Bell Telephone of Canada, in a relationship similar to that between Western Electric and AT&T in the United States. Northern Telecom became a wholly owned subsidiary of Bell Canada in 1962; Bell Canada remains a major shareholder.

Government Institutions

Canada was perhaps the first country to institute a ministry specifically for communications. The Canadian Department of Communications (DOC) was established by the Trudeau government in 1969, with responsibility for development of national policies to ensure the orderly evolution and operation of Canadian communications as key elements in the achievement of Canada's social and economic goals. The DOC's roles and responsibilities were to formulate policies relating to development of the national communications network; to conduct research on new technologies and services; to promote access to and use of new telecommunications and information technologies; and to manage utilization and development of the radio spectrum.[33]

In mid-1993, under a federal government restructuring initiative, the Conservative government abolished the Department of Communications, assigning telecommunications policy and program responsibilities to an expanded department of industry called Industry Canada. Broadcasting and cultural activities were moved to a new Department of Canadian Heritage. The DOC, and its successor, Industry Canada, have funded numerous initiatives to extend the benefits of telecommunications to remote areas and social services, including distance education and telemedicine. In recent years, the government has sought to increase involvement of the private sector in such initiatives, for example, through jointly sponsored "information superhighway" projects and an Information Infrastructure Task Force (see chapter 19).

The Canadian Radio-Television and Telecommunications Commission (CRTC) was established in 1968 to regulate broadcasting and cable television; since 1976, it has also exercised authority over the federally regulated telecommunications carriers. Following a 1989 Supreme Court ruling, most telephone companies now come under the jurisdiction of the CRTC. The mandate of the CRTC is to insure that rates charged are just and reasonable, and that carriers do not engage in any unjust discrimination or afford any undue preference or advantage in the provision of telecommunications services or facilities.

The Major Carriers

The major telephone companies are privately owned except those operated by the provincial governments of Alberta, Saskatchewan, and Manitoba. The largest companies are Bell Canada (once affiliated with the U.S. Bell System, but now totally separate) and British Columbia Telephone (BCTel), a subsidiary of General Telephone and Electronics (GTE) of the United States. A few local telephone companies are owned by city councils.

Stentor (formerly Telecom Canada) is an association of the largest telephone companies and Telesat, the national satellite operator. Stentor's members jointly provide domestic and Canadian-U.S. long distance public voice services, public switched data services, switched messaging service, integrated network services, leased circuits, and electronic message and mail services, and other enhanced telecommunications services. Stentor also operates a national cellular mobile phone service, CellNet. To position itself for competition, Stentor has formed an alliance with MCI.[34]

Unitel, formerly known as CNCP Telecommunications, is a joint venture between CNCP (originally a subsidiary of the Canadian National and Canadian Pacific railroads) and Rogers Communications, the country's largest cable TV operator. Unitel originally was allowed to offer only public switched data services and other enhanced services. However, since 1992, it has been allowed to compete in the provision of public long distance services. AT&T acquired a 20 percent interest in Unitel in 1993. In return, Unitel received AT&T network software and switching and transmission equipment, and gained access to AT&T expertise and research and development.

Since 1992, other carriers have been allowed into the long distance market. Resale is legal, and several U.S. carriers have become active resellers in Canada. For example, Lightel, a Toronto-based reseller, is partly owned by Rochester Telephone. Also, the United Kingdom's Cable & Wireless leases lines and offers discounts of 20 percent or more below telephone company rates.[35]

More than 75 percent of international traffic is with the United States, and is handled by regional telephone companies, long distance companies, and resellers. Teleglobe, a privatized monopoly, is the wholesaler of facilities-based overseas services, handling traffic to destinations other than the United States. Teleglobe has stated that it will not seek renewal of its exclusive mandate, due to expire in 1997, but proposes establishment of a "sustainable competitive environment," with access to the U.S. market and modifications in foreign ownership restrictions to allow up to 49 percent ownership, rather than the current 20 percent, in order to facilitate strategic alliances and infusions of foreign capital.[36]

Canada's Domestic Satellite System

Canada became the first country to use a geostationary satellite for domestic communications with the launching of Anik A-1 in 1972. (The U.S.S.R. was already using satellites in nongeosynchronous polar orbits for domestic telecommunications.) The name Anik, meaning "brother" in Inuktitut, the language of the Inuit, was chosen to symbolize the commitment of the government to use satellites for improving communications in the far North.

Telesat Canada was established by an act of parliament in 1969 to own and operate Canada's communication satellites. The act stated that Telesat was to be owned one-third by the government, one-third by the carriers, and one-third by the public. Telesat was never considered profitable enough to issue public shares. Instead, the government share was sold in 1992 to Alouette Communications, a consortium of Canadian telephone companies and Spar Aerospace. This privatization effectively eliminated the possibility of any competition between Telesat and the telephone companies.

Satellite broadcasting in Canada is intertwined with cable television. Canadians were early adopters of cable technology as a means of receiving U.S. channels. By 1961, there were already 200,000 subscribers and 260 cable systems. Canada now has one of the world's highest cable penetration rates; some 1,800 cable systems serve a total of 7.8 million subscribers, or 76.2 percent of Canadian households. Canadian cable grew in the 1960s and 1970s to meet the demand for reception of U.S. television networks. Cable systems also now provide specialized news, music, sports, and movie channels distributed by the domestic Anik satellite system.

However, perhaps the greatest success in the Canadian satellite experience to date has been in native broadcasting. Radio broadcasting in the remote North involves three elements: the Canadian Broadcasting Corporation's (CBC) network, which is distributed across Canada; the CBC Northern Service, which produces regional programs in English, French, and native languages, primarily originating in northern production centers; and community radio stations operated by local people in the most remote settlements. The satellite feeds a mixture of the CBC national and regional programming to the communities, which then add their own locally originated programs. This model builds on existing facilities at the regional and local level, and demonstrates that a satellite does not have to be only a central distribution system for national programming.

Canadian satellite policy is frequently driven by new U.S. satellite services that can be received in Canada. In 1980, the CRTC authorized Canadian pay television channels to counter the threat of U.S. satellite-delivered programming. In the mid-1990s, the CRTC licensed Canadian direct-to-home (DTH) services in response to the DirecTV service available in the United States. However, as in many other instances, Canadian policy makers have sought to balance cultural and economic concerns, in this case to ensure access to Canadian content without threatening the Canadian cable industry, which has called the direct broadcasting satellites "death stars."

Mobile Services

There are about 1.9 million cellular subscribers in Canada, that is, about 6.5 per 100 residents. In 1983, the government licensed mobile cellular service as a duopoly of two competing national systems. One network, Cellnet, was licensed to a consortium of telephone companies, Telecom Canada (now Stentor). The other license is held by Rogers Cantel, a wholly owned subsidiary of Rogers Communications, the country's largest cable television operator. Rogers also offers paging, voice mail, and mobile data services. Telesat Mobile International (TMI) operates satellites with the American Mobile Satellite Consortium (AMSC) to provide mobile communications throughout North America.

Restructuring Initiatives

Canada has taken a more cautious approach to deregulation and competition than the United States. It has a large geographic area, fewer concentrated markets, and many small, isolated communities. There was a strong lobby to retain the status quo of public network monopoly, based on the assumption that Canada's population was too small and scattered to support competition, and that the result would be an increase in basic rates and prices of services for remote areas. Bell Canada estimates a cross-subsidy of US$1.7 billion per year from toll to local service, equal to C$20 per local line per month.[37]

However, market forces created pressure for fundamental changes in the structure of Canadian telecommunications. The threat of competition from U.S. carriers through traffic routed via the United States, and threats from aggressive resellers and big business users sped up the introduction of new services and competition. Customer premises equipment and value-added services were liberalized, but the CRTC initially decided to limit competition to two long distance carriers, Stentor and Unitel. However, under the new telecommunications regulatory framework adopted in 1994, virtually all services are open to competition, including the local market which is now open to cable television and wireless operators. Carriers are also allowed to offer information services.

To rebalance local and toll tariffs, the CRTC has authorized an increase of C$2 per year for local service for three years. In 1998, the CRTC will shift from rate of return to price cap regulation for "utility services," defined as services which have a monopoly or dominate the market, and are likely to be primarily local exchange services. The CRTC has pledged to keep local service charges among the lowest in the world; they are the fifth least expensive among industrialized countries, according to the OECD. Also, the CRTC has capped the price of calls from Canada to anywhere in North America for three years to ensure that consumers do not pay higher rates that would finance long distance reductions for big business.

Free Trade

In January 1988, Canada and the United States signed a free trade agreement which came into force in 1989, removing a wide range of tariffs and trade barriers. Bilateral trade with the United States amounted to C$193 billion in 1990, far more than any other pair of trading partners in the world.[38] The North American Free Trade Agreement (NAFTA), which was signed in 1993, includes Mexico as well.

Basic services such as local and long distance services are not affected by free trade, but there is now open competition in the enhanced services market. The CRTC will continue to regulate enhanced services provided by federally regulated common carriers, but will not regulate other domestic or foreign suppliers. The United States and Canada will work together to establish compatible standards in such areas as technical specifications, technical regulations, and certification systems that apply to goods, processes, and production methods. Equipment affected by the agreement includes central office switching apparatus, PBX switching, electronic key telephone systems, telephone handsets, and modems.[39]

AUSTRALIA

Overview

Australia is the thirteenth largest economy in the world, and the world's seventh largest foreign exchange market. About 80 percent of Australia's population of 17.8 million lives along its eastern and southeastern coast. The rest are scattered in a few cities and large towns, and on isolated farms, sheep stations, and aboriginal settlements.

Australia's telecommunications networks are extensive, with 49.6 lines per 100 population, of which about 38 percent are digital.[40] ISDN services are available for business customers in major cities. Several new fiber links are being built. Australia has its own domestic satellite, Aussat; there are over 350 satellite transmit/receive stations and 15,000 total satellite earth stations.[41] Mobile cellular service has grown dramatically, with a compound annual growth rate from 1990 to 1994 of 61.2 percent. There are more than 1.2 million cellular subscribers, or more than seven per 100 population.[42]

The Public Sector

Until 1991, telecommunications in Australia was provided by a public monopoly, following the PTT model. The Australian Telecommunications Commission (Telecom) was created in 1975 under the Telecommunications Act of 1975, which separated postal and telecommunications functions previously carried out by the postmaster general's department. Telecom became a statutory authority known officially as the Australia Telecommunications Corporation in 1989. The Overseas Telecommunications Commission (OTC Australia) was established in 1946 as a statutory authority. In 1989, it became a public company called OTC Limited, wholly owned by the Australian government.

The Department of Transportation and Communications (DOTAC) was created in 1987 from an amalgamation of the Departments of Transport, Aviation, and Communications. The government has since reorganized departments so that telecommunications is now under the Department of Communications and the Arts (DOCA).

Another key agency is the Department of Industry, Technology, and Commerce (DITAC), whose role is to produce policy aimed at development of efficient and competitive industry. It places particular emphasis on services and equipment manufacture, and includes industry policy for information technology.

The Australia Telecommunications Authority (Austel) was established under the Telecommunications Act of 1989 as an independent authority for the telecommunications industry in Australia. Austel's mandate is to supervise competition between the carriers and transition to full competition. It consists of a board chairman, two members, and such associated members as the minister may appoint (currently three). It administers (and does not set) government policy, and has three advisory committees: the Consumers Advisory Committee, Standards Advisory Committee, and Law Enforcement Advisory Committee.[43]

Satellite and Cable Services

Like Canada, Australia recognized that a domestic satellite system could reach isolated communities that were difficult and expensive to serve with terrestrial technologies. Aussat was formed by the Australian government in 1981 to own and operate Australia's national satellite system. The first two satellites were launched in 1985; the third followed in 1987, with a replacement in the mid-1990s (the first replacement was destroyed when a Chinese Long March rocket exploded). However, the established carriers and broadcasters saw Aussat as a threat. When Aussat was launched, Telecom was already committed to building a digital microwave system in the Outback. Also, the broadcasters did not want urban viewers to have the option of receiving programs directly via satellite rather than from local stations.

Aussat failed commercially as a competitor to Telecom because of inadequate funding and regulatory restrictions that served to protect Telecom and the broadcasters. For example, Aussat was originally allowed to provide only private line and video services. Since April 1991, Aussat has been permitted to provide public switched telephone and data services. Aussat also delivers a package of television programs to Outback communities and homesteads. However, new broadcasting entities licensed to serve the Outback via Aussat have found that it is very difficult to survive because the small population is not attractive to advertisers.

As in Canada, the most innovative satellite applications have been by indigenous people. An aboriginal communications organization called Imparja succeeded in obtaining a license to serve the Northern Territory, and now transmits a mixture of aboriginal and commercial programming from Alice Springs. Another aboriginal group has established a videoconferencing system called the Tanami Network, which links several aboriginal communities with each other and with Alice Springs and Darwin via satellite. The Tanami Network is used for community exchanges as well as for distance education and videoconferencing for the Northern Territory government.

Unlike Canada, Australia has not faced the spillover of neighboring broadcasting and satellite signals. Because of this isolation, and resistance from its powerful broadcasters, Australia was one of the last industrialized countries to introduce cable television. After several flawed attempts, the federal government finally issued licenses for cable television, and allowed the two carriers to build infrastructure for cable distribution.

Restructuring the Telecommunications Sector

Reform of the sector was difficult for the Labour government because of strong opposition by trade unions and the Left to any reduction in government ownership. Thus, initial moves to privatize either Telecom or OTC had to be dropped in favor of a compromise under which Aussat was privatized as a basis for a private sector second carrier.[44] Restructuring involved the merging of international carrier OTC with domestic carrier Telecom Australia, licensing of a second full-service carrier, further mobile telephone licenses, and introduction of competition over a full range of domestic and international services. The immediate advantage of restructuring was the likelihood of lower costs through increased efficiencies, higher quality network and service delivery, sharper customer focus, cost containment, and product innovation.

In September 1990, the government decided to reform the telecommunications monopoly, merge Telecom and OTC, and sell off Aussat. As part of the government's reform process, Telecom merged with OTC to become Australian and Overseas Telecommunications Corporation (AOTC), now known as Telstra. In July 1991, seven new telecommunications acts came into force, which provided for the sale of Aussat to form the basis of duopoly competition with the merged carrier Telecom/OTC. Among the main provisions were:

- the previous service-based monopolies were replaced with a network facilities-based duopoly;
- domestic and international resale were permitted;
- provision of mobile service was initially restricted to Telstra and Optus; Vodafone was later licensed as a third mobile operator.

Public Access Cordless Telephone Service will be opened to full
competition following development of a licensing system;

- prices for certain services are subject to price control;
- arrangements and procedures designed to prevent abuse of market power are administered by Austel;
- carriage of telecommunications services are to be open to full competition from July 1997.[45]

The winning bidder for the second carrier license was Optus, in
which BellSouth and Cable & Wireless each hold 24.5 percent, and
Mayne Nickless, an Australian transport group, holds 20 percent.
Optus stated that it would invest US$3 billion over five years to pro-
vide competitive long distance and international access to all Aus-
tralians, and digital cellular service to 80 percent of the population.
The dramatic growth in cellular subscribers since 1990 is a response
to the duopoly policy, which stimulated investment in cellular networks and some competition in services and pricing.

Thus, Australia chose to introduce competition in basic services
without privatization. However, the Conservative government elected
in 1996 pledged in its campaign that it would privatize Telstra. It
did not elaborate on how it would manage this transition while also
opening the duopoly to further competition in mid-1997.

NEW ZEALAND

Open for Competition

New Zealand was the first country to open fully its telecommunications market to competition. Until 1987, virtually all telecommunications services were provided by the post office. The reformist
Labour government split the post office into three sections: Telecom,
New Zealand Post, and Postbank, which were to be privatized. In little more than three years, New Zealand telecommunications under-
went a major structural change from a monopoly government
department to a private company in a competitive and largely unreg-

ulated environment. The current regulatory environment is the result of a series of decisions made during the years 1985-90, with the overall goal of making the New Zealand economy more internationally competitive, and in recognition that an efficient and competitive telecommunications infrastructure was essential to achieving that goal.[46]

The restructuring was done in three phases. In phase one, during the years 1985-87, the initial impetus to reform was the former Labour government's desire to improve efficiency and performance of the public sector. It separated commercial from noncommercial functions, and reorganized commercial functions as businesses. The government corporatized the telecommunications business activities of the New Zealand Post Office as Telecom New Zealand in 1987. The Telecommunications Act of 1987 ordered partial liberalization of telecommunications services, including customer premises equipment and value-added services. In phase two, the Telecommunications Amendment Act of 1988 authorized full removal of restriction to entry to the telecommunications market. The Radiocommunications Act of 1989 provided for management of some parts of the radio spectrum based on a legal scheme of tradable rights. In phase three, in 1990, Telecom New Zealand (TCNZ) was privatized.

Regulation

The Ministry of Commerce assumed responsibility for providing policy advice to the government on telecommunications following the breakup of the old post office into three state-owned enterprises, including Telecom Corporation of New Zealand in 1987. The Communications Division has a general manager responsible to the secretary of commerce, and provides policy advice to government through the minister of commerce. The government did not establish a detailed ad hoc regulatory regime and special industry regulatory authority, but instead relied on a recently strengthened antitrust statute, the Commerce Act of 1986, which prohibits monopoly conduct designed to deter competitive entry, and is overseen by the Commerce Commission and enforceable in the courts. The commission has the power of investigation and the right to seek pecuniary penalties in court. Consumer rights in telecommunications are protected under the Fair Trading Act of 1986, which is the general consumer protection law of New Zealand.[47]

The Major Carrier: Telecom New Zealand

Telecom New Zealand (TCNZ) provides local, long distance, and international services as well as many enhanced services. Since 1987, TCNZ has gone through substantial restructuring. Telecom New Zealand was sold in September 1990 for US$2.6 billion to a consortium of Bell Atlantic, Ameritech, and Freightways Holdings Limited (a New Zealand transport firm). The sale and purchase agreement entered into between the government and TCNZ's buyers specified that the shareholding by any single foreign entity be limited to 49.9 percent, and that at least NZ$500 million of TCNZ stock be released on the New Zealand market. In July 1991, Bell Atlantic and Ameritech sold 27 percent of their shares on the New York, London, and Asian exchanges. The sale was very successful, with shares selling at a 25 percent premium. Bell Atlantic and Ameritech were given three years to reduce their holdings to 49.9 percent, and completed the sell down in 1993.

The government retains a "Kiwi" share, which ensures that the company adheres to foreign ownership restrictions and that it meets certain service obligations which are specified in the Articles of Association. These articles require that free local calls remain available to residential subscribers (although TCNZ may introduce optional tariff packages), that residential line rentals do not rise faster than the consumer price index (CPI) (unless the overall profitability of TCNZ's regional operating companies is unreasonably impaired), that line rental to residential subscribers in rural areas will not be higher than in city areas, and that residential services will remain as widely available as they are at present. There is a provision within the articles for amendments to principles and modifications, with the agreement of the Kiwi shareholder (the government). The Telecom (Disclosure) Regulations of 1989 require that TCNZ release information to ensure that it does not cross-subsidize its competitive areas from areas where effective competition has not yet emerged.[48]

TCNZ was reorganized and downsized, and facilities were upgraded to prepare for deregulation and competition. TCNZ reduced its employees from 24,500 in 1987 to fewer than 15,000 in 1991, and adopted a holding company structure which reduced central management, decentralized decision making, and made each regional business directly accountable for customer satisfaction and financial performance. TCNZ rebalanced tariffs to reduce cross-subsidies and move to cost-based pricing, and invested $1.46 billion to modernize the network, including replacing old exchanges so that nearly 90 percent of the lines were linked to digital exchanges.

The New Competitors

The competition for provision of network services began in early 1991, when Clear Communications Ltd., a consortium of local and North American interests, began to offer long distance service. Clear is a consortium of New Zealand-based Todd Corporation, TV New Zealand, which has a national digital microwave network, New Zealand Rail, which owns a fiber-optic cable network, MCI, and Bell Canada Enterprises. Clear provides a full range of national and international switched and nonswitched services. It uses fiber owned by New Zealand Railways and optical cable installed by the Electricity Corporation across Cook Strait between the North and South Islands.[49] Clear now offers competitive telecommunications services over 80 percent of the country.

Netway Communications Ltd., a joint venture between TCNZ and Freightways Services Ltd. of New Zealand, provides electronic message services, data transfer, voice, and data network management. Another competitor is Optus, which offers trans-Tasman service between Auckland/Wellington and Australia via Aussat.

Both TCNZ and BellSouth operate cellular services. BellSouth paid NZ$25 million for TAC A digital cellular frequencies in 1990. Telstra of Australia has also acquired spectrum rights to provide GSM digital services.[50] In broadcasting, commercial activities in television and radio were corporatized and entry restrictions into broadcasting were abolished. Commercial news and sports channels are available via satellite and cable.

Auctioning Spectrum

The Radiocommunications Act of 1989 radically overhauled New Zealand's spectrum management policies. The intent was to create an open market in spectrum so that radio frequencies could be bought and sold like other commodities. The law authorized the government to define and auction long-term property rights to radio frequencies, creating a two-tiered system of spectrum property rights consisting of "management rights" to own a nationwide band of frequencies for 20 years and "license rights" that specify uses and users of the band.

The initial tenders generated NZ$45.6 million. The government initially chose a controversial "sealed bid, second price" tender, in which the highest bidder obtained the license, but paid the amount of the second highest bid. This approach resulted in some huge disparities between first and second bid prices. It was not unusual for first and second bid prices to differ by a factor of ten, and in extreme cases, bids differed by factors of 1,000 and more.[51]

While traditional sealed bidding has been used for subsequent tenders, more fundamental issues concern the difficulty in implementing major reform because of political pressure from incumbents, uncertainty and complexity of the transition, and competition policy concerns raised by the small number of commercial entities that could take on spectrum management. Nevertheless, the New Zealand experiment pioneered auctions as an alternative to conventional methods of allocating frequencies; auctions have now been adopted for some frequency bands in the United States and are being considered by several other countries.

The New Zealand Model

New Zealand's approach of relying on existing antitrust and consumer protection legislation exclusively will be important to monitor as a potential model for other countries (compared with, for example, the United Kingdom and Australia, which set up new telecommunications regulatory bodies). While this approach appears to be generally viable, the result is that any significant dispute must be resolved by the courts, where proceedings can be lengthy and judges are likely to lack expertise in telecommunications.

Clear, the new long distance carrier, has resorted to two legal proceedings in New Zealand High Court, alleging abuse of the dominant position by TCNZ in issues of interconnection in local and toll markets. Users filed a complaint with the Commerce Commission on leased circuit prices, also alleging abuse of TCNZ's dominant position. There have also been concerns about delays in decisions from the Commerce Commission and the cost of court disputes.

UNRESOLVED ISSUES

How Much Competition?

Competition means more than opening the door to new entrants. In industrialized countries where one carrier is already in place, it becomes critical to establish rules of the game such as interconnection, unbundling of facilities, and equal access. There need to be safeguards to prevent potential abuse of business information that the dominant carrier obtains from connecting carriers, and disclosure by the dominant carrier of technical information, such as network design and traffic data that other carriers need in order to plan and operate their networks. Without these safeguards, new entrants will be at a significant disadvantage. For example, the president of one of the major Japanese NCCs has stated publicly that "There is no real competition in Japan."[52]

While telecommunications markets may be legally open, there is no assurance that there will be multiple competitors in every market segment. New Zealand's market is completely open, but its small population makes it less attractive for multiple operators than more densely populated countries. Canada and New Zealand allow competition in the local loop, but there are few competitors. Canada is also unlikely to see many carriers vying to serve its remote areas. However, neighboring Alaska does have two long distance carriers that serve its villages as well as its cities, demonstrating that rural areas need not be unattractive.

Countries that have adopted the duopoly model may find that two operators are not enough to generate many competitive benefits. In cellular services, for example, two operators are likely to offer only small variations in price or service. Duopolists may also invest in duplicating facilities even when demand does not justify the capacity, as appears to be the case in parts of Australia. Opening markets further to full competition may actually result in the formation of new monopolies, such as alliances of cable and telephone companies to offer video dial tone. As the strongest competitors drive out or swallow up smaller players, strategic alliances among telephone, cable, and wireless operators are likely to result in new oligopolies to provide broadband services in industrialized countries.

Competition vs. Overinvestment

In the initial years of the British duopoly, Mercury gained a very limited foothold, concentrating on the urban business market, primarily the City, London's financial district. Now that the United Kingdom has opened the entire telecommunications industry to competition, will there really be effective competition among local and long distance services? If so, will the proliferation of network provision lead to wasteful duplication, overinvestment, and the creation of uncertainty that would hold back development of a fully effective, modern infrastructure? The British regulators think not. Oftel contends that duplication appears to be offset by the benefits of competition through innovation, better marketing, larger volume, and improved efficiency. Secondly, concerns about the level of investment tend to underrate the natural caution of investors in conditions of unlimited competition. Also, reliance on investors for development decisions has introduced a factor of stability.[53]

The verdict is still out on the extent to which Oftel's predictions will be borne out. Other countries will also have to weigh the benefits of fully open competition with the risk of overinvestment in facilities. Australia has witnessed enormous investment in optical fiber as both of its duopoly carriers built fiber rings around the continent. Carriers in the United States also went on a building frenzy after the divestiture of AT&T, as the competing long distance carriers invested in nationwide fiber networks, building their own redundant capacity to ensure reliability.

ISSUES AND CHALLENGES

Liberalization without Privatization: Keeping the Cash Cow

France, Germany, and Australia have shown that it is possible to introduce some competition without privatizing the government monopoly. For all three countries, this appears to be a transitional strategy, with plans now in motion for privatization. Still, telecommunications administrations that remain government monopolies are vulnerable to being used as cash cows by their governments. For example, France Telecom had a net income of more than $1.6 billion in 1994. The French government has forced France Telecom to invest in other firms, including two state insurers, and Groupe Bull, an ailing computer conglomerate.[54] Germany has used the profits of Deutsche Telekom to help finance the modernization of eastern Germany. Deutsche Telekom will have spent some DM 60 billion ($35 billion) renovating eastern Germany's telephone system; the company's debt rose to DM 140 billion by 1995.[55]

Industrial Policy

France's strategy in the 1970s and 1980s was to use investment in telecommunications facilities to stimulate not only economic growth, but also its own telecommunications industries, and to generate consumer interest and demand for telecommunications and information services. Although this strategy has enabled France to create one of the most modern digital networks in the world, access to services is still much more limited than in the competitive U.S. environment. Also, the Minitel, which brought French consumers into the computer age and spawned online services, is now an electronic antique. Have the French locked themselves into obsolete technology, or will they be able to translate their investment in computer literacy into widespread utilization of the Internet and other new interactive services?

Other countries, such as Canada, are facing the duel challenges of preserving domestic industries while facing global competition. Toughening their own players by opening the doors to competition seems a better long-term strategy than protectionism, which is now being challenged under international trade agreements. Perhaps ironically, the United States, which in the 1980s shunned industrial policy as harmful and unnecessary interference with the marketplace, is now embracing it in its telecommunications and electronics policies. The so-called "information superhighway" is intended to give all Americans access to interactive data and video services. But the model is not one of government investment or subsidy; rather, it serves as an incentive for industry to upgrade facilities, and for projects to demonstrate health, education, and community applications (see chapter 18).

The Gorilla Next Door

While policy makers and regulators in many countries are struggling to keep up with technological change, Canadians face an extra challenge: new technologies and services available in the United States can easily cross the border. Satellite signals are an obvious example; the footprints of U.S. fixed and broadcasting satellites cover much of Canada. But pricing differences can also force policy changes. For example, Canadians found that it was cheaper to call from Vancouver to Toronto by routing calls through the United States than by sending them across Canada. An industry sprang up that would forward Canadian calls across the border where discount U.S. carriers would pick up the calls and transmit them to the nearest border point, and a transborder carrier would deliver the call back into Canada. Faced with significant transcontinental bypass, Canadian regulators were forced to consider allowing more long distance competition to reduce domestic long distance rates.

France and Germany also face a gorilla in their backyards; in this case, the European Union, to which they belong, has mandated separation of regulators and operators and introduction of public network competition by 1998. EU policies are driving members to act faster than some governments or powerful unions would have otherwise permitted. Australia and New Zealand, however, with much more distant neighbors, have charted more independent courses. Interestingly, their choices have been very different on some issues. Australia set up a regulator, established a duopoly without privatizing Telstra, and was one of the last countries to introduce cable and satellite television. New Zealand threw the whole sector open to competition, auctioned off spectrum, and did without a regulator.

The Need for Regulation

The countries examined above demonstrate that the characterization of deregulation is largely a misnomer. While open competition may not require regulation, issues of standards, interconnection, and consumer protection may be even more important in a competitive environment. The U.S. experience shows that predatory pricing and unequal treatment of competitors are very real dangers in the transition to a multiprovider environment. Most industrialized countries have chosen to set up an arm's length regulator, such as Oftel in the

United Kingdom and Austel in Australia (the United States and Canada have a much longer history of commission-style regulation). New Zealand has chosen a different model, applying consumer protection and antimonopoly laws to the telecommunications industry through existing government agencies. This approach seems to be working in New Zealand. However, using the courts as the only vehicle to resolve carriers' disputes has been cumbersome, and some analysts have complained about the courts' lack of technical and economic expertise in deciding telecommunications cases.

It may be appropriate to use existing consumer protection laws and agencies to oversee consumer issues, although the concern about lack of specialized expertise also applies. Consumer protection is generally much more limited in the United Kingdom and other European countries than in the United States. Many countries are just now introducing itemized billing, and subscribers often face disconnection if they fail to pay a disputed bill. Waiting periods for installation are much longer than in the United States, and there are typically no sanctions if a carrier does not provide adequate quality of service. Competition may improve quality of service, but consumers are still likely to need safeguards. State regulators in the United States have found that most consumer problems, including billing and service disputes, do not disappear with competition.

CHAPTER 7

The European Union: Telecommunications Policies and Initiatives

"According to EU estimates, by the year 2000 more than 60 percent of all jobs in the European Union will be strongly information based and therefore relate strongly to telecommunications."

—HERBERT UNGERER, European Commission[1]

EUROPEAN TELECOMMUNICATIONS

The telecommunications sector in Europe shares many of the challenges highlighted in the previous chapters. These include technological changes, new markets, especially in data communications and in value-added services, and a greater role for market forces in regulation and standardization. However, European nations face two other important factors: the role of the European Union in shaping a "Single Market," and the political and economic transformation of Eastern Europe. This chapter addresses the European Union's role; Eastern European issues are addressed in chapter 11.

The European Economic Community was established with six founding member states in 1957. The Community has been enlarged several times since its foundation, to a current membership of 15 countries with a total population of over 320 million, namely, Austria, Belgium, Denmark, Finland, France, Germany, Greece, Ireland, Italy, Luxembourg, the Netherlands, Portugal, Spain, Sweden, and the United Kingdom. With the addition of Eastern Europe, the total European population is over 470 million.

Under the framework of the Maastricht Treaty, which came into effect in 1993, the member states of the European Community are in the process of creating a broader European Union that will remove monetary as well as trade barriers, and harmonize many social and economic policies. There are pressures to expand the Community because of the success of the single market program and the collapse of communism in Central and Eastern Europe. To join the EU, a state must be both European and democratic, and able to integrate fully into the economic activities of the Community after a relatively short transitional period of adjustment.[2] Several of the young democracies in Eastern Europe intend to apply for membership as soon as their level of economic development makes this possible. Poland, Hungary, the Czech Republic, and Slovakia hope to join by the year 2000. Peripheral nations, including Turkey, Cypress, and Malta, also want to join.

Telecommunications is an important sector in the European economy; it represented just under 3 percent of the Community's GDP in 1990, and is expected to grow to 5 to 7 percent by the end of the century.[3] Seven of the world's 15 largest equipment suppliers are European. The EU estimates that more than six out of ten jobs depend on information and communication technology, directly or indirectly;[4] the telecommunications sector alone provides about 1 percent of industrial employment.[5] Yet, access to telecommunications facilities and services ranges widely across Europe, with telephone densities ranging from more than 68 lines per 100 population in Sweden, to about 35 in Portugal and Ireland, and fewer than 20 per 100 inhabitants in most of Eastern Europe.

The European Union sees telecommunications not only as an important industry in itself, but as a vital component of the expansion and integration of the European economy. "Telecommunications are a major factor in the completion and success of the single European market, because they form the backbone of the information revolution: they stimulate industrial productivity, promote the growth of a service economy, and provide the link between industry, services and the market. As technology evolves, it is making telecommunications one of the driving forces behind market growth and the future prosperity of Europe."[6]

The EU market for telecommunications equipment and services has grown rapidly in new and competitive services such as value-added services and mobile communications. The market is becoming more competitive as a result of restructuring, as well as the arrival of new operators such as IBM, AT&T, GE Information Services, and EDS (Electronic Data Services). There are growing VAS information retrieval and closed-user groups such as financial transactions or electronic trading, using Electronic Data Interchange (EDI), and electronic messaging. Some information companies are establishing niche markets such as news services, financial services, and airline reservations systems. Others are setting up global networks with a wide range of services.

In 1989, the European Community had a trade deficit of $326 million in telecommunications equipment, primarily resulting from terminal equipment imports from Asia, such as telephone sets and facsimile machines. By 1992, it had reversed this trend, generating a $1 billion surplus. The largest net exporters were Germany and France, now joined by Sweden and Finland in the expanded 15-member European Union. Major exports are transmission and line and switching equipment, much of it destined for developing economies in Eastern Europe, Asia, and Latin America that are investing heavily in infrastructure.[7]

Europe has faced a double challenge to its telecommunications industries, from both the convergence of telecommunications, data processing, and broadcasting, and the opening of markets first in the United States and then elsewhere in the world. The pace of technological change no longer allows for the costly duplication of research and development efforts of "national champions" nor differences between standards. For example, there were six incompatible mobile communications systems in Europe before 1990, and three videotex systems. Also, national monopolies have stifled innovation and deprived suppliers of markets, because more than 95 percent of equipment contracts were reserved for national suppliers.[8] These constraints have also affected users, who want more innovative services and internationally competitive telecommunications prices in order to access the information they need to compete in global markets.

TOWARD AN INTEGRATED MARKET

Initial Steps

The European Union agency primarily responsible for telecommunications policy is the Directorate General of Telecommunications, the Information Market and Exploitation of Research (DG XIII) of the European Commission. Directorates of industry and of competitiveness are also involved, particularly in efforts to open the telecommunications sector to competition. The EU has established three goals for telecommunications:

- promotion of an advanced telecommunications infrastructure;
- stimulation of a homogeneous regionwide market for services and equipment;
- encouragement of greater competitiveness among European industry.

To reach these goals, a Community policy in the telecommunications sector was launched in 1984, with three major objectives:

- to make available to users at minimum cost, with maximum efficiency and minimum delay, a wide range of services which will sustain the competitiveness of the European economy;
- to promote the competitiveness of the European telecommunications industry to ensure that it maintains a strong position at European and world levels;
- to enable network operators to confront, under the best possible conditions, the technological and industrial challenges they will face.[9]

The criteria adopted for achieving these objectives included transparency, equality, open competition, harmonization, and full consultation with all major players, particularly manufacturers and users.[10] Steps toward these goals are summarized in Table 7.1.

The 1987 Green Paper

In recognition of the rapidly changing telecommunications environment and the importance of telecommunications for the European Single Market, the Commission of the European Communities (EC) produced a Green Paper on Development of the Common Market for Telecommunications Services and Equipment in 1987. The Green Paper recognized that technically advanced and low-cost telecommunications throughout Europe would provide an essential infrastructure for improving the competitiveness of the European economy and achieving the internal market.

The Green Paper saw telecommunications as essential for the realization of the EC's single market program and proposed functional change in the organization of telecommunications as soon as possible.

TABLE 7.1
EUROPEAN UNION: STEPS TOWARD INTEGRATION AND COMPETITION

1987	Green Paper on Development of the Common Market for Telecommunications Services and Equipment
1988	Action Program
	ETSI (European Telecommunications Standards Institute) established
	Terminal equipment opened to competition
1990	Full competition in value-added services
	Green Paper on Satellite Services
1991	Full competition in data transmission
1992	EC Open Market (actually January 1, 1993)
1994	Green Paper on Infrastructure
	Green Paper on Mobile and Personal Communications
1998	Proposed date for full liberalization of all services including public voice telephony

The two main thrusts of the Green Paper were liberalization of terminal equipment and the partial opening up of telecommunications services; its major goals were:

- opening of terminal equipment markets to competition;
- progressive introduction of competition in services;
- separation of regulatory and operational activities;
- movement toward more cost-based tariffs.

It argued that PTTs should retain monopolies only over basic voice networks, and should be allowed to compete in all new enhanced services and equipment markets. Member countries were not requested to privatize. The emphasis was on organizational reform: to separate the regulatory function from business operations. Members could retain public operators' monopoly of infrastructure and basic (voice) service. The EC agreements permit a mixture of competition and cooperation. Under the notion of reserved services, the PTOs do not compete with each other at the national level; however, they actually do compete for business at the international level, e.g., in setting up hubs, and offering least-cost routing of international traffic.

In parallel, the EC began to develop a new scope of Open Network Provisioning (ONP) intended to define conditions under which the basic public network could be opened to rival private service providers. The aim of ONP is to provide a harmonized, intracommunity telecommunications infrastructure which allows nondiscriminatory and open access to reserved services.[11] ONP is similar to the U.S. Open Network Architecture (ONA), designed to ensure that access to network facilities is truly open to all service providers.

The Green Paper also recommended that public procurement contracts be opened to all European suppliers, including subsidiaries of foreign companies, on a nondiscriminatory basis. An agreement in 1990 progressively extended bidding rights to suppliers from all member states on contracts valued at 600,000 ECU and above. However, there is still resistance to allowing competition with national industries for contracts, and the market remains difficult for foreign suppliers to penetrate. For example, the EC agreement on open tendering allows tenders from EC-based companies to be chosen over cheaper non-EC based bidders if price difference does not exceed 3 percent.[12]

Terminal equipment was liberalized in 1988. There were also agreements on unified terminal approvals: equipment approved and certified in one country is valid for all member states. An agreement on services and ONP was adopted in 1989. It sanctioned full and rapid introduction of competition in all VAS, including electronic mail and database access and resale of basic data communications services from January 1993. However, it allowed the possibility of extending data monopolies to the beginning of 1996 for member states with undeveloped public data networks. Thus, Greece, Spain, and Portugal were allowed to delay competition in data transmission until 1996 in order to modernize their networks. All governments will be allowed to set license conditions to safeguard the general economic interest of their data networks.

In summary, priority actions of the 1987 Green Paper that have been implemented include:

- opening up the terminal equipment market;
- opening up the value-added services market;
- Open Network Provisioning (ONP);
- opening up public procurement contracts;
- mutual recognition of type approval for equipment.[13]

The Treaty of Rome

The Treaty of Rome, which established the European Community, provides an important basis for pan-European liberalization; the lines for admissibility of monopolies are extremely narrow. Article 85 of the treaty prohibits agreements on competition, but foresees conditions for authorization (exemptions); Article 86 prohibits abuse of dominant positions. These articles are somewhat analogous to U.S. antitrust laws. However, they apply to enterprises, as opposed to Article 90, which applies to states, and is the legal basis for the suppression of statutory monopolies.[14] To endorse proposals for competition in telecommunications, the Commission relied on Article 90, which gives the Commission power to curb monopoly organizations, and allows the Commission to issue directives binding on member states without their prior consultation. This approach was used for telecommunications to short-circuit debate in the Council on introducing competition. The European Court judged that the EC's use of Article 90 of the Treaty of Rome to issue directives was legitimate; its decision apparently ended drawn-out legal challenges.[15]

1992 and After

The European Community's 1992 goal of creating a regional open market (actually implemented in January 1993) was intended to remove technical and legal barriers to free movement of goods, capital, people, and services. The development of advanced communications networks and services is vital to the success of the single market. For example, with different regulations, crossborder data transmission has been expensive and error prone; in 1990, 90 percent of data traffic was within national boundaries.[16] And in 1992, it cost four times as much to make a call between two places 300 kilometers apart in different countries, as it cost to make a call over the same distance inside one member state.[17]

In 1992, the *Financial Times* observed: "In four years the EC has transformed the legislative landscape of the region's telecommunications."[18] However, there have been implementation problems. The European market is fragmented, with different and often incompatible national regulatory traditions, supplier relationships, and equipment standards. The PTTs are also resistant to change, and governments are reluctant to surrender sovereignty. Yet, there have been changes in thinking during the period. For example, former German Telecommunications Minister Christian Schwarz-Schilling defended monopoly in 1987, but in 1991 announced: "Competition is the rule; monopolies or restricted regulations are the exceptions to be justified."[19]

Toward 1998

The European Commission has prepared an Action Plan designed to hasten the construction of European "information highways." One of the major consequences of the drive to hasten information infrastructure initiatives has been accleration of the liberalization program for the telecommunications sector toward full-scale voice telephony and public network liberalization by 1998. Recognizing that future telecommunications services will be provided via a web of wire and wireless services, the Commission has focused on liberalizing wireless mobile communications that will be a vehicle for personal communications, satellite communications, which are necessary for regional and global systems, and cable television networks, which are seen as an entry into the multimedia world.

Satellite communications was liberalized by an EU directive adopted in 1994. A Green Paper on Mobile and Personal Communications was published in 1994, proposing market structures that should transform wireless-based services from premium services to mass-market deployment by the end of the decade. A draft cable directive proposes to implement the use of existing or licensed cable television networks for telecommunications, opening EU telecommunications to multimedia development.[20]

January 1, 1998, is the deadline for lifting all remaining exclusive and special rights in the sector, in particular for public voice telephony and network infrastructure. The European Union has also drafted competition safeguards in the form of ground rules concerning licensing, interconnection, and sharing of universal service obligations. These rules are based on principles that such mechanisms should be nondiscriminatory, transparent, and as little restrictive as possible.[21]

In a transitional step to full competition, in July 1996, new operators in Europe were permitted to lease lines from cable operators, railways, and utilities, bypassing the monopoly operators. Several ventures were poised to take advantage of this opportunity to bypass the monopolists, including Hermes Europe Railtel, a consortium of

railway companies backed by financier George Soros that is building a
$1.2 billion trans-European backbone network.[22]

Satellite Policy

The 1987 Green Paper on telecommunications was followed in 1990
by a Green Paper on a common approach in the field of satellite communications in the EC. The Green Paper on satellites proposed four
major changes in the regulatory environment to fully exploit the potential of satellite communications for Europe:

- full liberalization of the earth segment;
- free access to space segment capacity;
- full commercial freedom for space segment providers;
- harmonization measures required to facilitate provision of Europe-wide services.[23]

Specifically, the Commission recommended:

- full liberalization of the earth segment, including receive-only
  and transmit-receive terminals, subject to appropriate type approvals and licensing for transmit stations where justified to implement necessary regulatory safeguards;
- free (unrestricted) access to space segment capacity subject to
  licensing procedures; access should be on an equitable, nondiscriminatory, and cost-oriented basis;
- full commercial freedom for space segment providers, including direct marketing of satellite capacity to service providers
  and users subject to above licensing procedures and Community
  law;
- harmonization measures to facilitate provision and use of
  Europe-wide services such as mutual recognition of licensing
  and type approval procedures, frequency coordination, and coordination of services to and from countries outside the Community.

However, progress in implementing these recommendations has been slow. A 1992 Commission paper, "The European Community and Space: The Next Steps," pointed out that "the underdevelopment of the European market and the weakness of European industry is the consequence of a regulatory regime that has restricted the potential growth of the satellite communications markets, together with a European internal market which is still fragmented."[24] While the EU has abolished restrictions in provision of satellite equipment (such as VSATs), restrictions remain on use of satellites for voice telephony and some aspects of direct broadcasting, as national telecommunications administrations and broadcasters resist the satellite-based competition.

RESEARCH AND DEVELOPMENT

The European Community has invested heavily in research and development (R&D) in information and telecommunications technologies (ITT). Among the priorities of European R&D have been:

- developing the skills of the workforce;
- increasing user friendliness of ITT systems;
- improving ITT standards;
- developing new ITT applications;
- promoting the use of R&D in knowledge management.

There has been a shift in recent years, especially in Europe, to support the introduction and efficient use of ITT applications—with less emphasis on the supply side, and more on the demand side of ITT. As some of the member countries are too small to manufacture, the EU has also emphasized diffusion policies for information technology, education and training, and research and development in information services, especially software development. The EU has also allocated funds to upgrade telecommunications in rural areas and in the poorer regions of the Community.

From 1987 to 1991, the EC spent ECU 2.1 billion* on research of information and communications technology, or about one-third of its total research and development budget. For the years 1990–94, it allocated ECU 3.4 billion, an increase of 62 percent.[25]

Research Initiatives

The RACE (Research on Advanced Communications for Europe) program is the research and development part of European telecommunications policy. Its aim is to promote "precompetitive" R&D, enabling Europe to set up an integrated broadband communications (IBC) network during the years 1995 to 2000 in order to provide broadband services that cannot be offered using ISDN. The objective is to have operators, users, standards institutions, manufacturers, and governments agree to a standards reference model and develop basic technologies for commercial use of IBC by 1995. Private MANs (metropolitan area networks) and flexible access systems are being installed for business areas.

The objectives of the RACE Program include:

- promoting the Community's telecommunications industry;
- developing the competitiveness of European network users;
- creating a single European market in broadband equipment and services;
- developing the poorest regions so that they will be able to benefit from advanced telecommunications.[26]

New standards are also developed through cooperative R&D projects, such as "precompetitive" RACE programs where corporations based in Europe are organized into pan-European R&D consortia for projects formulated under the overall direction of the EU. Standards formulated in the process of R&D in one project are made available to

---

* An ECU (European Currency Unit) is worth about US$ 0.80.

participants in other projects under the program, and will eventually be offered to other firms outside the program.

RACE supports crossborder cooperation between manufacturing and network operations in R&D and pilot testing. The EU also sponsors research and pilot projects in various fields of telematics, including social and administrative services involved with establishing the internal market and transborder activities; transport services; health care; distance learning; libraries; linguistic research and engineering; and services for rural areas.

For example, the DELTA Programme (Development of European Learning through Technological Advance) sponsors studies and projects aimed at using telecommunications and information technologies for distance education. The IMPACT Programme sponsors shared-cost projects to produce interactive multimedia information services aimed at a wide consumer and professional market.

Rural Projects

Half of the European population lives outside major conurbations. The quality of rural life and employment opportunities are threatened; there is a decline in the quality of education, provision of health care, and commercial services. There are also restraints on agricultural production, increasing agricultural productivity, and a decline in manufacturing, coupled with growth in new industries with increased reliance on information and telecommunications.

Telematic systems linked to better telecommunications can facilitate economic growth by bringing jobs to people and enabling more cost-effective support and services to be provided to rural residents and businesses. However, rural areas may lag in the introduction of new communication and information technologies and services. A Commission report on the future of rural society highlighted the important role of new technologies in rural development. Research and development on telematic systems for rural areas received unanimous support of the Council of Ministers, which allocated ECU 14 million for research.

The result was ORA (Opportunities for Rural Areas), a program designed to address some of the challenges of rural development through telematics. ORA funded R&D on telematic systems for rural areas, including pilot projects, infrastructure planning, and implementation strategies over a four-year period from 1992 to the end of 1995. Projects address applications of telematics for teleworking, agriculture, tourism and leisure, education, and social services.[27]

Another program designed for rural areas is STAR,[28] which is designed to reduce the disadvantages of remoteness by providing telecommunications facilities and services in rural areas of Greece, Ireland, Portugal, Italy (Mezzogiorno), the United Kingdom (Northern Ireland), France (Corsica and overseas departments), and regions of Spain. STAR had three main lines of action:

- helping to set up infrastructure offering advanced services such as digitalization, ISDN, links with advanced networks;
- support for projects designed to stimulate demand;
- encouraging use of advanced services, particularly by small- and medium-sized enterprises (referred to as SMEs).[29]

TOWARD A EUROPEAN INFORMATION SOCIETY

A 1993 White Paper by the European Commission on "Growth, Competitiveness, Employment"[30] acknowledged the importance of the information revolution to the future of European society, stressing that information and communication technologies have the potential to promote steady and sustainable growth, to increase competitiveness, to open new job opportunities, and to improve the quality of life for all Europeans. Following the adoption of the White Paper, the EU Council mandated Commissioner Martin Bangemann to investigate the means by which the market could meet the network requirements of the information society. The Bangemann report emphasized the need to ensure that a coherent and comprehensive network of networks would be accessible to all Europeans. It concluded that the private sector, in new forms of partnership with the public sector, could develop these networks. The most effective role for the EU would be to establish stable political and economic conditions within which this investment can occur. The report also proposed a single regulatory authority to oversee the development and management of these networks within a "Common Information Area."[31]

Responding to this vision, in 1994, the European Commission produced an action plan that proposed regulatory and legal frameworks and addressed technical and service issues on "Europe's Way to the Information Society."[32] The action plan reiterates many of the themes in previous Commission reports on telecommunications, stressing the need to open communication monopolies to competition, bring tariffs in line with rates in other industrialized regions, ensure interconnection and interoperability of networks, and establish a European regulatory framework. As part of its "Information Highways" initiative, the plan prioritizes projects in interconnected, high-speed networks, general electronic services such as electronic mail and interactive video, and telematics applications in teleworking, teletraining, and telemedicine.

To complement the information highway initiatives, the EU has earmarked nearly one-third of the available funds of the fourth Framework Programme for Research and Technological Development (1994-1998) to Information and Communication Technologies (ICT). Some 3.4 billion ECU will be used to fund research on telematics, an Advanced Communication Programme linked to RACE, and to establish an integrated digital broadband network within the EU, and information technology projects, particularly in software.[33]

INDUSTRIAL POLICY

The European Commission's strategies of mandating competition and funding research and pilot projects in new technologies and services are part of a deliberate industrial policy to provide affordable access to telecommunications for all Europeans and to improve the global competitiveness of European communications and information industries. The Bangemann report stresses that the biggest barrier to the development of a European information society is neither technical nor financial, but pyschological: Europeans need to understand how important information and communication technologies are to the EU economy.[34]

The Commission states: "Community industrial policy should promote permanent adaptation to industrial change in an open and competitive market. It is based on the principle of free trade and on the competitive functioning of markets around long term industrial and technological perspectives."[35] To implement these principles, the Commission proposes that the Community and member states:

- improve the functioning of the internal market;
- improve the functioning of the world market; and
- pursue positive adjustment policies aimed at building a favorable economic environment for private initiative and investment in the Community.

Strategies aimed at the internal market include developing Europe-wide standardization; opening up public procurement; abolishing national quotas on imports; providing a coherent legal framework for business, including mergers and intellectual property; and building trans-European networks, including telecommunications. The Commission recognizes that the creation of trans-European networks will also stimulate demand for information technology and electronic equipment.[36]

The Commission proposes that the Community's actions on behalf of the telecommunications equipment industry should have four main aims:

- establishing a true internal market within the EC, including liberalization of services and harmonization based on coherent and dynamic European standards;
- supporting technological research and development to provide

the European economy with advanced services and maintain global competitiveness in the industry;
- improving the position of companies providing terminal equipment;
- seeking conditions for fair competition in the world market.[37]

It is interesting to note that opening markets is the goal of the first and fourth aims, which in turn could help to stimulate demand and competition that might accomplish the second and third aims of maintaining global competitiveness and improving the position of companies' marketing terminal equipment.

While championing the private sector and competition, the EU has made significant public sector investments in research and development, subsidizing European initiatives and supporting companies that have not been commercially viable. Its approach is in marked contrast to the United States, which has promoted a private sector, competition-driven approach to building "information highways" and developing information services. Increasingly, however, the EU is seeking new private-public partnerships in order to develop its "Common Information Area."

SETTING STANDARDS IN EUROPE

Standards are important to ensure interoperability of networks and to facilitate the growth of markets for communications equipment.European standards for ONP are designed to ensure transparent, nondiscriminatory access for users and service providers in leased lines, packet-switched data services, and voice telephony. The European digital cellular standard, GSM (Global Service Mobile), has been adopted not only across Europe, but in much of Asia and Latin America. The growing importance of voluntary market-driven standards constitutes one of the major developments in Europe since 1987.

European Standards Organizations

The European Standardization System replicates the global system to a considerable extent (see chapter 16). Three bodies cooperate in several areas. The European Committee for Standardization (CEN) dates back to 1961. It is a regional organization under Belgian law, and the analog of the International Standards Organization (ISO). The European Committee for Electrotechnical Standardization (CENELEC) has been in operation for more than 20 years. The European process in developing standards through CEN and CENELEC is similar to that of the ISO. Standards are voluntary; however, once approved, they must be adopted and published without deviation in participating European countries.

The major regional telecommunications organization in Europe for more than 30 years was the European Conference of Posts and Telecommunications Administrations (CEPT), which was founded in 1959 to provide a forum for Western Europe in the fields of posts and telecommunications. Its original members have grown to more than 30 with the addition of Central and Eastern European countries since 1990.[38] CEPT's working group structure parallels that of the ITU, but has evolved as a consequence of structural reform in many countries. The main committees are separated into regulator and operator committees. The Radiocommunications Committee of CEPT was reformed as the European Radiocommunications Committee (ERC) in 1990. The ERC produces CEPT Recommendations and is the forum for coordination of radio policy. The ERC has three working groups: Frequency Management, Radio Regulatory, and Spectrum Engineering. ERC established the European Radiocommunications Office (ERO) based in Copenhagen. ERC is committed to offering wider consultation and greater emphasis on long-term planning.

CEPT typically decided standards through compromise (for example, the CEPT videotex standard is a combination of U.K. and French standards), or the lowest technical denominator of existing technical standards. CEPT has also been criticized for being overly bureaucratic and protective of the interests of its members, the European telecommunications administrations. However, to respond to the changing environment, CEPT has established a Commercial Action Committee (CAC), which aims to bring member administrations into commercial and competitive orientations.

Standards Setting in the European Union

The European Union is taking a new approach to technical regulations. European standards must conform to European Directives, laws which define the essential requirements necessary to protect public health, safety, and the environment. The goal is to establish a portfolio of European standards to replace national standards, a "global" European approach to testing, inspection, and certification.

Despite the existence of CEPT, the European Community decided that there was a need for an EC standards-setting body. The apparent reasons were not only to formulate standards and policies specifically for EC member states, but also to develop a more innovative and efficient mechanism for standards setting to respond to the rapidly changing technological environment. The EC established the European Telecommunications Standards Institute (ETSI) in 1988. ETSI's headquarters are in the high-technology region of Sophia-Antipolis in southern France. It is an open forum involving all interested parties, who work together on common problems in the telecommunications sector. ETSI's members include not only PTTs, but private network operators, manufacturers, users, and research organizations as well (see Table 7.2).[39] There are now more than 260 members from 32 countries. New members in the 1990s include the Czech Republic, Hungary, Poland, Russia, and Slovakia.

ETSI has three major fields of interest, namely telecommunications, the interface between telecommunications and information technology, and the interface between telecommunications and broadcasting. ETSI works with CEN and CENELEC on standards related to telecommunications and information technology, and with the European Broadcasting Union (EBU) on standards related to telecommunications and broadcasting.

The primary goals of ETSI are:

TABLE 7.2
ETSI MEMBERSHIP[40]

Type of Member	Percentage of Members
Manufacturers	49.5%
Public Network Operators	15.8
Administrations	9.8
Associate Members	6.0
Users	6.5
Research Bodies, Public Sector Providers, Others	12.4

- to speed up the standardization process;
- to provide greater transparency to the process;
- to boost the level of participation of all involved parties.[41]

The basic objective of ETSI is to set uniform technical standards for Europe that will be adopted by each member country, thus linking national networks and services and ensuring interoperability of equipment.[42] ETSI's structure is open, accessible, and nonbureaucratic. The objective is to grant any organization with relevant interests free access to the work in progress. This approach facilitates dialog about technical specifications between private and public operators and manufacturers of network facilities and terminal equipment. Compliance is voluntary.

"ETSI's approach to standards making is innovative and dynamic. It is ETSI's members that fix the standards work programme in function of market needs. . . . the fact that the voluntary standards are requested by those who subsequently implement them, means that the standards remain practical rather than abstract."[43] ETSI's General Assembly is responsible for the overall rules and budget. The Technical Assembly supervises arrangements for production and approval of standards: it puts topics on the work program, and assigns target dates for completion.

ETSI works through technical committees, subtechnical committees, and project teams for rapid creation of standards which are highly product-specific. It can also issue interim standards for topics not yet mature enough for formulation of a long-term standard. ETSI's innovative structure and procedures have many advantages for operators, manufacturers, and users. User participation alerts industry to user needs; users, in turn, have an opportunity to influence standards, and to get advance information on new technical developments.

European Standards and Spectrum Initiatives

The standards prepared within ETSI and approved by the ETSI Technical Assembly are called European Telecommunications Standards (ETS). ETSI has also adopted ITU standards such as ISDN. Its goal is to take the best from any source and avoid duplication of effort and proliferation of competing standards. After its first three years in existence, ETSI had approved 170 standards and about 300 standardization projects were underway.[44]

Mobile communications is one of the fastest growing telecommunications services in Europe, with an annual growth rate of 30 percent. Europe will need both common standards and a framework of cooperation among frequency experts to find room for growth of mobile services, as it is estimated that half of the phones may be cordless by the year 2000.[45] The European Commission is developing a close liaison with Central and Eastern European countries, which are investing in wireless systems to upgrade their urban networks and to extend service in rural areas.

ETSI has formulated two major pan-European wireless standards that have been adopted in other parts of the world:

- GSM (Groupe Special Mobile) is the group that originated in CEPT and continued work on Public Land Mobile Networks (PLMN) in ETSI. In recognition of its potential as a global standard, GSM now also stands for Global System for Mobile Communications. GSM PLMN is a digital cellular standard that ensures compatibility in access for any subscriber in any CEPT country. It includes facilities for automatic roaming, locating, and updating mobile subscribers plus accounting, licensing, subscriptions, and subscriber directories. GSM has been widely adopted in Asia and Latin America.

- DECT (Digital European Cordless Telecommunications) is a set of service standards needed for cordless communications. It is primarily for voice traffic, and is envisaged for residential, neighborhood, and business environments. The DECT standard has also been adopted in some Asian countries.

The European Commission has passed a resolution on the strengthening of a Europe-wide frequency coordination. The Commission is concerned with provision of harmonized frequency bands for new mobile systems; it therefore issued directives making it legally binding to assign the same frequencies in each member state for GSM, ERMES (European Radio Message System), and DECT.

An important component of the proposed trans-European networks (TENs) is a common standardized Euro-ISDN network. A Memorandum of Understanding (MoU) was signed by European network operators to enable European ISDN services to be offered across Europe. Standards issues include standards for a common range of services, user-network interfaces and protocols to ensure terminal interchangeability, and interconnecting national systems to provide international services.

ISSUES AND CHALLENGES

While the EU has made enormous strides in promoting competition, funding research, and streamlining standards setting, there are still significant challenges facing European telecommunications.

Industrial Policy

Western European nations have supported their own telecommunications companies, largely through preferential considerations on bidding for PTT contracts. While advocating competition, the EU has also channeled research funds to these "national champions." This emphasis on national champions has met with mixed results: some European analysts feel they are too protected, and not competitive; others respond that without government help, there would be very few telecommunications and information technology firms in Western Europe today. Yet with protectionist barriers coming down, European companies will have to be more innovative and efficient if they are to hold on to their own markets, let alone compete with foreign suppliers in other regions. The strategy of depending on former overseas colonies for export revenue is quickly becoming outmoded.

The EU's heavy investment in selected technologies has also had mixed results. While ISDN has been touted in Europe, it has not been widely adopted by users. The EU is funding research for development of integrated broadband communications (IBC) as a solution to the shortcomings of ISDN. The EU also made a major investment in research on high definition television (HDTV), hoping to develop a system that would become a global standard in the face of competition from Japan and the United States. However, the EU has abandoned the development of an analog standard for HDTV, in recognition that its efforts have been overtaken by developments in digital technology, led by U.S. companies. The EU has been more successful in promoting digital mobile communications; its GSM standard has been adopted widely throughout the world, except in North America.

While many of the EU's research projects are aimed at developing applications of telecommunications technologies that might not be developed in the marketplace, such as transborder exchange of medical records and imaging systems for libraries and museums, many of the others appear to be either usurping functions that the marketplace could fulfill, or ignoring other factors necessary for them to have any significant benefit. For example, the EU has sponsored research on EDI, aircraft information systems, and electronic publishing;[46] in the United States, research in these fields has been the responsibility of the private sector, in anticipation of a market for these applications. A major problem is the lack of private European venture capital to support start-ups; American venture capital firms have been the source of support for many of the most innovative information and communication technology companies in the United States.

The EU's rural communications projects aim to provide more diversified employment and more balanced economic activities in rural areas, and to "ensure that the introduction of information and communication technologies into rural areas does not contribute to further centralization of economic and administrative activities and the loss of cultural and economic diversity in rural parts of Europe."[47] Yet many additional factors besides communications infrastructure will influence these goals, such as labor costs and skills, transportation facilities, and financing, as has been found in North America and other regions.[48]

Telecommunications in Developing Countries

CHAPTER **8**

The Role of Telecommunications in Socioeconomic Development

"We believe that by the early part of the next century virtually the whole of mankind should be brought within easy reach of a telephone and, in due course, the other services telecommunications can provide."
—THE MAITLAND COMMISSION, *The Missing Link*[1]

INTRODUCTION

The past quarter century has been marked by dramatic technological developments in computers and telecommunications, and the growing importance of information in all aspects of human life. Access to information, and to the facilities to produce, store, and transmit information, is now considered vital to development, so that the classifications "information rich" and "information poor" may mean more than distinctions based on GNP or other traditional development indicators.

This chapter examines the role of telecommunications in socioeconomic development, what progress has been made toward bringing the "whole of mankind within easy reach of a telephone," and what problems remain. It then proposes strategies to achieve the goal of universal access to reliable and affordable telecommunications services.

THE INFORMATION GAP

In 1984, the ITU's Maitland Commission noted that telecommunications was a "missing link" in much of the developing world.[2] A decade later, policy makers were calling for a "Global Information Infrastructure" that would link everyone into a worldwide network, or more specifically, network of networks. Yet a majority of individuals and whole communities in the developing world still have no access to telecommunications. And where telecommunications facilities exist in developing regions and remote parts of the industrialized world, limited capacity and unreliability may leave users decades behind their better-equipped counterparts who are able to take advantage of converging technologies and new services.

The "information gap" between industrialized and developing countries is reflected in their access to telecommunications resources. Although there has been a dramatic increase in telecommunications investment in the past decade, there are still enormous gaps between the developed and developing world in accessibility to telecommunications, and within the developing world, between urban and rural areas. While there are now almost 50 lines per 100 people in high-income industrialized countries, there is still an average of only one line per 100 in the poorest countries.

Of course, telephones are not evenly distributed throughout the population in developing countries. Not only do wealthier people have greater access to telephones, but the gaps are even greater between urban and nonurban areas. There are almost three times as many telephone lines per 100 in the largest city of lower-middle-income countries as in their rural areas, and more than seven times as many lines per 100 in the largest city of low-income countries as in their rural areas. These gaps are even more significant given the fact that more than 50 percent of the population, and as much as 80 percent in the poorest countries, lives in rural areas (see Table 8.1).

TABLE 8.1
ACCESS TO TELECOMMUNICATIONS

Country Classification	Teledensity (lines/100 pop)		
	Entire Country	Largest City	Rest of Country
High-Income Countries	48.8	51.7	48.5
Upper-Middle-Income Countries	12.9	21.9	10.6
Lower-Middle-Income Countries	8.1	19.0	6.8
Low-Income Countries	0.9	5.2	0.7

Derived from ITU, *World Telecommunication Development Report*, 1995.

Even in most middle-income countries, including those with access to satellites for domestic communications, the infrastructure is still extremely limited. Brazil and Mexico, which now both have their own domestic satellites, have an average of seven and nine telephone lines per 100 population respectively; Indonesia, the first developing nation to have its own domestic satellite, has only 1.3 telephone lines per 100. Rural densities are much lower. Several developing countries have also leased Intelsat transponders for domestic use. Yet these satellites cannot close the gap without an investment in "the last mile" of the network to reach the population (see Table 8.2).

THE TECHNOLOGICAL CONTEXT

Promising New Technologies and Services

An analysis of strategies for increasing access to telecommunications in rural and developing regions must be placed in technological context. Telecommunications technologies have changed dramatically in the past decade, and many recent innovations offer promising solutions for extending services at lower costs than were generally thought possible. Perhaps the most telling evidence of change is the cover of the Maitland Commission report itself, which showed two rotary dial telephones. This is not to imply that digital switching did not exist by 1984, but rather that it was not considered necessary or

TABLE 8.2
THE LAST MILE PROBLEM:
DEVELOPING COUNTRIES WITH INTELSAT DOMESTIC LEASES

Country	Teledensity (liner/100 pop)
Algeria	4.1
Argentina	14.1
Brazil	7.4
Chile	11.0
China	2.3
Colombia	9.7
Côte d'Ivoire	0.8
India	1.1
Indonesia	1.3
Libya	4.8
Malaysia	14.7
Mexico	9.3
Morocco	3.8
Mozambique	0.4
Nigeria	0.3
Sudan	0.2
Thailand	4.7
Venezuela	10.9
Zaire	0.1

Derived from Intelstat, *Annual Report,* and ITU, *World Telecommunication Development Report,* 1995.

perhaps even appropriate for developing regions. A second indicator is that the commission specifically identified only telephone service, and proposed access "in due course [to] the other services telecommunications can provide." Today, many of those services could be available as soon as telecommunications service is provided.

Among the recent innovations that can make telecommunications services more reliable and cheaper to provide in developing countries are advances in radio technology such as cellular radio and rural radio subscriber systems. Cellular networks can be used for pay phones and other "fixed" services as well as for mobile communications. "Wireless local loop" technologies can link subscribers to the network without the need for laying cable or stringing copper wire. Small satellite earth stations (very small aperture terminals or VSATs) are proliferating in developing regions, usually for distribution of television signals. However, VSATs can also be used for interactive voice and data, and for data broadcasting. Satellite terminals can also serve as hubs for wireless local networks.

These technologies can provide access to services that may have many applications for developing regions, such as:

- **Data Broadcasting:** The flow of information within the developing world has been hampered by the cost of distribution and by the lack of access to telecommunications facilities in rural areas. VSATs now make it possible for wire service information to be disseminated to virtually any location. Wire service copy is transmitted by satellite from a hub earth station which may be shared with other data, voice, and video customers. Reuters uses this VSAT technology for news service feeds to Latin America. In Asia, China's Xinhua News Agency transmits news via AsiaSat.
- **Electronic Transactions:** Computers combined with telecommunications enable organizations to conduct business from virtually any location. Banks can transfer funds internationally using the SWIFT network; airlines can book reservations from ticket offices, airports, and travel agencies. Brokers and traders can buy and sell coffee, soybeans, copper, or petroleum electronically; agricultural cooperatives can use computer terminals to find where to get the best prices for their crops; and tourist lodges in rural areas can book reservations.
- **Electronic Messaging:** Facsimile transmission and electronic mail may be particularly viable alternatives to sending hard copies of correspondence and documents through the mail where service may be slow or unreliable. Managers and researchers located in different cities may exchange information; these technologies can also be used to link project staff in the field with each other and with headquarters.
- **Electronic Meetings:** Managers, development experts, and project staff may now stay in touch electronically rather than having to travel for face-to-face meetings. Audioconferencing allows

participants at several sites to participate in the same meeting, while computer conferencing allows for interaction among group members at their convenience by reading and contributing to a discussion stored on a host computer. These electronic meetings do not offer the richness of face-to-face interaction, but they may be particularly important to supplement travel to meetings where transportation costs severely strain limited travel budgets.

- **Virtual Telephone Service:** Voice-mail systems can do much more than replace analog answering machines. TeleBahia in northeastern Brazil is using voice messaging technology to offer "virtual telephone service" to people who are still without individual telephone service. Customers can rent a voice mailbox for a monthly fee. Callers can leave messages in their mailboxes, which the subscribers can retrieve from a pay phone. (A similar approach has been used in the United States in some homeless shelters so that job-seekers can be contacted by prospective employers, and in camps for migrant farm workers so that their relatives can stay in touch with them.)

- **Distance Education:** Audioconferencing over terrestrial or satellite circuits can enable isolated students taking correspondence courses to participate in tutorials with their instructors and interact with other students. Students may also communicate with instructors and fellow students using computer conferencing, and send assignments back and forth via electronic mail. Televised courses, often disseminated by satellite, can reach students in classrooms and adults in community centers or their workplaces.

- **Access to Databases:** Using the Internet, researchers in developing countries can access databases anywhere in the world, such as those at the Food and Agriculture Organization (FAO) in Rome and the National Library of Medicine in Bethesda, Maryland, as well as databases for agriculture and energy in India and for development project management in Malaysia. Alternatives to online searching are CD-ROM databases, which offer full text of journals, video images, and other graphics searchable using a personal computer. The advantages of CD-ROM include vast storage potential, low cost, durability, and ease of use; users do not have to pay for connect time to online services. Of course, the disks must be frequently updated to keep information current.

- **Dissemination of Information:** Journalists, development agencies, and nongovernmental organizations (NGOs) can use telecommunications to collect and disseminate information for rural areas. For example, development workers and reporters in the field could send in reports by electronic mail or facsimile; these materials could then be edited and published in newsletters in the city. Newsletters could be transmitted electronically either directly to the communities or to regional centers for duplication and dispatch to communities or schools and clinics in their territory. Information obtained from various sources, such as news services, databases, and teleconferences, can also be disseminated to development workers throughout a country or region via facsimile or electronic mail.

THE IMPORTANCE OF INFORMATION

Information is critical to the social and economic activities that comprise the development process. Information is obviously central to education and training, but it is also important to health services, where providers need advice on diagnosis and treatment of cases beyond their level of expertise or the capacity of local facilities. As discussed in chapter 3, information is also critical to economic activities ranging from agriculture to manufacturing and services.

If information is critical to development, then telecommunications, as a means of sharing information, is not simply a connection between people, but a link in the chain of the development process itself. The role of telecommunications in transmitting information can be particularly significant in rural areas, where alternative means of obtaining and conveying information, such as personal contact, transport, and postal services, are likely to be less accessible.

Distance represents time, in an increasingly time-conscious world. In economies that depend heavily upon agriculture or the extraction of resources (lumber and minerals), distance from urban markets has traditionally been alleviated only with the installation of improved transportation facilities, typically roads. But transportation links still leave industries without the access to information that is becoming increasingly important for production and marketing of their commodities.

Another disadvantage faced by many developing countries is economic specialization. As they strive to diversify their economies, timely access to information becomes even more critical. For example, in a manufacturing environment that increasingly depends upon outsourcing of component parts and just-in-time delivery to assembly plants, the farther the production facilities are located from markets, the more some other economic characteristic (for example, lower wages) must compensate for the time/distance penalty.

As developing countries also join the global market by attracting multinational corporations, establishing joint ventures, and developing service industries, they soon recognize the need for a reliable and modern telecommunications network. Telecommunications is also vital to the emerging information sectors in developing regions. The great distances between the major research institutes and development centers and the vagaries of postal services and expense of airfares mean that experts are isolated from each other as well as from the people they are trying to help. For example, the National Research Council points out that for Africa sharing information is vital if Africans are to contribute to finding solutions to their own development problems:

> Economic development in Africa will depend heavily on the development of the information sector. Countries will need the ability to communicate efficiently with local and overseas markets to determine where they may have comparative advantages for supplying their products to consumers or to purchase essential imports, based on current prices and services. Many of the economic development problems facing African countries have scientific and technological components that will require solutions to be developed in Africa by African scientists. . . . Lack of information is a critical constraint.[3]

Advances in communications technology now make it possible to extend reliable communications to any village or camp, whether in the desert or the jungle or on a remote island. But the progress in taking advantage of these technological advances to meet the needs of rural people in the developing world has been painfully slow. Until recently, telecommunications was considered a luxury to be provided only after all the other investments in water, electrification, and roads had been made—and after all the demand for telecommunications services in the cities had been met. Instead, telecommunications should be considered a vital component in the development process—a complement to other development investments—that can improve productivity and efficiency of rural agriculture, industry, and social services, and can enhance the quality of life in developing regions.

The importance of information in social, cultural, and economic development was recognized by UNESCO's MacBride Commission, which stated: "There can be no genuine, effective independence without the communication resources needed to safeguard it."[4] While characterized initially in terms of news flow, the imbalance in access to information includes access to the tools of telecommunications and information, and the necessary expertise to utilize them effectively.

Research conducted during the past decade has shed considerable light on the ways and extent to which telecommunications contributes to development. Several studies sponsored by the International Telecommunication Union (ITU) in collaboration with the Organization for Economic Cooperation and Development (OECD) and by the World Bank, as well as by other development agencies, have documented the indirect benefits of telecommunications for economic activities and social services. These studies show ratios of benefits to costs of telecommunications usage from 5:1 to more than 100:1 from improved efficiency in managing of rural enterprises, time savings in ordering spare parts, savings in travel costs and time, etc.[5] These studies have been augmented by recent research on rural telecommunications in the United States.[6]

The above studies indicate that telecommunications can contribute to economic growth, not only of rural areas, but of developing nations as a whole. Yet for many years, it was difficult to find tangible evidence of this contribution at the national level. Numerous studies have demonstrated a strong positive correlation between telephone density and economic development measured by GNP per capita or a similar statistic. However, this simple correlation did not explain what appears to be a chicken-and-egg relationship: does investment in telecommunications contribute to economic growth, or does economic growth lead to investment in telecommunications?

However, a model developed by Andrew Hardy[7] has shown that investment in telecommunications can make a statistically significant contribution to economic growth as measured by Gross Domestic Product (GDP). Hardy's methodology is significant in providing for the first time a means of disentangling the chicken-and-egg relationship between telecommunications investment and economic growth. Hardy's analysis shows that the causality runs in both directions: while economic development leads to more investment in telecommunications, telecommunications also contributes to economic development. And the economic impact is comparatively greater in regions with low telephone density—which are likely to be rural and remote areas.

What kinds of information do people in developing nations need?

- In the South Pacific, a village health worker wonders when emergency assistance will arrive to combat an outbreak of cholera on her typhoon-ravaged island.
- In Southeast Asia, a farmer needs help to combat a fungus that is destroying his crop.
- In South America, a rural cooperative wants to know where and when to sell its soybeans to get the best price.
- In the Caribbean, a teacher puzzles over a correspondence course she is taking to upgrade her skills.
- In East Africa, a mother in the village worries about her sons who have gone to the city to look for work.

All of these people share a need for information. Without telecommunications, they may have to wait for days—or even weeks—to send and receive a message, or they may have to travel themselves to get the information. In many cases, they simply have to do without it.

These are not isolated incidents. They happen all too frequently in the developing world. Each is a source of frustration, delay, or despair for the people involved, yet together their impact is much greater—the inability to send and receive timely information may virtually undermine the development process.

INDIRECT BENEFITS OF TELECOMMUNICATIONS

Information has unusual economic properties: it can be shared without being transferred, and its benefits may extend to others besides those directly involved in the information transaction. The benefits of a telephone call may accrue both to the caller and the person called, but also to others who are not involved in the information transaction. For example, a farmer would benefit if an agricultural extension agent could contact an agronomist to find out how to eliminate a crop fungus, and a patient would benefit if the health worker at a rural clinic could get advice from a doctor at a regional hospital. But even more important in terms of development is the fact that the society as a whole will benefit from these uses of telecommunications, both in economic terms and in improved quality of life. Access to a physician can reduce mortality rates, and allow more people to be treated locally without having to be transferred to a hospital. Consultation with agronomists and veterinarians can improve crop yields and livestock production.

The ability to communicate instantaneously can facilitate the development process in three major ways, by improving:

- *efficiency,* or the ratio of output to cost;
- *effectiveness,* or the quality of products and services;
- *equity,* or the distribution of development benefits throughout the society.

Among the benefits of telecommunications for improving efficiency and productivity are the following:

- Price Information: Producers such as farmers and fishermen can compare prices in various markets, allowing them to get the highest prices for their produce, to eliminate dependency on local middlemen, and/or to modify their products (such as types of crops raised or fish caught) to respond to market demand.
- Reduction of Downtime: Timely ordering of spare parts and immediate contact with technicians can reduce time lost due to broken machinery such as pumps, tractors, and generators.
- Reduction of Inventory: Businesses can reduce the inventories they need to keep on hand if replacements can be ordered and delivered as needed.
- Timely Delivery of Products to Market: Contact between producers and shippers to arrange scheduling for delivery of products to market can result in reduced spoilage (for example, of fish or fresh fruit), more efficient processing, and higher prices for produce.
- Reduction of Travel Costs: In some circumstances, telecommunications may be substituted for travel, resulting in significant savings in personnel time and travel costs.
- Energy Savings: Telecommunications can be used to maximize the efficiency of shipping so that trips are not wasted and consumption of fuel is minimized.
- Decentralization: Availability of telecommunications can help to attract industries to rural areas, and allow decentralization of economic activities away from major urban areas.

Telecommunications has also made contributions to effectiveness in other areas, including distance education, where real-time instruction and tutorials are more effective than simply requiring students to study correspondence materials, and telemedicine, where information about the patient transmitted electronically can help distant specialists to diagnose conditions and recommend treatment.

Telecommunications contributes to equity by enabling the disadvantaged, including isolated rural people, the disabled, and the poor, to access information that would otherwise be very difficult or impossible to obtain.

Examples from the Developing World

Health Services

Telecommunications is used for three different functions in support of health care delivery:

- *consultation* gives advice to rural health workers, or directly to isolated patients;
- *training* of health care workers;
- *education* of target populations including expectant mothers, mothers of young children, groups susceptible to contagious diseases, etc.

Telecommunications plays an important role in health care delivery in many developing countries. Studies in India, Costa Rica, Egypt, and Papua New Guinea, all showed that about 5 percent of rural phone calls were for emergencies and medical reasons. The indirect benefits of these calls in terms of saved lives and reduced suffering are highly significant. For example, in the South Pacific, the experimental PEACESAT satellite network has been used to summon medical teams during outbreaks of cholera and dengue fever, and to coordinate emergency assistance after typhoons and earthquakes.

Many developing countries now rely on paraprofessionals for delivery of basic health services, particularly in rural areas. These health workers receive basic training in treatment and prevention of common health problems, but need supervision and assistance in diagnosing and treating uncommon diseases and coping with serious health problems. Telecommunications links between village clinics and regional hospitals or health centers can be used for consultation and supervision.

In Alaska, village health aides are in daily communication via satellite with physicians at regional hospitals. More than 100 Alaskan villages are equipped with earth stations that are used for the dedicated medical network, long distance telephone service, and television reception. The medical network is a shared audioconferencing system that allows heath aides from several sites to participate, and can also be used for in-service training.[8]

In Guyana, rural health workers called "medex" use a two-way radio network to communicate with headquarters in Georgetown to check on delivery of drugs and supplies, and to receive advice on major health problems. They may also request emergency evacuations and follow up on patients referred to hospitals. The Georgetown training staff run refresher sessions and "grand rounds" over the radio. At night, chatting over the radio helps medex reduce their sense of isolation and boosts morale.[9]

Similar health communication networks are found in other parts of the developing world. Flying doctor services in several East African countries (including Kenya, Tanzania, and Malawi) use two-way radio networks to link nurses at rural clinics with headquarters, and to coordinate the aircraft that transport doctors to the clinics and evacuate seriously ill patients.

SatelLife of Cambridge, Massachusetts, operates two store-and-forward satellites, Healthnet I and Healthnet II, for medical communications in Africa. Field reports from the Gambia cited improved efficiency of collecting epidemiological data from vaccine trials using Healthnet instead of an employee from the Ministry of Health traveling 500 kilometers every week to pick up data. Similarly, in Cameroon, Healthnet is used for logistics coordination, administration, and communication, instead of an administrator traveling from province to province.[10] African medical faculty and students can search remote databases and retrieve articles from sources such as *The New England Journal of Medicine.* They can also exchange information with colleagues across Africa via e-mail to share research findings and treatment strategies.

Distance Education

Use of telecommunications for distance learning may reduce student dropout rates and staff turnover at remote locations. The University of the South Pacific operates a satellite-based audioconferencing network linking its main campus in Suva, Fiji, with its agricultural college in western Samoa and extension centers in nine Pacific island nations. The system is used for administration of extension services activities and courses, tutorials for students taking correspondence courses, and outreach services to bring the resources of the University to the people of the region—through consultation, in-service training, and seminars by the United Nations and other development agencies. The benefits of this experimental network have been significant. The savings in travel time and costs resulting from conducting meetings over the network rather than bringing a representative from each location to Fiji have been at least ten times the cost of using the network. Dropout rates of correspondence students in courses with effective satellite tutorials have also been reduced.[11] Distance education also enables students to stay in their home countries rather than leave their families and jobs to come to Fiji. The University of the West Indies (UWI) also uses telecommunications for seminars and tutorials with students at extension centers throughout the Caribbean. China delivers continuing education to the workplace using satellite communications to distribute televised courses nationwide. Thailand and Indonesia operate "open universities" with satellite-delivered lectures.

The Monterrey Institute of Technology and Higher Studies (ITESM) in Mexico is using a satellite-delivered, compressed digital video system to enable faculty in Monterrey to reach students around the country. The network links 26 campuses using Mexico's Morelos satellite. As the Institute expanded its number of campuses, it was no longer able to supply qualified instructors at every site. Using the network, an instructor teaches a class in Monterrey and also reaches 2,000 students at campuses around the country. Students at the receiving site can communicate via computer with their instructor during class. The network can also be used by Mexican businesses; for example, banks with multiple branches can provide training on-site rather than sending their staff to Mexico City.[12]

Agriculture and Fisheries

Telecommunications can also help farmers to market their crops. In Sri Lanka, small farmers were able to use newly installed rural telephones to obtain prices of coconut, fruit, and other produce in Colombo. As a result, instead of getting 50 to 60 percent of the Colombo price for their products, they were able to get 80 to 90 percent of the urban price.[13] A farm in Mexico's Baja California produces "designer vegetables" that command high prices in the United States, with production and shipping coordinated from the San Francisco Bay Area. Kenyan farmers export fresh flowers to Europe, taking orders by fax for overnight deliveries to flower markets in London, Paris, and Frankfurt.

Improved coordination of transportation is important in the marketing of perishable products. In the Cook Islands, agricultural officers notify the shipping agent of the amount of fresh fruit ready to be picked up from each island. The shipper then sets the schedule and notifies the farmers when to have the fruit ready for the ship's arrival. Without this information, farmers risk spoilage if the fruit is picked too soon, and the shipper risks major delays if the boat must wait at each port for the fruit to be delivered.[14]

Native fishermen in northern Canada have found similar benefits from using two-way radios to call bush pilots when they have a catch ready to be flown to market. Previously, the pilot would return on a specified day, with the risk that a large catch might have spoiled or a small catch would not amount to a full load.[15]

Transportation Savings

Telecommunications can also reduce the need for travel. In India, the benefits to villagers of using long distance public telephones were about five times the cost of the call, taking into consideration bus fare and time lost from work in traveling to town to deliver the message.[16] A study conducted for the ITU in Yemen estimated that the use of telecommunications to facilitate organization of the transport and storage of goods would save about 15 percent of all transportation costs, in addition to reducing delays caused by equipment breakdowns by a factor of five to eight.[17]

Related to travel savings is the ability to make more efficient use of transportation. This benefit could be extremely important for developing countries dependent on imported fossil fuels that are a major drain on their foreign exchange. Transportation accounts for 10 to 20 percent of the total energy consumed in the low-income developing countries. A study for the ITU estimated that through better telecommunications, oil-importing developing countries could save about $18 billion in scarce foreign exchange used to import oil.[18]

Rural Industry

Lack of adequate telecommunications can hinder the growth and efficiency of industries in developing countries. A study of businesses in Kenya estimated that the losses incurred as a result of poor telecommunications were on the average 110 times higher than the total cost of providing adequate telephone service, and amounted to an average of 5 percent of total turnover. The businesses ranged from a hotel and travel agent, to a freight forwarder, a biscuit maker, and exporters of vegetables and fresh flowers. Among the cost savings expected to be derived from better telecommunications were savings in managerial time, lower inventory levels (from reduced uncertainty of delays in reordering), fewer and shorter production outages due to ability to order spare parts, and better vehicle scheduling with higher load factors because of the ability to organize back loads.[19]

In Peru, businesses generated about one-third of the calls made over a rural network; business users estimated that each call saved them about $7.30 compared to alternative means of communication.[20] Similar effects can be achieved at the national level. For example, since the installation of Intelsat satellite earth stations in the South Pacific, officials have saved on procurements by sending out tenders and receiving bids by telex and facsimile, rather than placing standing orders with one supplier.

The manager of an entrepreneurial village cooperative in China summarized the benefits of telecommunications for rural businesses in a discussion with the author. The cooperative raises chickens, builds prefabricated houses, and assembles printed circuit boards. The village now has a telephone, whereas previously the nearest telephone was more than an hour's bicycle ride away. When asked if the telephone had made a difference to the cooperative, he responded: "We have a saying: 'When the telephone rings, business is coming.'"

WHO ARE TELECOMMUNICATIONS USERS?

People who need to communicate quickly or frequently for their work range from entrepreneurs and project managers to health care providers. But individuals may also use telecommunications facilities for many purposes. In Egypt, researchers found that better educated individuals were more likely to make calls to major cities and administrative centers, whereas those with little education tended to call only to nearby villages and towns. However, the most important characteristic of telephone users is thirst for information. Village chiefs without formal education may use the telephone to talk to other chiefs. Villagers who do not speak the national language or have limited education may rely on intermediaries, such as extension agents, cooperative managers, or other officials who in turn will use the telephone to obtain the information they need.

Thus, although telephone users tend to be better educated and more involved in the market economy than nonusers, literacy and "modernity" are not prerequisites for telecommunications use. Information seekers may be traditional people concerned about their families, their work, or problems in their community. They are likely to use whatever tools are available—from two-way radios to satellite circuits—to find the information they need.

There have been several studies examining characteristics of telephone users and purposes of use. A Costa Rican study of rural public call offices (PCOs) found that the villages that benefited most from the PCOs tended to be larger and better off economically, with relatively better educated populations engaged in more progressive agricultural techniques. The PCO users themselves tended to be employed, but their incomes were not higher than average. In fact, telephone users included a substantial number of low-income residents, although the most frequent callers had higher than average incomes.[21]

In developing countries, Hardy found that residential telephones appear to contribute more to economic development than business telephones.[22] The reason may be that residential phones are often used for business activities, and are available 24 hours per day, whereas business phones are available only during work hours. There may also be a difference between public and private sector use, with many businesses using their proprietors' home telephones.

Research has shown that the economic benefits of telecommunications are related to distance and density, so that benefits are proportionately greater where telephone density is low and communication alternatives are expensive and/or time consuming. In much of Latin America and India, for example, villagers must travel many hours by bus to the town or city. In some parts of the South Pacific, communication is by mail that arrives on a monthly boat. In many parts of the developing world, villages are isolated for weeks during the rainy season, when flooded roads become impassable. The telephone or two-way radio becomes a lifeline—not only for emergency assistance, but to keep up the contact necessary to administer government services and manage development activities, and to reduce the sense of isolation.

Where telecommunications services are available, rural people often use them more heavily and spend more of their disposable income on telephone calls and telegrams than do city dwellers. In the Australian Outback, "chatter channels" on two-way radios were at one time busy all day long with messages in many aboriginal languages. Now, terrestrial telephone links and satellite conferencing are helping people in the Outback to keep in touch. In northern Canada, Indians and Inuit spend more than three times as much as their urban counterparts on long distance telephone calls, even though their average income is generally much lower than that of urban Canadians.

The only alternative means of getting a message through quickly is to take an expensive ride on a bush plane, since there are no roads in the remote north.

In many of these remote communities, telecommunications authorities have had to activate extra circuits in village satellite earth stations much sooner than they had anticipated because of the growth in telephone use. The number of long distance calls in some villages in northern Canada increased by as much as 800 percent after satellite earth stations replaced high frequency radios. In Alaska, the installation of small satellite earth stations in villages also sparked tremendous growth in telephone use. When local telephone exchanges were installed in some villages, long distance telephone traffic spurted again by up to 350 percent.[23]

Telecommunications as Complement and Catalyst

The research findings outlined above show that telecommunications can contribute significantly to socioeconomic development. However, many other factors may influence whether and to what extent telecommunications may make an impact. Generally, certain levels of other basic infrastructure, as well as organizational activity, are required for the indirect benefits of telecommunications to be realized. For example, a well-managed, decentralized organization, such as a manufacturing enterprise, a tourist development, or a health service, will derive more benefits from telecommunications than a poorly managed or understaffed operation.

Telecommunications may also serve as a catalyst at certain stages of the development process, becoming particularly important when other innovations are introduced, such as improved farming practices, lines of credit, incentives for decentralization, and diversification of the rural economic base. Evidence from India suggests that telecommunications becomes more important when rural modernization begins; for example, when improved farming practices are introduced or credit is made available, or when an integrated development plan is being implemented. Other researchers suggest a step function of levels of development, so that telecommunications investment is likely to be most beneficial when certain levels of economic activity are reached. These might be considered takeoff points, where telecommunications investment is likely to have greatest impact.

Communication is necessary for the development process, but investments in telecommunications alone to facilitate communication are not sufficient for development to occur. Thus, telecommunications may be seen as a complement to other prerequisites for development such as transportation, electrification, and a clean water supply.

Cultural Development and Quality of Life

The research findings outlined above document the indirect benefits of telecommunications that can be quantified. However, research that focuses on economic benefits tends to overlook some of the important social benefits which, although perhaps difficult to measure, are nevertheless quite real. Telecommunications links throughout a country can be used to strengthen the bonds of national identity and cultural cohesion. Access to reliable telecommunications in rural areas may help to reduce disparities with cities that contribute to urban migration and high turnover in staff at rural posts. In short, telecommunications services can contribute to improved quality of life in the developing world.

THE CONTRIBUTION OF TELECOMMUNICATIONS

Several important conclusions can be derived from research on the indirect benefits of telecommunications:

- Investment in telecommunications contributes to economic growth.
- The indirect benefits of telecommunications generally greatly exceed the revenues generated by the telecommunications network.

- Use of telecommunications can improve the quality and accessibility of education, health care, and other social services.
- Benefits of investment in telecommunications may be greatest in rural and remote areas where distances are greater and telephone penetration is lower than in urban areas.
- Intangible benefits of telecommunications such as the fostering of the sense of community and strengthening of cultural identity, while difficult to measure, contribute to the development process.
- Telecommunications can be considered a complement in the development process; that is, other conditions must exist for maximum developmental benefits of telecommunications to be achieved.

PLANNING TELECOMMUNICATIONS TO SUPPORT DEVELOPMENT

In the past, telecommunications planning and economic development planning seemed to be "two solitudes," using different terminology and administered through agencies that often have little contact with each other. To forge links between these distant cultures, various approaches to planning and policy may be adopted that cut across institutions. The following models suggest methods for integrating telecommunications planning and/or policy with development goals and activities that can be applied in developing countries as well as in rural areas of industrialized countries.

Integrated Planning

Where telecommunications services and regional development are both government responsibilities, a coordinated plan may be proposed to enable infrastructure investment and new services to support development goals. In Asia, Singapore is investing in information and telecommunications technologies to become an "intelligent island" in order to expand its information-intensive economy. The government of South Korea has targeted telecommunications and information technologies as part of its industrial and socioeconomic development strategies. South Korea has not only become a major exporter of electronic equipment, but has also set ambitious domestic goals for these technologies. Having achieved its goal of one television per household in the early 1980s, South Korea has now reached the goal of an average of one telephone per household, and aims for an average penetration of one personal computer per household by the late 1990s[24] (see chapter 11).

Basic Infrastructure Support

Governments may also play a major role in implementing their telecommunications plans; for example, they can underwrite all or part of the cost of installing and upgrading telecommunications facilities in high-cost or high-risk areas such as rural and very poor regions. In the United States, the state of Alaska appropriated $5 million for the purchase of 125 satellite earth stations to provide basic telephone service to Alaskan villages. The earth stations were installed and operated by Alascom, the intrastate carrier.[25] At a national level, the Rural Utilities Service (RUS), formerly the Rural Electrification Administration, is part of the U.S. Department of Agriculture, reflecting the New Deal commitment to providing utilities in rural areas to improve rural quality of life. The RUS provides low-cost loans to rural telephone companies for installation and upgrading of facilities.[26]

The European Union's STAR and ORA programs are designed to develop telecommunications infrastructure and networks in its underdeveloped, primarily rural regions. Although the members of the EU are primarily highly industrialized countries, many contain regions that are economically disadvantaged. In addition, approximately 50 percent of the Union's population live in rural areas. STAR and ORA are intended to aid infrastructure investment in these regions, as well as to carry out trials and demonstrations of new technologies and services that could contribute to their socioeconomic development.[27]

Identifying Needs and Testing Solutions

Government agencies may support innovative projects designed to test the suitability of new technologies for development applications. Industrialized countries, including the United States, Canada, Japan, and the members of the European Union, have all funded pilot projects and demonstrations to evaluate the technical performance and applications of new technologies and services, or to test existing facilities in new environments. Recent projects emphasize high bandwidth "information superhighway" applications (see chapter 18).

The Swedish government has funded a number of "telecottages," rural sites with computer and telecommunications facilities that are used by community residents. Some are strictly educational; others have spawned small businesses. The project is part of a program called "Let All Sweden Live," which is intended to create social and economic incentives for people to live in rural Sweden. Telecottages are also found in other Scandinavian countries. The ITU is planning a pilot project to introduce similar village facilities called "telecenters" in developing countries.

The U.S. Agency for International Development's (USAID) Rural Satellite Program sponsored pilot projects using existing satellites to provide health and education services in rural areas of Indonesia and Peru, and to link university campuses and extension centers in the West Indies.[28] Intelsat also provided transponder capacity on its satellites for distance education and other development applications through its Project SHARE and Project Access.[29]

Several development agencies are supporting initiatives to help developing countries access and utilize the Internet. Canada's International Development Research Centre (IDRC) is sponsoring a project on Pan Asian Networking (PAN), which is assisting developing Asian countries in setting up Internet service providers.[30] USAID has set up the Leland Project to assist African countries with Internet access.[31] The World Bank is undertaking a new initiative to support developing country telecommunications projects. Its Information and Development Initiative (InfoDev) is designed to provide financing for unbiased expertise in support of government initiatives in the information sector. The intent is that projects will involve partnerships between government, the private sector, the international development community, and the World Bank. Activities that could be supported include assessments and plans for reform of the telecommunications sector, information infrastructure plans, regional and intersectoral workshops, and demonstration projects such as applications of telecommunications for education, health care, and public sector financial management.[32]

User Involvement

Although some industrialized countries such as the United States, Canada, and Australia have mechanisms for public input to telecommunications planning through hearings, public meetings, and presentations from consumer and other user organizations, many industrialized and developing countries have no procedures for involving users in telecommunications planning. Yet the users remain the most important element of any plan; without an understanding of their needs and constraints, telecommunications services may be inappropriately designed or priced. Why are these individuals and organizations so often silent? They may not have the technical expertise usually expected in planning activities; they may also be unaware of how and when to get involved.

In a sense, telecommunications planners have to act like extension agents in getting out and meeting with users, such as the staff of schools, libraries, hospitals, development agencies, and small businesses, to learn about their needs, and help them to translate these needs into requirements for facilities and services. This is a new role for telecommunications carriers worldwide, yet it is a particularly important function in developing countries where resources for new facilities are limited, and failure to meet user needs can hinder the economy as well as limit the carriers' projected revenues.

Users must also organize to be heard. Businesses and other large users can form telecommunications user groups to monitor policies and advise government agencies. Consumer organizations can represent the concerns of individuals and community groups. Public service organizations of teachers, librarians, and health care providers can also advise planners on their communications needs and priorities.

IMPLEMENTING THE VISION

The above strategies, which may be termed "development-based approaches to communications planning," are based on the assumption that communications planners must consider the socioeconomic implications of communications policies. In other words, they should include assessments of the indirect benefits of telecommunications investment and utilization to the economy and the society in framing their policies and regulations. This development-based model requires coordinated communications planning:

- telecommunications administrations must be informed about national priorities and development plans;
- national planners must be made aware of the importance of telecommunications infrastructure to national development;
- resources for extension and improvement of facilities must be allocated to the communications sector, and resources for training and utilization of facilities must be included in the sector budgets;
- potential users must be made aware of the services available and how they could benefit from them.

This approach assumes a broadening of the definition of "public interest" beyond the simple assessment of pricing of services. It involves an analysis of the potential benefits of access to education and social services, of the impact of geographical as well as income-related disparities, and of the potential economic benefits of affordable access to information for both individual and commercial activities. Many of the steps in implementing the vision will need to come from other entities: the telecommunications industries themselves, government agencies that can fund pilot projects, and users who can identify needs and develop strategies to aggregate demand and share costs. Still, policy makers must take a prominent role, both in the agenda-setting process and in devising incentive-based strategies to achieve national communications goals.

9

Telecommunications Planning for Developing Regions: Extending the Infrastructure

"We will use this {Global Information Infrastructure} to help our respective economies and to promote health, education, environmental protection and democracy. . . . {The ITU} adopted five principles for a GII which the nations of the world have been putting into practice: Private investment. Market-driven competition. Flexible regulatory systems. Non-discriminatory access. And universal service."

—ALBERT GORE JR., U.S. Vice President[1]

THE CHANGING POLICY ENVIRONMENT

The previous chapter pointed out some of the technological changes since the publication of the Maitland Report that appear particularly promising for developing countries. The global policy environment has also changed dramatically, with the introduction of privatization and competition in most industrialized countries. The developing world is also now looking to market-based strategies to attract investment and improve the efficiency of its telecommunications sector. This chapter examines these trends and identifies promising models for restructuring the telecommunications sector in developing countries and for attracting investment to upgrade and extend telecommunications networks.

Today, about 80 percent of the world's population has no access to reliable telecommunications, despite the availability of VSATs and other relatively low-cost technologies such as VHF and UHF radio, rural subscriber microwave systems, and cellular networks. One reason, of course, is lack of investment. It is estimated that at least an additional $4 billion per year must be invested in Third World telecommunications if at least minimal worldwide access is to be achieved by the year 2000.[2] Another related problem is that in most cases, the telecommunications administration (PTT) acts as a gatekeeper or bottleneck that prevents customers from obtaining equipment and services. Thus, the government-operated utility model, which was adopted to protect the public interest, now acts as a constraint to retard the growth of the telecommunications sector and, as a result, the economy as a whole.

Building an infrastructure requires certain preconditions. According to the ITU, investment as a percentage of revenue generated should meet a minimum threshold of 40 percent. The level of pretax profitability is higher in lower-income countries (average 38 percent of revenue) than in upper-income countries (average 26 percent) or OECD countries (average 13 percent), implying that there is not really a shortage of capital. The telecommunications sector must not be used to subsidize other sectors of the economy and starve telecommunications of investment funds. Revenue streams in terms of revenue generated per line must be adequate to support the investment program. For example, if the average investment cost per line is $1,500, and the surplus of revenues over operating expenses is 40 percent, then the revenue generation to pay off the investment within five years is $750 per line per year.[3]

To increase access to voice and data communications in developing countries, it is necessary to eliminate bottlenecks wherever possible, and to create incentives for carriers to meet customers' needs and to raise capital for expansion. Two major strategies are required. The first involves restructuring to introduce autonomy and privatization. The second involves liberalization, or introduction of competition in various facilities and services. The following sections focus on the issues that developing countries face in evaluating options and strategies for restructuring their telecommunications sectors.

RESTRUCTURING THE TELECOMMUNICATIONS SECTOR IN DEVELOPING COUNTRIES

Reasons for Privatization

Developing countries have generally emulated the European PTT model, in many cases retaining the structure established by colonial administrations. There is little incentive for innovation in such an environment. In many cases, the telecommunications revenues are used to subsidize the postal service; in others, these revenues go directly to the national treasury. Telecommunications officials cannot count on using their revenues to upgrade or expand their networks. Since demand from unserved customers is often enormous, there is constant pressure to meet urban needs, and few resources are committed to extending services in rural areas. Many developing countries have begun to restructure their telecommunications sectors; examples from Eastern Europe and Russia, Asia, and Latin America are examined in the following chapters. However, others have rejected these approaches, at least for the near term, because they fear loss of monopoly control by the government-operated PTT.

Privatization is only the first step in a larger process of deregulation and liberalization. "The distinguishing features of this first wave (the privatizations of the United Kingdom and Japan) were the commitment to reducing the government's role in the economy and establishing the preeminence of markets over regulation as the guiding principle of economic policy. The current trend is to view privatization as the tool and, indeed, the catalyst for fundamental restructuring requiring substantial investment."[4] The new wave of privatizations appears to be motivated by governments' need to reduce public debt and attract investment funds to modernize the sector. Privatization is also seen as a means to obtain technical and managerial expertise.

Legal reforms such as privatization and market reforms such as the introduction of competition, either together or alone, correlate with increased levels of infrastructure. The British consulting group

Analysys concludes from its research that prosperity (GDP) is a necessary, but not sufficient, condition for a high level of telecommunications infrastructure.[5] An Analysys study estimated that if (West) Germany had followed similar privatization policies to those carried out in Britain from 1984, its infrastructure score (a combination of several variables) would have been 15 percent higher, with 55 instead of 48 main lines per 100 population, or 4.2 million more telephone lines.[6]

However, privatization in itself is not a panacea. Governments should introduce telecommunications privatization as part of their overall economic policies, and ensure that they have avoided pitfalls such as undervaluing their assets or providing no effective enforcement of licensing terms. Political economist Eliana Cardoso analyzes the issues surrounding privatization in the Latin American context:

> The case for privatization is twofold: It enables governments both to unload loss-leading companies and to bolster efforts to rationalize economic relations. Nonetheless, except for unprofitable public enterprises (parastatals), which can be simply liquidated, the fiscal benefits of privatization are not clearly defined. If efficient parastatals are sacrificed to cover current budget shortfalls, the state may be faced with future cash flow problems. Similarly, hasty privatization might merely transfer monopolistic entities from public to private control unless such entities are regulated. Additionally, privatization carries social costs that derive from the redistribution of income and potential variation in employment patterns implicit in ownership change. Finally, privatization is not an automatic process; it requires a stable regulatory and economic environment to encourage private investors' participation. The rules of the game must be clear enough to instill sufficient confidence that privatization will not be reversed or be subject to bureaucratic meddling.[7]

Corporatization: Management Autonomy

The first strategy for creating incentives to improve efficiency and innovation in the telecommunications sector is to create an autonomous organization operated on business principles. The end goal in many cases may be privatization as part of a national strategy to turn government-operated enterprises over to the private sector, or simply as a means of freeing the sector to raise its own capital and to introduce management policies that are rarely tolerated in a public enterprise.

Many countries see the autonomous, government-owned corporation as simply an intermediate step. For example, Fiji, Sri Lanka, Nigeria, and the Gambia have each established government-owned telecommunications corporations, as an intermediate step toward privatization. Malaysia intends to fully privatize the autonomous corporation it established in 1987. Singapore has begun to privatize highly profitable Singapore Telecom.

The difficulty with a government-owned corporation is that it may still be subject to many political and bureaucratic constraints, even if granted considerable autonomy. For example, Malaysian telecommunications still suffers from government and union influences that reduce incentives. In India, Metropolitan Telephone, a government corporation established to serve Delhi and Bombay, has improved service substantially, but the government still dictates many policies that prevent the corporation from acting efficiently to meet customer needs.

Steps to Privatization

There are several strategies for privatization. The national operator can be privatized through an initial public offering (IPO), as was done with BT in the United Kingdom and Singapore Telecom in Singapore, if local markets are large enough to absorb the offerings. Governments can also bring in a strategic investor which takes a minority interest in terms of equity, but assumes a majority control of operations, achieved through different classes of stock, as in Mexico. The strategic investor raises the value of the carrier by improving productivity; tranches are then sold on the capital markets. This has the effect of increasing the value of the government's retained interest.[8] As noted in chapter 4, many governments choose to retain a "golden share" so that they have some means of ensuring that the privatized carrier serves the public interest. However, a viable mechanism for enforcing licensing terms and conditions is likely to be a better guarantee that the operator will be responsive to demographic and geographic needs for telecommunications.

Many countries have allowed foreign carriers to invest in their national operators as a means of attracting needed investment and improving management and operations. The Bell Operating Companies, Telefonica, and France Telecom have invested in several operators in Latin America; European and American companies are investing in Eastern European carriers; Cable & Wireless owns a major share in many national telecommunications operators in the Caribbean and the South Pacific.

Another approach is subscriber investment through share capital. Brazilian subscribers, for example, must become shareholders in their regional telephone companies. This strategy has raised much of the capital needed to construct the network. But the experience in Brazil shows that mandatory subscriber investment alone is not a solution. Brazilian shareholders have no say in the policies and priorities of the telephone companies, as shareholders would (if organized) in a regular corporation. Also, the government has not allowed the regional companies to function autonomously, but instead has impounded their revenues and allocated annual funds, rather than letting them reinvest their own profits. The result has been deterioration of the plant in the highly populated areas and very limited expansion in rural areas (see chapter 13).

Introducing Competition

The goal of many countries is to fully privatize their telecommunications sector. However, as Cardoso points out, privatization is unlikely to generate major gains in efficiency unless accompanied by other re-forms: "Macroreform is an essential prerequisite to successful privati-zation. Privatization can only achieve its ends if the marketplace is a stable, unfettered economic system. Thus, the mechanism of reform necessarily includes liberalization. Moreover, buyers of public enter-prises need to know the economic environment in which they will have to operate, what tax structure they will have to face, and how the trade regime will affect profits."[9]

Many developing countries are unwilling to introduce competition with the public switched network, fearing loss of control as well as revenue, and/or waste of scarce resources in duplicating facilities. However, most agree that competition can be successfully introduced in customer premises equipment (CPE). But lack of capacity is a seri-ous problem in many developing countries and cannot be solved sim-ply by selling telephones and other equipment on the open market. Facilities-based competition is often acceptable for new services such as cellular mobile networks and value-added services. The ITU has concluded: " . . . it is unlikely that the PTO would lose any potential revenue from competition with mobile services: if anything, mobile services are likely to generate more traffic."[10] Many developing coun-tries including Malaysia, Thailand, the Philippines, Mexico, Ar-gentina, and Chile allow competition in mobile communications. Satellite services such as data communications may also be offered through one or more private licensed carriers. Again, this approach is likely to get service installed much more quickly than through the PTT. For example, private banking networks using VSATs have now been authorized in Brazil, and VSAT networks link remote and off-shore petroleum operations in Mexico and Indonesia.

Both satellite and cellular technologies may be considered a form of bypass, although they may introduce services previously unavail-able. One advantage of both of these technologies is that they provide services that can be important for economic development. For exam-ple, cellular radio may be used for fixed communications to provide public call offices and community telephones where they were not previously available, as well as to provide service to more affluent business users. VSATs may provide data links for important national industries such as petroleum, banking, and tourism.

Other Restructuring Strategies: Local Companies

Although most countries have a single carrier that provides both local and long distance services, it may be appropriate to delineate territories that can be served by local entities. In the United States, the model of rural cooperatives fostered through the Rural Electrification Administration (now the Rural Utilities Service) has been used to bring telephone service to areas ignored by the large carriers. Local enterprises are likely to be more responsive to local needs, whether they be urban or rural. An example of this approach in urban areas is India's Metropolitan Telephone Corporation, established to serve Bombay and Delhi. Local companies also provide telephone service in Colombia and the Philippines. Cooperatives have been introduced in Hungary and Poland. A disadvantage of this approach is that local expertise to operate the system is likely to be in particularly short supply in many developing countries.

A related policy to speed coverage of currently unserved areas is to open them up to private franchises. Large carriers may determine that some rural areas are too unprofitable to serve in the near term. However, this conclusion may be based on assumptions about the cost of technologies and implementation that could be inappropriate. A more innovative carrier, willing to use more appropriate technology such as rural radio systems or small earth stations, and to cut costs by hiring local people rather than bringing in outside staff, may be able to run the system at a profit. In Alaska, small telephone companies provide local service in villages where earth stations were installed by Alascom, the statewide long distance carrier. Some American states also authorize competitive franchising of territory that the large carriers claim is prohibitively expensive to serve.[11] Indonesia and India are licensing operators to serve areas that have not been covered by the national operator.

Investment Requirements

Regulators may encourage investment in unserved areas or upgrading of equipment by making such investment a condition of a franchise or other concession. In the United States, several states, including Michigan, Tennessee, and Texas, have required accelerated upgrading of rural networks as a condition for reducing price regulation of telephone companies. In the Philippines, companies receiving licenses for cellular and international services are required to install several hundred thousand lines in a specified area of the Philippines currently without telephone service.

Resale

Carriers with excess capacity, such as new optical fiber links or satellite transponders, can be required to resell capacity at wholesale prices to new carriers in order to prevent duplication of backbone networks. Third parties can also be permitted to lease capacity in bulk and resell it in units of bandwidth and/or time appropriate for business customers and other major users. Resale helps to prevent costly duplication of facilities that can be a drain on foreign exchange reserves in developing countries; resale also generates additional revenue by creating pricing incentives for customers to increase their usage.

THE NEED FOR REGULATION

Setting and Enforcing the Rules

For privatization to be successful, the rules of the game must be clear, enforceable, and impervious to political influence. As Cardoso notes: "A credible and stable legal and economic environment is key to the ultimate success of the privatization process."[12] Thus, restructuring the telecommunications sector may not achieve its desired objectives without some form of regulatory oversight.

The above range of options available for restructuring the telecommunications sector is often referred to as "deregulation." However, in many developing countries, restructuring creates a need for regulation where no form of oversight previously existed. Not only is there a need to separate regulatory from operational functions, but in developing regions, the regulatory function itself must be created from scratch. Institutional structures must be developed, and laws must be introduced to spell out the responsibilities and authority of the regulatory agency. Perhaps most importantly, competent staff must be found or trained to take on these roles. In countries where there is already an acute shortage of professionals, including engineers, economists, and lawyers, establishing an effective regulatory body may be one of the most difficult challenges facing a government intent on restructuring the sector.

The chairman of New Zealand's Commerce Commission summarizes the current situation in that country, where there is deregulation, but only limited competition so far. His analysis could serve as a warning for other countries restructuring without introducing effective regulation:

> The resulting picture, in the Commission's view, is not that of an industry subject to "light-handed" regulation. In the absence of competition (the best regulator of all), the gap is filled by self-regulation. More precisely, in telecommunications, in relation to most important segments and most of the critical inputs, Telecom [Telecom New Zealand] is the *de facto* regulator. Telecom owns or controls the key factors and so Telecom makes the rules, and other parties to the industry, by and large, play by them.[13]

Protecting the New Carrier: The Pitfalls of Exclusivity

Policy makers may make concessions to attract investors and ensure the short-term profitability of the newly privatized carrier: ". . . the new strategic partnerships often mean that deregulation, competition and user choice are postponed for the sake of safeguarding the financial viability of the new telecommunications entity."[14] Mexico granted Telmex's new owners a six-year exclusivity period to provide basic services. Argentina granted two carriers geographic exclusivity and granted each group a ten-year exclusivity period to provide basic services. Jamaica awarded its newly privatized carrier a 25-year exclusive monopoly on domestic and international service. However, potential gains from privatization can be reduced when industry is overly protected. For example, consumer complaints have increased since divestiture of Telmex in 1990.

If the market is open to competition, the franchise value of the existing operator will be reduced, threatening successful privatization. However, protection will mean a potential loss of technological and economic benefits arising from greater competition. "The question is how to find an adequate balance between protecting the newly privatized company in the short term and establishing a competitive market that will foster economic prosperity in the long term." [15] As Richard Beaird points out: "The overarching question we must ask ourselves . . . is how to strike a balance between short term investment goals and long term questions of consumer sovereignty. Unless we can find a bridge between these two pressures, the vehicle of privatization may become a politically expedient way of replacing one form of operator based monopoly with another." [16]

Pricing Issues

From the users' point of view, their access to technology involves not only availability of equipment, but affordability. From the carriers' point of view, providing access to users is not cost free. They must determine how much to charge for use of their facilities. If they decide to offer the services at reduced prices or for free, they must justify this decision.

An ITU official has stated that if revenue per line is below a certain threshold (for example, $400 per line per year if line costs are $750 per year), then tariff reform through raising subscription and/or local calling rates should be a priority. [17] However, simply raising rates without understanding costs and considering service quality is not an appropriate solution. As technological changes drive the industry worldwide toward competition, prices must be driven toward costs.

Typically, such prices will be lower than previously, rather than higher. Revenues can also be increased by lowering prices to stimulate demand. This is the model used by the U.S. long distance networks and by many carriers through incentives such as off-peak pricing and bulk discounts. To avoid creating more bottlenecks, the network must have adequate capacity to handle increased usage. In addition, rates must be tied to quality of service. In some countries with newly privatized networks, customers have become more frustrated than they were before privatization because rates have increased without comparable improvements in quality of service. Such frustration creates incentives to bypass the network.

THE NEED FOR INVESTMENT

Reasons for Underinvestment

Telecommunications users implicitly recognize that the value of the information exceeds the price they pay to transmit or receive it. However, in the telecommunications sector, justification for providing services has been on a cost recovery basis, ignoring the indirect benefits to users and to the economy as a whole. Investment decisions are more likely to be based on the anticipated revenue to be derived from the telecommunications network. Further, it is often assumed that in rural and sparsely populated areas, the costs of providing and maintaining service will substantially exceed the revenues.

Instead, telecommunications networks could be considered basic infrastructure like roads, water mains, and electrical power grids. Planners do not expect these services to make money directly, but rather to facilitate the development process. However, unlike other utilities, a telephone network can generate its own revenues, and thereby become a profitable investment for a country. Yet this potential profitability is almost an Achilles' heel from a development perspective; planners tend not to look beyond the revenues to determine what greater benefits might be gained from the telecommunications investment. Furthermore, they may expect the revenues from each part of the network to cover their costs, so rural areas may be considered inherently unprofitable, and therefore not deserving of investment of scarce resources until urban needs are satisfied. This approach continues to overlook the important developmental value of the indirect benefits of telecommunications.

In general, there appear to be many reasons why the funding for rural telecommunications has not met even minimal needs:

- The revenue-producing nature of telecommunications has led planners and investors to concentrate on improving services that will generate profits, such as interurban and international links;
- The indirect benefits of telecommunications for socioeconomic development have not been widely understood or appreciated, so that telecommunications often receives a low priority in development plans;
- Telecommunications has been seen as an urban luxury rather than an essential component of infrastructure that should be included in integrated development plans;
- The increased demand and competition in the industrialized world has resulted in product innovation, but suppliers have concentrated on the markets in industrialized countries;
- The lack of adequate training of technical and managerial personnel has often contributed to service problems that result in limited service and thereby limited benefits;
- As most developing countries do not have indigenous telecommunications manufacturing industries, they must buy their systems and spare parts abroad with scarce hard currency, which is often needed for necessities such as imported food, medicine, and fossil fuels;
- The telecommunication sector's profits may be used to subsidize the postal service or to contribute to general government revenues, rather than being available for expansion and improvement of the network;
- Financial institutions and equipment suppliers may be reluctant

to extend credit for telecommunications purchases if they perceive excessive risk due to the above factors or to political or economic instability in the country.

Indicators of Pent-Up Demand

Lack of telecommunications cannot necessarily be attributed to lack of demand or purchasing power. In many developing countries, television sets are much more prevalent than telephone lines. In industrialized countries, both TV sets and telephone lines are almost universally accessible. However, in lower-middle-income countries there are almost two and a half times as many TV sets as telephone lines, and in low-income countries, there are more than 13 times as many TV sets as telephone lines (see Table 9.1).

The problem is apparently a bottleneck in provision of telephone service rather than lack of sufficient disposable income to pay for telephone calls. It appears that where television is available, a significant percentage of families will find the money to buy TV sets. These numbers indicate a potential pent-up demand for other communications services, and the availability of sufficient disposable income to pay for the service if it is deemed important.

Most countries do not have a specific policy intended to provide greater access to television than telephones; however, people can buy TV sets if they can afford them, whereas if telephone lines are not

TABLE 9.1
ACCESS TO TELEPHONE LINES AND TELEVISION SETS

	Telephone Lines/100	TV Sets/100	Ratio TV Sets/ Telephone Lines
High-Income Countries	48.8	59.7	1.2
Upper-Middle-Income Countries	12.9	24.1	1.9
Lower-Middle-Income Countries	8.1	19.8	2.4
Low-Income Countries	0.9	11.8	13.1

Derived from ITU, *World Telecommunication Development Report*, 1995.

available, there is no possibility of obtaining telephone service. Some socialist countries do intentionally foster access to mass media, while limiting access to interactive communications, which could be used for political as well as social and economic purposes. Regardless of whether there is a deliberate policy, a country that provides television service without telephones not only deprives people of the means to get help in emergencies, but of the means to have a voice in their own development.

Bypass: The Users' Response

When users are unable to obtain the capacity they need, or to afford available services, they look for alternative solutions. In the past (and even today in some parts of the developing world), they turned to high frequency (HF) radio. HF is frustrating in terms of its signal quality and varying reliability, but the price is right. If the users own their radios, they can use them whenever they want without paying a carrier. Now, satellites offer a more reliable bypass option.

Two examples of public service satellite bypass illustrate the problems that small users face in developing regions. The PEACESAT network was founded at the University of Hawaii in 1971, and linked universities and development centers throughout the Pacific using the "experimental" ATS-1 satellite operated by the National Aeronautics and Space Administration (NASA), until ATS-1 finally drifted out of orbit in 1985. By that time, most of the island nations were linked to each other and the rest of the world via the commercial Intelsat system. However, the PEACESAT members were unable to afford the use of Intelsat. Therefore, they searched for funds and a "free" satellite to reestablish their network. In 1987, the U.S. National Telecommunications and Information Administration (NTIA) received a congressional appropriation to restore the PEACESAT network using the GOES-3 satellite (Geostationary Operating Environmental Satellite), a meteorological satellite operated by the U.S. National Oceanic and Atmospheric Administration (NOAA).[18] Thus, although commercial satellite service is now available throughout the Pacific, PEACESAT, with U.S. government support, turned to another stopgap experimental satellite because it was free.*

SatelLife, another nonprofit development organization, has gone even further, and launched its own satellite for medical communications in the developing world. Its "microsatellite" provides store-and-forward data communications to small terminals in developing countries. As noted in chapter 3, these physicians raised money for their own satellite because, despite modern technology, telecommunications facilities in the poorest regions were either unavailable or unaffordable.

These applications may seem rather inconsequential in terms of usage and revenues lost to carriers, but commercial users are turning to bypass on a much larger scale. In Latin America, banks, brokerages, and oil companies, among others, are establishing their own private networks on PanAmSat. Businesses in Asia now have the same opportunity using PanAmSat or AsiaSat, a regional satellite operated by interests based in Hong Kong. Increasingly, commercial users are choosing to bypass the national carriers in order to get the services they need and to obtain better prices. (These competitive satellite systems are discussed in chapter 15.)

Another bypass service known as "callback" enables users to obtain much cheaper international rates than are offered by most developing country administrations. Callers dial a service in countries with cheaper rates, such as the U.S. The service calls them back and sets up the call at the cheaper rate (see chapter 17).

STRATEGIES TO ENCOURAGE INVESTMENT

The extension and upgrading of telecommunications facilities is highly capital-intensive. In the past, the approach to financing of

---

* Some PEACESAT members went back to using HF radio, nearly twenty years after ATS-1 demonstrated that satellites could reliably and affordably replace HF radio in Alaska and the Pacific!

PTT facilities was bureaucratic, with allocations from the national budget that typically were unrelated to revenues generated or efficiency of operations. Therefore, the PTTs overstated needs and overdesigned systems; they overbuilt whenever possible, and spent all of their funds quickly to avoid returning any to the national treasury. Today, internal generation is likely to be a main source of funds; as a result, PTTs need to be efficient, increase traffic, and maintain economic tariffs. They will therefore need to improve their tariff policies, set better investment priorities, reduce costs, and market their services effectively.

Internal generation alone is not likely to produce sufficient profits to finance the investments in infrastructure required in many developing countries. Many PTTs rely on high profits from international traffic as a major source of funds for extending their infrastructure. However, the growth of competition in international telecommunications and opportunities for users to bypass the operator through "callback" services are likely to drive international rates toward costs. Also, many governments are moving away from the PTT model by corporatizing or privatizing the former PTTs, and introducing competition in some services. These strategies are designed to attract foreign investors with the necessary capital and expertise to provide the service more quickly than it could be offered through the PTT.

In developing countries, attracting sufficient investment capital for telecommunications is likely to remain a challenge because of lack of investment incentives and the demand for capital for other sectors. Direct foreign investment is desirable because it provides equity and may bring foreign management expertise. Many forms are possible, such as foreign purchase of share equity in existing companies, new companies to provide new services such as mobile communications and data networks, joint participation in new ventures, joint purchase of facilities, and so on. Among the strategies used to encourage foreign investment are:

- Joint ventures: partnerships of foreign companies with the operator to offer new services, or subsidiaries of the operator with overseas partners;

- Revenue sharing: foreign investors receiving a share of the revenue in return for investing in the operator;
- Build-Operate-Transfer (BOT): a form of investment offset; private investors build the facility, own and operate it for a negotiated period to earn back their investment, then turn it over to the carrier;
- Build-Transfer-Operate (BTO) or revenue sharing: similar to BOT, but requiring immediate transfer of the project to the carrier after it is built by private investors; the carrier operates the network, and shares revenue with investors for a negotiated period;[19]
- Debt equity swaps: used creatively in some Latin American countries to trade unpaid national debt obligations for equity in industries such as telecommunications;[20]
- Local alliances with international manufacturers to produce and sell telecommunications equipment.

Rather than turning to foreign investors, operators may raise money through domestic outside financing, for example, with subscriber bonds or subscriptions. This approach has been used in Hungary, as well as Brazil and South Korea.

Other sources of capital are the development banks. The development banks offer loans at below-market rates or with significant grace periods, and consider themselves "lenders of last resort" to provide financing when private capital has not been sufficient. The development banks also typically provide technical and managerial expertise, and can advise the government on restructuring of the telecommunications sector. The major multilateral development lenders are the World Bank and regional development banks including the African Development Bank, the Arab Development Bank, the Asian Development Bank, the European Bank for Reconstruction and Development (EBRD), the European Investment Bank (EIB), and the Inter-American Development Bank (IDB). Telecommunications lending by the six largest multilateral development agencies was about 6 percent of their total disbursements in 1992, ranging from about 20 percent for the EBRD to no telecommunications lending during that period by the Inter-American Development Bank.[21] Central and Eastern Europe have been very successful in obtaining multilateral financing, with project financing of over $400 million in one year.

Another international organization that provides funding and technical assistance in telecommunications is the United Nations Development Program (UNDP). The UNDP may implement its own telecommunications projects or work through the ITU. The ITU does not provide funding for facilities, but does offer technical assistance and training in telecommunications planning, operations, and management through its Telecommunication Development Bureau (BDT) (see chapter 16).

A new organization designed to raise funds for telecommunications projects in developing countries is WorldTel, an initiative launched by the ITU. WorldTel, which describes itself as a buyer-based mobilizer of funds, is headed by Sam Pitroda, an Indian telecommunications entrepreneur who returned to India from the United States in the 1980s to advise then Prime Minister Rajiv Gandhi on developing India's telecommunications and electronics industries. Pitroda founded C-DOT, an innovative center that designed electronic switching equipment for Indian towns and villages. The goal of WorldTel, which was first proposed in the Maitland Report, is to provide a revolving fund for telecommunications investment in developing countries.

Another financing source is bilateral aid, which often takes the form of "tied aid," requiring the borrower or grantee to buy equipment manufactured by companies from the donor country. Japan is the largest bilateral donor in telecommunications, with funding going primarily to Asian countries. Japanese aid appears closely related to its trade policy, benefiting not only the recipients, but Japanese vendors. The EU provides grants and loans to developing countries through the Lomé Convention, under which countries in Africa, Latin America, and Asia can become associate members of the Union. Sweden, Canada, and Australia also provide some development assistance for telecommunications infrastructure and training.

The United States provides very limited funding for telecommunications equipment, but sponsors projects using telecommunications and information technologies to support other development sectors such as health, education, and agriculture through the U.S. Agency for International Development (USAID). The United States also offers training for telecommunications professionals from developing countries through the U.S. Telecommunications Training Institute (USTTI), which is jointly sponsored by the government and the telecommunications industry.

Supplier credits are another source of funds for operators unable to borrow sufficient funds from other sources. Supplier credits have possible hidden costs: they can sidetrack the recipient from its own priorities, and they may include higher equipment or servicing prices than could be negotiated through outright purchase. Buyers may become locked into one vendor for spare parts and upgrades. Conversely, if the operator relies on supplier credits from several vendors, it will face problems in maintaining too many different types of equipment.

Small carriers can increase their bargaining power by aggregating demand with other carriers to obtain better prices and financing terms. The Caribbean Association of National Telecommunication Organizations (CANTO) has negotiated volume discounts worth several million dollars with equipment vendors since its founding in 1988. In East Africa, telecommunications operators are pooling their tenders into a single order for one million lines to gain increased bargaining power.[22]

CHALLENGES FACING THE LEAST DEVELOPED COUNTRIES

The least developed countries face daunting challenges in obtaining the investment required to build the infrastructure they need to develop their economies. "Low-income countries" as they are identified by the World Bank, range from the giants of China and India, which together account for nearly 40 percent of the world's population, to small countries such as Bhutan and Guyana, with fewer than 1 million people each. They include some of the most densely populated countries in the world, such as Bangladesh, as well as some of the most sparsely populated, such as Mongolia and the countries of the African Sahel.

All share very low per capita GDP (averaging only $415), and very limited infrastructure (averaging only 1.5 lines per 100 population). But averages do not tell the whole story: telephone densities are less than 0.2, or one telephone per 500 people in Afghanistan, Guinea, Liberia, Niger, and Somalia, and less than 0.1, or one line per 1,000 in Cambodia, Chad, and Zaire.[23] Since most of the facilities are in cities, rural residents in the poorest countries generally have no access to telecommunications.

Problems and Opportunities in Africa

Approximately 60 percent of the low-income countries are in Africa. Teledensity south of the Sahara is only 0.8 lines per 100 population. As much as 80 percent of Africa's population is rural, thus, most Africans have no access to telecommunications. Africa's telecommunications networks are more inefficient and less reliable than networks in any other region, with an average of only 44 lines per employee, compared with more than 200 in high-income countries, and an average of 110.1 faults per 100 main lines per year, more than 11 times the rate in high-income countries.

African countries face many challenges to meet the enormous demand for telecommunications. There is a shortage of management and technical expertise, and many promising employees sent for training are lured away by higher paying jobs in other sectors or in the Middle East. Most African countries have limited or no manufacturing capabilities in telecommunications, and must rely on imported foreign equipment. Most of these countries also have an acute shortage of foreign exchange, which they rely upon to import necessities such as fuel and medicine. The ITU estimates that about $8 billion would need to be invested by the year 2000 in order for the continent to reach a teledensity of one line per 100 inhabitants. Of this amount, only about $5 billion could be funded internally, leaving a shortfall of $3 billion.

A microwave network called Panaftel (the Pan-African Telecommunications Project) is an important means of regional communications, but its capacity is limited, and it is vulnerable to outages if towers are damaged by storms or generators run out of fuel. Some West African countries such as Benin, Senegal, and Côte d'Ivoire appear to have kept up their portion of the network, but poorer countries have reverted to relying on links to France to transport their international traffic.[24]

African nations rely on Intelsat for their intercontinental communications, and for much of their communications within Africa, although these connections may also require double satellite hops through European gateways. African administrations have explored the possibility of a regional satellite system through a project called RASCOM (Regional African Satellite Communications System), supported by the ITU and the Pan-African Telecommunications Union (PATU). However, both the politics of establishing and managing a regional system and the financing required have proved daunting. As an interim strategy, some countries are sharing Intelsat capacity, which several countries lease to provide domestic links between their major cities and towns. In 1997, Intelsat, in cooperation with RASCOM, will launch two satellites to provide extensive African coverage.

Other new initiatives may offer attractive alternatives for communication within the region. As part of its global communications system, PanAmSat now covers Africa. In Latin America, PanAmSat provides domestic and regional television distribution and VSAT networks that link banks and other private networks. These services are also now available in Africa. Competition between Intelsat and PanAmSat should result in lower prices for satellite services than previously offered in Africa. PanAmSat may also be used to provide rural telephony through VSATs or small earth stations serving as hubs for terrestrial wireless access.

Another option for regional networking is Africa One, a consortium led by AT&T that plans to ring the entire continent with an undersea fiber-optic cable. Countries along the coast would tap into the cable to connect with each other and with international networks. Landlocked countries would need to negotiate access through coastal countries. A potential benefit of Africa One would be to minimize the outflow of international transit fees, which now amount to some $400 million per year.[25] Other regional fiber-optic initiatives include Alcatel's West Coast Africa cable and Siemens' Afrilink.

Pricing remains a major problem for African users. Annual subscription rates for telephone service in Africa's poorest countries average 24.6 percent of per capita GDP, compared to an average of 0.7 percent in high-income countries.[26] As discussed earlier, the medical consortium SatelLife decided to invest in a small satellite for store-and-forward data communications because communications in Africa were either excessively expensive or nonexistent. Uninet, a Zambian e-mail service provider, pays $150,000 per year for an Internet link that would cost $5,000 in the United States.[27]

Tariff rebalancing is resisted by most operators because African countries have used international revenues to cross-subsidize domestic operations; some countries that run telecommunications as part of a government department may simply siphon off foreign exchange revenues to the national treasury. Restructuring is also viewed with skepticism in some countries because privatization and liberalization will lead to loss of government revenues and civil service jobs in the telecommunications sector. Yet restructuring must be seen as more of an opportunity than a threat. Governments cannot count on internal revenue generation or foreign aid to provide the enormous investment needed to build telecommunications networks. And investment in telecommunications is likely to lead not only to more jobs in the telecommunications sector in new and competitive services, but also contributions to national economic development.

Some 41 percent of African countries have telecommunications facilities run by government ministries, while another 28 percent are government corporations. Most countries, such as Namibia, Tanzania, and Zambia, have separated post and telecommunications services, and others are taking the next step to corporatize their operations. A few countries have begun to privatize; operators in Ghana, Uganda, Côte d'Ivoire, Cape Verde, and Guinea have issued tenders for partial privatization. Of the 23 countries with cellular services, at least 11 have joint ventures between national operators and foreign operators.

Ghana, Nigeria, and Namibia, allow competition with the national operator. Some foreign investors are realizing that African countries offer excellent business prospects. For example, Millicom says its operation in Ghana is one of its three most profitable in the world.[28]

South Africa is considering partial privatization of Telkom South Africa, as recommended in a 1995 Green Paper. South Africa is also in the process of setting up a telecommunications regulatory agency. The Congress of South African Trade Unions remains opposed to moves to restructure the sector. Meanwhile, Telkom is adding 5 million new lines through a Build-Operate-Transfer (BOT) scheme to avoid adding new debt.

Although South Africa's economy and infrastructure are much more developed than other African nations, initiatives in South Africa may lead to more investment in other countries. South Africa plans to spend more than $1.6 billion to build more than 1 billion lines in mostly black rural areas. Telkom South Africa received bids from 22 consortia, many led by foreign companies that may also be interested in entering other African markets.[29]

Some of the technical and managerial techniques developed for rural telephony in South Africa may also be transferrable to other African countries. For example, an entrepreneurial approach has helped black townships to get access to telephone service. Cellular operator Vodacom works with local entrepreneurs to operate "phone shops" in black townships and squatter settlements. The phone shop consists of a specially outfitted shipping container with several metered cellular phones.

Similarly, in Rwanda, pay telephones are installed in kiosks where local entrepreneurs also sell soft drinks, candy, and newspapers. Many also offer facsimile services. The kiosk owners retain a percentage of the telephone revenue. These models are attractive because they use entrepreneurial incentives to provide access, without necessarily requiring restructuring of the telecommunications sector. Governments not yet ready for privatization can retain control over telecommunications operators while allowing the private sector to participate in provision of services.

Botswana is an example of an African country that has improved its telecommunications infrastructure significantly without privatization. Botswana operates a completely digital network with a teledensity of 3.1 lines per 100 subscribers, nearly four times the average in sub-Saharan Africa. It also has packet-switched data and paging networks. Botswana's economy is based on diamonds and other natural resources, although tourism is becoming a major sector. The Botswana Telecommunications Corporation (BTC) is a government-owned, autonomous corporation managed by Cable & Wireless. BTC maintains a high quality of service, with some 80 percent of faults repaired by the next working day, and sustains a high rate of investment. International traffic, fueled by Botswanans working abroad as well as multinational mining companies and tourism, contributes about one-third of the total revenue.[30]

Botswana has advantages over the poorest African countries, with its small population, natural resources in strong demand, and favorable treatment from Commonwealth and other aid donors. Like many of the Commonwealth countries in the Caribbean and the South Pacific, Botswana relies on Cable & Wireless to manage its network. While the result has been a high quality of service, it will be important to ensure that both management and technical skills are transferred so that Botswanans can run their own telecommunications enterprises.

THE DEVELOPMENT CONNECTION

The strategy of using telecommunications to contribute to national development requires an active government policy to ensure that telecommunications plans and services are designed to meet national goals. It also requires flexibility and innovation in services, equipment, and pricing to respond to user needs. New technologies offer new opportunities, but they also pose additional challenges for telecommunications carriers. The most important challenge will be to plan and manage telecommunications not only as a source of revenue, but as a strategic resource for development.

The previous chapters have examined telecommunications issues facing developing regions in general. The next chapters examine the status and trends in the telecommunications sectors of selected emerging economies and developing countries in Central and Eastern Europe, Asia, and Latin America.

10

Telecommunications in Eastern Europe and Russia

"The telephone is indispensable for building a democratic state and introducing a market-oriented economy."

—TADEUSZ MAZOWIECKI, Polish Prime Minister[1]

THE ROLE OF COMMUNICATION IN POLITICAL AND ECONOMIC DEVELOPMENT

Having broken with the past—with its totalitarian regimes and centrally controlled economies—Eastern Europe and the former Soviet Union are struggling toward democratic societies and market-based economies. The goal in Eastern Europe is to move to a Western-style economic system, with a role for private enterprise, and a free capital market rather than state budget as the primary means of resource allocation, and, eventually, membership in the European Union. The transition is wrenching and without precedent: "Eastern Europe's transition from the command economy to the market economy is a premiere performance with no dress rehearsal. Every actor in the drama has to interpret his role himself."[2]

Two objectives important for economic modernization are restructuring of the economic system to a market economy and development of efficient manufacturing capacity. Meeting these goals will require modern telecommunications and full seamless interconnection with Western Europe and the world.[3] The per capita GNP of Central and Eastern Europe (CEE) in 1990 was about $2,500, and could be expected to grow to about $6,000 by the year 2000. World Bank planners estimate that telephone penetration to support this level of development should be about 30 lines per 100 population, requiring an annual growth rate of 11 percent.[4]

Democratization is also dependent on sharing information. Chapter 1 discussed the role of telecommunications in the fall of the Berlin Wall. In August 1991, communications within Russia and with the outside world also helped Boris Yeltsin abort the coup against then President Mikhail Gorbachev. The coup leaders, who became known as the Gang of Eight, did not attempt to shield from the outside world the press conference that they intended to use to explain and justify their actions. They took hostile questions from Soviet and foreign press, all of which were transmitted live over Soviet television and carried internationally via satellite. Soviet citizens could also listen to unjammed western broadcasts such as the BBC, Voice of America, and Radio Liberty.

The coup leaders also neglected to take effective steps to isolate Yeltsin electronically. They could have cut off international traffic at the Moscow international gateway switch, the Ostankino relay tower, or the Dubna Intelsat earth station. The Ministry of Defense ordered international communications to be shut down, but the staff did not obey the orders. According to one analyst: "One senior official at the Moscow Long Distance and International Operations Center said that during the crisis, he and his staff had seen their duty as maintaining the functioning of the network. Had any key telecommunications official collaborated with the plotters, the situation in the Soviet Union today might have been considerably different."[5]

From the Russian White House, Yeltsin could communicate with the West via phone, fax, telex, and electronic mail. Foreign business people brought him cellular phones. New private networks kept channels of communications open within the country and to the rest of the world. A woman working at the Moscow office of a trading company interviewed by telephone said that while the conventional network's international lines choked with traffic, "the Comstar [satellite] phone is working great." Citizens of the city of Togliatti credited MCI's telex-based Insight news service with keeping the legal government in power.

Apparently, the coup leaders underestimated the power of communications technologies as well as their ability to control them. Former totalitarian regimes that did recognize the threat of open communication faced a dilemma: " . . . either they had to try to stifle informational technologies and telecommunications with the result of falling further behind the new industrial revolution, or they had to permit microelectronics, telecommunications, and media technologies and see their totalitarian control eroded."[6]

In fact, before 1989, telex networks were relatively extensive in Eastern European countries such as East Germany and Poland. Was this a contradiction of the policy of limiting and controlling communications channels? Probably not. Telex was a major means of government communications. A printed message may have carried more authority than a voice message, as is true in many cultures. Telex networks were easily controlled and monitored, whereas facsimile, which also produces hard copy, but can be used over any public network, is much more difficult to monitor.

Governments of countries close to Western Europe also found television reception hard to control. East Berliners could watch television transmitted by West Berlin television stations. Now European satellite footprints also cover much of Eastern Europe, and there are now millions of viewers of satellite television in that region, watching programs transmitted primarily via the Luxembourg-based Astra satellite. In 1995, there were an estimated 2.3 million satellite antennas in Poland, 1.1 million in Hungary, 890,000 in the Czech Republic, and 660,000 in Slovakia.[7]

TELECOMMUNICATIONS IN
EASTERN EUROPE

While Eastern European governments now recognize the importance of telecommunications for building their economies as well as for sustaining democracy, they face formidable obstacles to upgrading and extending their facilities. Factors conditioning the pace of modernization include the availability of finance and willingness of foreign companies to build production facilities and train workers. Some countries need skilled jobs for former defense industry workers. In addition to capital investment, Central and Eastern Europe (CEE) needs professional and technical know-how, and organizational and managerial development to improve management of existing concerns and ensure that new businesses can operate efficiently.

A critical problem is the lack of free market experience among middle management and the difficulty in attracting, training, and retaining the best local technical personnel. This problem extends to telecommunications: "Unfortunately, the mindset of the typical Eastern European telephone company is even more resistant to consumer pressure than its Western counterpart, and until now there has been little incentive to improve efficiency or adopt new ideas."[8]

CEE telecommunications are further behind the West than many other sectors of their economies. They share many of the problems that developing countries are experiencing:

- Their telecommunications networks are expensive to install and operate: one line may cost one year's per capita GDP;
- They have state-run monopolies which are inefficient, and suffer from lack of investment;
- Large parts of the countries are unserved or poorly served.

Central and Eastern Europe have inherited underdeveloped, worn-out, and unbalanced networks. In 1990, the seven countries in the region (Bulgaria, the Czech Republic, Hungary, Poland, Romania, Slovakia, and the former Yugoslavia) had a population of about 100 million, and 11 million connected direct exchange lines (DELS), giving an average penetration rate of about 11 percent, compared to 37.5 in the European Community. Thus, there were only about one-quarter as many telephones per 100 inhabitants as in Western Europe; one-fifth to one-sixth as many facsimile machines and data terminals, and virtually no mobile communications.[9] Some rural regions had no communication facilities; for example, Romania had 3,300 villages without telephones.

On average, Eastern European telecommunications are 20 to 25 years behind Western Europe. Infrastructure is poor, with some exchanges dating to pre-1917 Russia, and too many lines hooked up to exchanges without sufficient long distance infrastructure, so that systems frequently fail with repeated call attempts. There was chronic underinvestment in telecommunications by previous regimes. "This lack of investment in public telecommunications infrastructure was based on political ideology: in Stalinist economics, services were not considered a form of production and consequently their provision was considered merely as a cost to the economy. Additionally, the rights of individuals and enterprises to communicate were restrained to ensure State control of information."[10]

The technology is also inferior partly because of the Eastern bloc goal of self-sufficiency. Eastern bloc (so-called Comecon) countries were 94 percent self-sufficient: they exchanged only 4 percent of goods with other Comecon countries, exported only 2 percent outside the region, and imported only 4 percent from outside the region. "Today it is no longer possible for a single country or region to develop the full range of top technologies and large network systems, or to independently finance the costs involved. Efforts of this nature now require an international division of labor and a global market place."[11]

TECHNOLOGICAL STRATEGIES
FOR MODERNIZATION

In Western Europe, the gross annual investment by public telecommunications operators (PTOs) is about one-third of their annual revenues or about two-thirds of 1 percent of GDP, whereas in Eastern Europe it is only one-third of one percent of GDP and lower on a per capita basis. Modern digital communications offers the most efficient and cost-effective solution for present and future needs. However, digitalization will require massive investment, as well as familiarity with the technology: "CEE telecommunications sectors must now move more rapidly than anyone has ever done before to catch up with the digitalization process which has been going on in the West for 15 years." [12] The investment required to modernize and extend the networks is enormous: estimates of the cost of increasing the number of main lines to 30 million range from $38 to $76 billion. [13] If the former U.S.S.R. is included, the total number of lines required is estimated at 120 million, with a projected cost of about about US$330 billion. [14]

Approaches to meeting the challenges include short-term strategies to provide rapid relief such as:

- installing new digital international switches to relieve bottlenecks and generate income;
- constructing digital overlay networks to relieve congestion in the national trunk network, provide high-quality service to large users, and build the skeleton for long-term modernization;
- licensing one or more cellular operators to provide service quickly to those able to pay, and to generate income from franchise fees;
- licensing or building packet-switched data networks for large data users. [15]

Strategies in CEE countries include overlay and digitalization in Hungary, the Czech Republic, Slovakia, and eastern Germany; digitalization is also being introduced in Poland and Romania. Poland, Hungary, and the former Czechoslovakia are on a fast track; the pace of modernization in Bulgaria and Romania is much slower. Cellular systems can be installed quickly as overlay networks. The customers are almost exclusively business and government users who can afford to pay relatively high rates, assuring financial viability of the enterprise. Mobile service also provides a limited area where competition between two or more systems, as well as foreign co-ownership, can be introduced quickly. Although mobile may be limited to capitals and surrounding areas at the start, it is still a prime candidate for joint ventures.

Another complication facing CEE operators is that telecommunications manufacturing industries have been slow to adopt or develop new technologies and to improve efficiency. However, governments are reluctant to see them collapse, and therefore pressure the operators to buy from domestic manufacturers. While there is strong interest in joint ventures and licensing agreements, there are also dangers of suboptimal equipment, tied procurements, and high costs. In addition, telecommunications enterprises have been unpopular, and are often regarded as inefficient, bureaucratic, and symbolic of old regimes.

FINANCING STRATEGIES

Investment Requirements

The volume of financing required to upgrade CEE telecommunications is enormous. For a growth rate of 10 to 12 percent per annum, CEE countries would need an estimated investment of from 0.5 to 1.5 percent of GNP. Some newly industrialized Asian countries, such as South Korea, Singapore, Malaysia, and Taiwan have achieved high telecommunications growth rates for sustained periods, but usually in the context of faster overall economic development. According to World Bank analysts: "For CEE to increase its telecommunications financing from less than 0.5 percent to around 1.5 percent during a period when GNP is growing by 1 to 2 percent will be difficult." [16]

Eastern Europe has relatively low revenues to finance expansion and development of technology, and a shortage of hard currency, which is needed to buy foreign components. Previously, about 90 to 95 percent of CEE telecommunications investment was paid for in domestic currency, to procure equipment from domestic and other Eastern bloc sources.

Now operators also need massive amounts of hard currency to import equipment. World Bank officials estimated that operators would have to import 40 to 50 percent of their equipment from abroad for five years from 1990 to 1995, and 30 percent for the next five years. [17] Possible sources of funds include earnings from international telephone traffic, barter transactions (such as oil, natural gas, and electrical energy), re-export of semifinished products, and contributions to development of software.

Financing Options

Direct foreign investment is desirable because it contributes equity and may bring foreign management expertise. Many forms are possible, such as foreign purchase of share equity in existing companies, investment in new companies for mobile and data networks, joint participation in new ventures, purchase of facilities, etc. Among the proposals for private investment in telecommunications operators (TCOs) are:

- different degrees and forms of equity ownership;
- joint ventures for particular activities;
- management contracts;
- conventional debt or mezzanine financing for all or part of the TCO;
- creation of a financial subsidiary to raise capital, own assets, and lease to the TCO;

- local telephone companies and cooperatives for rural areas, such as those in the United States and Finland;
- upgrading of dedicated networks that may also provide public services, such as networks operated by railroads, pipelines, and electricity suppliers.

A popular investment strategy involves joint ventures and alliances with international manufacturers. The Eastern European partner usually provides land and buildings, and the investor provides production equipment, which has to be financed. For example, Siemens has joint ventures to manufacture equipment in all the countries of Eastern Europe. Typically, the local partner makes land and buildings available, and Siemens provides production equipment. There are mobile joint ventures in Hungary, the Czech Republic, and Poland, and similar ventures are planned for Romania and Bulgaria. However, direct foreign investment is slow because of lack of political and legal clarity on ownership and transfer of assets, and outdated regulatory frameworks. Also, foreign investment raises issues of control; some countries are considering setting up separate subsidiaries which would raise capital in private markets, purchase telecommunications assets, and lease to TCOs.

Foreign outside financing options include international institutions plus tied credits, either from suppliers directly or a combination of suppliers and their governments. The European Investment Bank, the European Bank for Reconstruction and Development (EBRD), and the World Bank are lending funds to Eastern European nations to modernize and extend their telecommunications systems as part of assistance to help them move toward market-oriented economies. The EBRD is entrusted with the task of assisting Central and Eastern European countries which have embraced multiparty democracy in order to help their transition to market economies. The bank will include in its initial emphasis the creation or strengthening of infrastructure necessary for private sector development and the transition to a market economy, including institutional infrastructure (regulatory communications and services) and physical infrastructure (telecommunications, transportation, and energy).[18]

The European Union's PHARE Program provides grants for technical assistance in planning studies, training, and pilot projects. A similar program called TACIS provides assistance for Russia and the CIS. Since many of these countries have joined CEPT, that organization will also provide technical assistance in operations and regulation.[19]

Domestic outside financing is also an option, for example, with subscriber bonds or subscriptions. Hungary has used subscriber deposits, municipal contributions, and sale of bonds, and will expand into equity instruments and more general debt/bond instruments.[20]

Western Corporate Initiatives

Western companies have found that it is important to come into the country early, understand how its power structure works, and establish relationships with several levels of officials. As noted above, Siemens has expanded not only into eastern Germany, but has also established joint ventures throughout Eastern Europe. Other equipment suppliers such as AT&T, Ericsson, and Northern Telecom are also participating in Eastern European modernization. US WEST made its first ventures into the then U.S.S.R. in 1988. It now has an established cellular service in St. Petersburg, and a consortium in Moscow; it has also installed three international gateway switches in Moscow, St. Petersburg, and Kiev, which will improve the flow of traffic in and out of Russia and the Ukraine. US WEST had to invest only $18 million, recovering the rest of its costs through utilization of the network; it will get 16 percent of revenues over 15 years from calls funneled through its switches[21] (see Table 10.1.).

Several optical fiber projects are being undertaken by Western companies. A 1,300-kilometer optical fiber cable stretching from Copenhagen to St. Petersburg is being built, with a radio link to Moscow; this is a venture of two Danish companies, Great Northern Telegraph and TeleDanmark. There are plans for other international fiber links involving Cable & Wireless, U.S. Simplex Wire and Cable, and the Russian PTT.[22]

Bell Atlantic is part of a joint venture, building public data and cellular networks for the Czech Republic, while US WEST is in-

TABLE 10.1
CELLULAR JOINT VENTURES IN EASTERN EUROPE
AND THE FORMER SOVIET UNION

Country	Consortium	Ownership	License Award Date
Belarus	Belcel:		1991
	Commstruct International		
	(includes C&W, United Kingdom)	50%	
	Ministry of Posts, Telecommunications and		
	Informatics	50%	
Czech Republic	Eurotel Prague:		1990
	US WEST	24.5%	
	Bell Atlantic (United States)	24.5%	
	SPT Telecom	51%	
Estonia	Estonian Mobile Telephone Company (EMT):		1991
	Telecom Finland	24.5%	
	Telia (Sweden)	24.5%	
	Estelekom	51%	
Hungary (analog NMT & GSM)	WesTel:		1989 (NMT)
	US WEST	49%	1993 (GSM)
	Hungarian Telecommunications Company	51%	
Hungary (digital GSM)	Pannon GSM:	N/A	1993
	Telecom Finland		
	Telia (Sweden)		
	TeleDanmark		
	Norwegian Telecom		
	three Hungarian companies		
Latvia	Latvian Mobile Telephone Company:		1991
	Telia (Sweden)	24.5%	
	Telecom Finland	24.5%	
	Latvian Companies	51%	
Lithuania	Comliet:		1991
	Millicom International Cellular (Luxembourg)	29%	
	TeleDanmark	20%	
	Lithuanian Telecom	41%	
	Antene UAB	10%	

continued

TABLE 10.1
CELLULAR JOINT VENTURES IN EASTERN EUROPE
AND THE FORMER SOVIET UNION (*cont.*)

Country	Consortium	Ownership	License Award Date
Poland	Polska Telefonia Komorkowa (PTK):		1991
	Ameritech (United States)	24.5%	
	France Telecom	24.5%	
	Polish Telecom	51%	
Romania	Telefonica Romania:		1991
	Telefonica de España	60%	
	ROM Telecom	40%	
Russia	Moscow Cellular:		1991
	US WEST	22%	
	Millicom International Cellular (Luxembourg)	20%	
	Ministry of Posts and Communications	50%	
	Fydorov Eye Microsurgery Science and Technology Complex of Moscow	8%	
Russia– St. Petersburg	Delta:		1991
	US WEST	45%	
	St. Petersburg City Telephone Network Company	55%	
Slovak Republic	Eurotel Bratislava:		1990
	US WEST	24.5%	
	Bell Atlantic (United States)	24.5%	
	SPT Bratislava	51%	
Ukraine	Ukrainian Mobile Communications:		1992
	DBP Telekom (Germany)	16.3%	
	PTT Telecom (Netherlands)	16.3%	
	TeleDanmark	16.3%	
	Ministry of Posts and Telecommunications	51%	
Uzbekistan	Uzdunrobita:		1992
	International Communications Group (United States)	45%	
	Ministry of Communications (United States)	51%	

Source: ITU, *World Telecommunication Development Report*, 1994.

volved in similar ventures for Slovakia. In the Ukraine, the UTEL joint venture involves AT&T, PTT Telecom Netherlands, Deutsche Telekom, and the Ukraine's communications ministry. It has installed international switches which will route calls via satellite.[23]

TELECOMMUNICATIONS MODERNIZATION: EXAMPLES FROM CENTRAL AND EASTERN EUROPE (CEE)

Eastern Germany

When Germany was reunified, there were 1.3 million applications on the waiting list for telephone service in East Germany, and it took 18 years to get a telephone. Virtually all of the network and switching were still analog, while some 70 percent of local connections were 25 to 60 years old, and 30 percent of local connections dated from 1922 to 1934. As well as being outdated, the network was difficult to maintain because spare parts were no longer available.[24] These conditions resulted from decades of underinvestment.

West German legislation in 1989 created BDP Telekom (Deutsche Telekom) as a separate operating arm of the MPT (Ministry of Post and Telecommunications). Following reunification, West and former East German telecommunications operators were merged in October 1990. Deutsche Telekom was then a public enterprise owned by the government, but was privatized in 1996.

Good telecommunications are important to the economic development of eastern Germany, which is trying to modernize its industries and to establish new enterprises. For example, entrepreneurs cannot operate a modern company that requires complex production and decentralized management if they have to wait five hours to make a call or send a fax. To overcome the East-West gap as soon as possible, Deutsche Telekom established the Telecom 2000 program, which is to provide modern infrastructure throughout the country by the end of the decade. Deutsche Telekom will invest DM 55 billion in building up the network in eastern Germany between 1991 and 1997, and plans to install more than 7 million connections by 1997.[25] Telekom is the largest employer in eastern Germany, with 50,000 employees, so that as well as being a contributor to economic growth, it also creates jobs. The challenges Deutsche Telekom faces in undertaking such a massive infrastructure investment are discussed in chapter 10.

Telekom is using satellites such as Eutelsat and Kopernikus 2, the German satellite, primarily for text and data to augment terrestrial networks. Commercial users are also using private VSAT networks for voice and data communications where terrestrial facilities are inadequate. The MPT allowed private satellite and mobile communications companies to offer data services in the east, provided that Telekom was unable to supply a data link in a two-month period. The three-year limit on VSAT licenses was extended to six years early in 1991.[26] Satellite operators hope that this precedent to allow bypass will lead to permanent authorization of competitive networks as part of Germany's move toward liberalization.

Hungary

In December 1990, the Hungarian telephone density was only 9.6 lines per 100 population, despite the addition of 420,000 main lines during the Seventh Five-Year Plan. Fifty percent of the telephones were in Budapest, which had only 20 percent of the country's population. Many rural exchanges were manual and operated only six hours per day. There were 550,000 on the waiting list, with an average waiting time of 12 years, and this was only the official demand. Call completion rates were only 45 percent for local calls, 50 percent for long distance, and 35 percent for international subscriber-dialed calls.[27] The old equipment also had to be replaced because it had been installed in the 1930s and 1940s, with technology dating from the 1920s.

Hungary's telephone density of less than 10 lines per 100 people in 1990 approximated the situation in the Republic of Ireland in 1974. By 1990, Ireland had 27 lines, the goal for Hungary by the end of the decade.[28] To achieve this goal requires network growth comparable to the growth in the French network, which expanded from 12 lines per 100 in 1974, to 30 in 1980, and 42 in 1985. However, compared to Western nations that have well-established, market-based economies, convertible currencies, and advanced banking systems, Hungary faces severe difficulties. Despite these problems, Hungary is the furthest along of Eastern European countries in restructuring its telecommunications sector along Western lines. Three factors have contributed to Hungary's reforms: the worldwide change in the telecommunications sector in technology, structure and market; the radical political and economic changes in Hungary (and Eastern Europe); and its very limited and outdated telecommunications infrastructure.

In restructuring the telecommunications sector, Hungarian officials decided that the state should be only a regulator, and should not be able to intervene in market processes directly; the monopoly of the Hungarian Post should cease; and the exaggerated protection of the manufacturing industry should be stopped: it should supply quality goods at world market price in a competitive environment.[29] Therefore, in 1989, operation and regulation were separated, with regulation transferred from the Hungarian PTT to a Telecommunications Supervisory Board in the reorganized Ministry of Transport, Communications and Water Management. The Frequency Management Institute allocates terrestrial and satellite uplink frequencies. The Hungarian PTT was divided into three independent companies: Hungarian Post, Hungarian Telecommunications Company (HTC), and Hungarian Broadcasting Company. HTC joined CEPT in 1990. Cross-subsidization of loss-making postal activities was also eliminated in 1990.

The Telecommunications Act was amended in 1989 to make privatization possible; private investors can own a minority stake in service provision, and state ownership must be at least 51 percent.[30] In July 1991, Hungary took the first step toward privatization with the transfer of the Hungarian Telephone Company (HTC) from the Ministry of Transport, and Communications and Water Management to the State Property Agency. The government is likely to keep a 51 percent stake in HTC, while foreign investors will likely be limited to 25 percent. Current regulations prohibit foreign majority ownership of basic telecommunications facilities and operations.[31] The monopoly of HTC has been broken in two respects: all kinds of terminal equipment can be sold with type approval, and the Ministry can license other state-owned companies for nonwired telecommunications services. For example, VSATs, paging, and mobile communications can be provided in competition with HTC. New and competitive services will be split off into "daughter companies."[32]

A three-year development program from 1991 to 1993 was to install 1,215,000 main lines, including reconstruction of old facilities, with a 75 percent increase in the number of main stations. The goal for business customers was satisfaction of the majority of professional demand by 1993 "within a period that corresponds with customer expectations," while for the general population, public pay phones were to be available in every locality to make domestic and international trunk-dialed calls, with at least 95 percent of localities having 24-hour service.[33] (Previously, about 1,200 settlements had manual service from 8:00 A.M. to 2:00 P.M. on work days.)[34]

To meet these targets, Hungary planned to build a nationwide digital transport network by the end of 1993 that would connect all 56 network nodes; by the same time, all secondary nodes and county seats were to be provided with digital transit and local switches.[35] HTC's three-year investment project included a digital overlay network to connect 450,000 to 500,000 lines, mostly for business subscribers; installation of about 60 large digital switches in main cities; and connections with fiber and digital microwave, including access to packet-switched data and videotex services. For mobile service, Hungary is introducing the European GSM system. It will use wireless local loops in villages and as a quick fix overlay in cities.[36]

Financing of these projects was through a combination of domestic and foreign loans and direct investments (40 to 45 percent), and network revenues (55 to 60 percent).[37] HTC has a second telecommunications credit agreement from the World Bank. Foreign direct investments include an HTC joint venture with US WEST in mobile communications called WesTel, which started operation at the end of 1990. Hungary has also used domestic outside financing through subscriber deposits, which raised approximately 7.9 percent of the total value of capital expenditure from 1986 to 1989. Municipal contributions raised 9 percent, and sale of bonds 8 percent.[38] HTC will establish investment companies to develop local networks and plans to be a minority partner in them. Where HTC or local government does not initiate local companies, such as in economically disadvantaged areas, HTC will provide public pay phone service.

Poland

Poland is the largest country in Eastern Europe, with a population of 38.4 million. The Polish government and society are in a period of fundamental systemic change; the objective is to set up a market economy and political system akin to that found in industrially developed countries. The economic program adopted at the end of 1989 calls for two types of action: stabilization of the economy and transforming the economic system. Political changes in Poland in 1989 had a fundamental influence on decisions concerning modernization and accelerated development of the telecommunications sector. The first non-Communist prime minister, Tadeusz Mazowiecki, stated: "The telephone is indispensable for building a democratic state and introducing a market-oriented economy."[39]

Poland's telephone density in 1989 was 8 per 100, with 11.4 lines per 100 in urban areas, but only 0.2 per 100 in rural areas. Up to 60 percent of the network was made up of pre-World War II, step-by-step technology. The network was highly congested, as the emphasis had been on local distribution as opposed to the intercity network. In 1989, the average waiting time for residential service was 13 years, but some waited as long as 20 years. The waiting list was estimated at 3.5 million.[40]

The development of telecommunications is one of four key priorities of the Polish government. Short-term solutions are to unclog the network in order to facilitate economic activity, followed by long-term plans to modernize and put Poland on par with the rest of the industrialized world by 2010. The target for telephone density for the year 2000 is a total of 20.9 lines per 100 subscribers, of which urban is to be 28.1, and rural 7.9 lines per 100 population.

In 1990, telecommunications was classified as the priority sector of the economy in the use of foreign credit sources. The Law on Communication, which came into force in 1991, created the Ministry of Communication. The Polish PTT (PPTT) was under the Ministry of Communications, and was administered through 49 districts that covered the country. The PPTT was transformed into two separate enterprises: the Polish Post, a public services enterprise, and Polish

Telecommunications SA, initially a joint stock company with the state treasury as sole shareholder, and with the possibility of privatization in the future.

The Polish government decided to privatize the telecommunications sector in 1991, splitting the phone network into three parts: international toll, which is 100 percent Polish-owned and operated by the Ministry; national toll, with up to 49 percent foreign ownership; and local phone networks that offer the country's 49 regions access to the national network and can have 100 percent foreign ownership. The act distinguishes between services and facilities provision; facilities may be provided through Polish-foreign joint ventures. A new law on foreign investments passed in 1991 provides various guarantees to foreign investors, and allows free transfer of profits from foreign investments.[41]

In the short term, Poland requires foreign capital and technology, but it also needs to modernize Polish industry to meet national needs. The Polish modernization strategy includes upgrading to a fully digital system, technology transfer, reliance on government-guaranteed foreign credits, and introducing new services commonly used in Western countries. First phases of modernization included introduction of digital switching and new networks using optical fiber, microwave radio, and satellite links. New local exchanges were to add capacity of about 1 million lines. Poland has also introduced a paging system and a packet data network known as Polpak. To improve international communications, fiber has been installed under the Baltic Sea to Copenhagen; there are plans for more fiber links with Denmark and Germany. Poland is a user of Intelsat, Intersputnik, and Inmarsat, and has joined Eutelsat, CEPT, and ETSI.

In 1990, part of the funds acquired by selling food aid from the European Community in Poland was used to help the development of telephony.[42] The World Bank and European Investment Bank granted the PPTT a credit of $200 million for implementation of the most pressing tasks from 1991 to 1993. Now the Poles use low-interest European Union credits for this purpose. Poland has also received foreign credits for exchanges and transmission equipment from Spain, France, Italy, Switzerland, and South Korea.

Several foreign companies are involved in modernizing Polish telecommunications. Joint ventures with foreign partners to produce equipment include Alcatel, Siemens, and the Austrian KAPSCH company.[43] AT&T won a $100 million project to modernize the telecommunications system: the goal is to triple existing lines by the end of the century. Nortel (Northern Telecom) is building digital switches in Poland as part of a contract with Polmaik, a Polish telecommunications company.[44] Ameritech and France Telecom won the license to set up a cellular system, in a joint venture in which PPTT owns 51 percent of the shares.

TELECOMMUNICATIONS IN RUSSIA AND THE CIS

The Need for Investment

The countries of the former Soviet Union together rank third in the world after the United States and Japan, in terms of the number of telephone lines, with 40.5 million, of which Russia has 24 million. However, telephone density in Russia is only 16.2 per 100, compared to 60.2 in the United States and 48.0 in Japan.[45] In the mid-1980s, the Soviet Union had about the same ratio of telephone lines to total GNP as the United States; however, its total GNP was only about one-seventh of that of the United States.[46] Like many developing countries, the former Soviet Union faced the problem of a high cost of service provision in relation to per capita GDP. For example, the cost of implementing a telecommunications network is estimated at up to $2,500 per line, while the per capita GDP in the former U.S.S.R. in 1993 was $2,218.

To achieve a target penetration rate of 25 lines per 100 people by the year 2000, Russia needs to invest more than $3 billion per year in telecommunications. By 1995, foreign investors and multilateral financial institutions had pledged more than $1 billion.[47] Enormous investment is required both in new plant and in upgrading of existing facilities, as most of the central office switches still use crossbar technology, and will need to be replaced with digital equipment. Russian engineers are working to reduce the cost of upgrading and expanding their networks through development of their own electronic switches and the use of wireless technologies in some regions. Several U.S. and European telecommunications companies are involved in installing cellular and satellite bypass systems. However, these facilities are currently affordable only for business and government users.

Since early 1993, cellular licenses for 58 regions have been awarded through competitive tenders. GSM licenses were awarded to US WEST International, M-Bell (a consortium of BCE Telecom International and Moscow Local Telephone Company, which is also building a digital network in Moscow), Krakor (a Russian commercial firm), Telecom Finland, SMARTS (a Russian radio-telecommunications entity), and Bashtelecom (an international telecommunications company). US WEST already operates analog cellular networks in Moscow and St. Petersburg using the Nordic Mobile Telephone standard.[48] US WEST financed and operates three international gateway switches in Moscow. Previously, the former Soviet Union had a single international gateway in Moscow with 1,200 circuits. The new switches add 15,000 circuits in Moscow and 4,000 each in St. Petersburg and Kiev.[49]

An example of digital overlay is the Russian Digital Overlay Network project (RDON), which will require an investment of more than $40 billion over the next decade. RDON plans to double the current network capacity of 22 million lines by installing digit transit switches in at least 70 Russian cities, to be linked by some 70,000 kilometers of fiber-optic cable. Participants in RDON include Rostelecom (the Russian long distance and international telephone company), Deutsche Telekom, France Telecom, and US WEST.[50]

From 1992 to 1995, the Ministry of Communications awarded licenses for international service to more than 60 foreign and domestic companies. Bypass systems have been installed to ease the bottlenecks in international communications. Sovintel is a consortium of the Moscow PTT, GTE, and Sovam Teleport, which is building and operating a network linking hotels and business centers in Moscow with the West and Asia. The system uses digital microwave links to satellite terminals.[51] Andrew Corporation has formed a joint stock venture company with a Russian partner to build and operate a fiber system between Moscow and St. Petersburg.[52] Until 1992, Russia had only one major public international exchange, located in Moscow. Now there are two digital switches in Moscow and one in St. Petersburg. There are also 15 international satellite gateways and several VSAT networks for international business communications.[53]

The regional operating companies (previously known as GPSI Rossvyazinform) have been privatized and separated from their postal and broadcasting units. Although standards, frequency management, and licensing are still federal responsibilities; the regional companies control the scope and pace of network development. Their networks are old and congested, with only 10.5 percent of the switches electronic or quasi-electronic.[54] In the past, Russia's equipment market was largely controlled by the "state orders system" in which production was geared toward specific local or regional telephone companies. Inadequate production capacity caused the networks to be built in small and inefficient increments. Now the equipment market is being liberalized, so that any domestic or foreign firm can sell approved or certified equipment into the Russian market.[55]

In 1995, President Yeltsin signed the Law on Telecommunications, which was previously passed by the Russian Parliament. The new law seeks to formalize liberalization and make the market more predictable and transparent for both domestic and international investors. However, the law remains ambiguous in several key areas, including interconnection agreements, spectrum management, tariff regulation, and dispute resolution procedures.[56]

Satellite Systems

The Soviet Union was the first nation to establish its own domestic satellite network, using a series of polar orbit satellites. Molniya 1 was launched in 1965. (Molniya means "lightning" in Russian; a colloquial meaning is "news flash.")[57] A further 16 Molniya-1 type satellites were launched into elliptical polar orbit before an improved version, Molniya 2, was introduced in 1971. These satellites form the space component of the Orbita domestic communications system. The elliptical polar orbit allows the signal to reach the extreme northern latitudes of the former Soviet Union, with an apogee of 500 kilometers and perigee of 40,000 kilometers. (Nearly one-quarter of the former Soviet Union's territory is above the 60th parallel.) Each satellite is visible for approximately six hours, at which time the signal is transferred to another satellite entering its polar apogee. Two antennas are needed at each site to execute a synchronized hand-over or hand off as one satellite passes over the horizon and another emerges.

The Molniya system transmits television from two primary stations in Moscow and Vladivostok. In addition to one television channel, there is capacity for some telephony, telegraphy, and facsimile transmission. The Molniya system operates with more than 100 12-meter tracking earth stations. In 1974, the U.S.S.R. launched its first Raduga satellite, designated Statsionar-1.[58] Unlike the Molniya system, the Raduga satellites are geostationary, and are used for telephone and telegraph services, television and radio distribution, and remote printing of newspapers. The Molniya and Raduga systems together give complete coverage of the former Soviet Union.

Soviet satellite and launching technology was developed not only for civilian communications, but primarily for the military. Since the end of the Cold War, the former Soviet Union has been looking for markets for its surplus satellites and customers for its launching facilities. Some satellites have been sold to Western entrepreneurs and moved to orbital locations over the Pacific Ocean to serve Asian and Pacific Island customers. One of these new ventures is a company called Rimsat, which has moved a Russian satellite to an orbital location reserved by Tonga to provide satellite services in Southeast Asia and the South Pacific.

POLICY ISSUES

As discussed in chapter 4, restructuring requires a regulatory framework and policies to protect the public interest. "Rules of the game covering franchising arrangements, interconnection, tariffs, taxation, standardization, frequency allocation, enforcement, and dispute resolution must be clarified—in law, regulation, and administrative responsibility."[59] "This is what governments—especially telecommunications ministries should be focussing their attention on—not choosing switching systems or making joint venture deals with foreign companies. . . ."[60]

Increasing demand is putting pressure on Eastern European governments to allow alternative providers to enter markets to provide services not available from the telecommunications operator (TCO). The TCOs are also forced to rethink their business and to redirect resources away from traditional network development toward targeted investment for special user groups, while at the same time facing intense popular and political pressure to build a general, nationwide public network. An inherent conflict arises if the operator is pressured to use surpluses from profitable services to extend the basic network, while competitors are allowed to "cream skim" the most attractive customers. A more equitable strategy would involve contributions from all providers to extend services to rural areas for example.

An innovative model that appears promising for other regions is the role of cooperatives in providing rural services in Hungary and Poland. These cooperatives are likely to be responsive to local needs because they exist to serve their member-customers. The U.S. government has provided technical and managerial assistance from rural telephone cooperatives in the United States. Formation of cooperatives and local private telephone companies could also help to meet demand for service in rural areas of Eastern Europe and the former Soviet Union.

While they are an attractive short-term solution to pent-up demand, overlay networks can set a dangerous precedent if they are installed only to provide reliable but expensive service to large businesses and government. One danger is that individuals and small entrepreneurs who have access only to antiquated, unreliable service or no access at all will continue to be excluded. Another is that the emphasis on high-priced service will perpetuate the notion that telecommunications is affordable only for large commercial enterprises and the well-to-do, rather than an important tool for the development of the entire country and its people. Thus, overlay networks should be implemented as part of a plan to provide reliable and affordable service to the entire population, with specified target dates for extending and upgrading national and local networks.

Another concern is the need for transfer of technological and managerial expertise. Foreign investment can create dependency on imported technology and expatriate experts. Governments need to ensure that foreign investors are required to transfer managerial skills as well as technology. While the CEE has well-educated people who will be able to master new technologies, it needs experienced managers. As a result, Western companies have had to second their own managers while training local professionals to take over. To contribute to the economic health and political stability of the region, Western companies need to be prepared not only to provide technology, operations, and management know-how, but to forgo short-term gains in the interest of helping to solve industry problems in Central and Eastern Europe.[61]

CHAPTER *11*

Asia: Demand and Diversity

"We have a saying: 'When the telephone rings, business is coming.'"
—Manager of a rural cooperative in China[1]

THE ASIAN REGION

Two factors differentiate Asia from other parts of the world: its immense size and its diversity, in terms of cultures, languages, and level of economic development. The contrasts are striking between ultra-modern Japan and Singapore, and Sri Lanka, which has only 100,000 telephones, or 0.7 per 100 population, and Papua New Guinea, where many people have never seen a telephone.

The Asian region is home to 59 percent of the world's population, accounts for 29 percent of the world's economy, and has 25 percent of the telephone lines.[2] More than 90 percent of the population of the Asia Pacific live in lower-income economies. China and India account for just 12 percent of the wealth and telephone lines, but have two-thirds of the region's population.[3] Despite widespread poverty in the least developed countries, the economic growth rate in Asia has surpassed other industrialized as well as developing regions. The Association of Southeast Asian Nations (ASEAN) has a higher per capita GNP and volume of exports than Eastern Europe.[4] The combination of an enormous market area (more than 3 billion inhabitants), considerable wealth (combined GDP of over $6.9 billion or a quarter of the global total), and untapped potential (overall telephone density is less than 5 main lines per 100 population), makes the area very attractive to telecommunications suppliers. Main line growth has averaged 8 percent per year, and among the four tigers or dragons (Hong Kong, Singapore, South Korea, and Taiwan) recent main line growth has been twice this rate.[5]

Whereas it took the developed world 30 to 40 years to increase teledensity from 10 to 30 lines per 100 population, some Asian countries are making that transition in a decade. Telecommunications is a high priority in Malaysia, Thailand, Taiwan, Indonesia, and the Philippines. There is an enormous need for investment. For example, in Indonesia, while there are 2.5 million telephone lines, this represents only one line per 75 people. At least 7 million lines are needed

TABLE 11.1
TELECOMMUNICATIONS IN ASIA

	Teledensity 1994	GDP/Capita 1993 (in US$)
The Giants		
China	2.9	$ 611
India	1.1	257
The Four Tigers		
Hong Kong	54.0	18,687
Republic of Korea	39.7	7,509
Singapore	47.3	19,214
Taiwan	40.0	10,280
Emerging Economies of Southeast Asia		
Indonesia	1.3	763
Malaysia	14.7	3,392
Philippines	1.7	817
Thailand	4.7	2,103

Derived from ITU, *World Telecommunication Development Report*, 1995.

just to meet current demand. In Malaysia, Syarikat Telecom Malaysia Bhd. invested $1.94 billion through 1995 to improve its system. Taiwan plans to spend $7.39 billion through 1996.[6] Thailand plans to install 2 million new lines; in Bangkok alone it needs to triple its lines, which reached only about one out of six people in 1990.[7]

The investment needed to maintain current levels of growth in Asia is at least $100 billion between 1993 and the end of the century. The investment required in lower-income countries to reach a teledensity of ten lines per 100 inhabitants would be almost $40 billion. Market analysts predict that beyond the year 2000, the Asian market will be comparable to what the European market is today.

Three major priorities are required for rapid expansion of Asian networks: innovative investment strategies, regulatory reform, and introduction of new services. Asian structural models include a range from the mixed private and public networks of the Philippines to the newly privatized PTTs in Malaysia and Singapore, to the highly competitive local and value-added services in Hong Kong, to completely state-controlled systems in Vietnam and Cambodia. In this chapter, we examine the changing telecommunications sectors of some of the major economic powers in the region: China, Hong Kong, Singapore, South Korea, and Taiwan. The following chapter focuses on emerging and industrializing economies including India, Indonesia, Malaysia, the Philippines, Thailand, and Vietnam. (Japan was discussed in chapter 5.)

REGIONAL COMMUNICATIONS

Asia-Pacific countries are linked by both satellite and cable networks. Six Intelsat satellites cover the Pacific and Indian Ocean regions, and there are at least eight regional and national satellite systems. The first satellite with pan-Asian coverage was AsiaSat, launched from China in 1990 and owned in equal shares by the British carrier Cable & Wireless, China International Trust and Investment Corporation (CITIC), and Hutchison Whampoa of Hong Kong. AsiaSat's footprint covers 2 billion people, or more than one-third of the world's population. Indonesia's Palapa satellites cover southeast Asia. More than half of Palapa's transponders are leased to other users in the region for various services. China and Australia have their own domestic satellite systems, while Thailand introduced a domestic satellite in 1994, and Malaysia's Measat was launched in 1996, both with Southeast Asian footprints. PanAmSat coverage of Asia began in 1995. (For more information on international and regional satellite systems, see country analyses below and chapter 14.)

New undersea fiber cables link countries in the ASEAN region and connect Asia with North America. The TPC-4 and North Pacific Cable fiber cables connect Japan to the United States. The HAW-4 and HAW-5 cables terminate in Hawaii and connect with PacRimEast to New Zealand and TPC-3 to Guam. Regional projects include PacRimEast and PacRimWest projects linking Australia, New Zealand, Guam, and Hawaii. PacRimWest will connect Guam to Australia and to New Zealand via the Tasman-2.

The Asean Optical Fiber Submarine Cable Network (AOFSCN) links ASEAN nations. A $53.5 million 1,500 kilometer optical fiber cable link between West Malaysia and East Malaysia (Borneo) will be part of AOFSCN. A new fiber-optic cable owned by the 38 carriers of the Asia Pacific Submarine Cable Project (APC) links Malaysia, Singapore, Hong Kong, Taiwan, and Japan. The 7,500 kilometer APC system entered service in 1993; Singapore Telecom is its largest investor.

Another regional fiber-optic cable, the Asia Pacific Cable Network (APCN), will connect Singapore, Indonesia, Malaysia, Thailand, Hong Kong, the Philippines, Taiwan, Japan, and South Korea. The APCN will be able to transmit 5 Gbps, or the equivalent of 660,000 voice circuits. Singapore Telecom is the major investor, with participation from carriers in each of the other countries. The cable is to be completed by the end of 1996.[8] For more information on international submarine cable networks, see chapter 15.

CHINA

The World's Largest Market

The People's Republic of China, with a population of 1.2 billion and a telephone density of only 2.9 per 100,[9] is potentially the largest telecommunications market in the world. However, there is still an acute imbalance between supply and demand for telecommunications services. Investment in telecommunications could contribute dramatically to China's economic growth. Chinese researchers calculated that 100 million yuan invested in posts and telecommunications would increase national income by 1.38 billion yuan in ten years, with an implicit internal rate of return of 45.1 percent.[10]

Growth in external trade, particularly through Hong Kong, is reflected in the growth in international traffic. International calls to and from China in 1992 numbered almost 207 million, of which 82 percent were to or from Hong Kong.[11] International fax transmissions grew from 34,000 in 1987 to 544,500 in 1992.[12] The compound annual growth rate in outgoing international traffic from 1984 to 1994 was 40.9 percent, one of the highest growth rates in the world.[13]

There are wide disparities in access to services: the coastal and southern provinces have relatively extensive and reliable services because they have foreign exchange available, approval from Beijing to invest in improving telecommunications, and ties to Hong Kong and the outside world for trade and investment, for which good communication is essential. However, telephone access in the rural areas of the north and the interior is extremely limited, with no service available at all to much of the rural population.

Factors that delayed development of telecommunications in China include withdrawal of Soviet science and technology experts when the relationship with the U.S.S.R. was broken in the early 1960s, and the policies of Mao Zedong from 1957 to 1976, when the development of the military took precedence over economic development. Under the leadership of Deng Xiaoping, new policies were introduced, aiming at rapid development of the economy, including the accelerated development of telecommunications.

During the 1980s, Chinese telecommunications developed rapidly as a result of political reforms and the opening up of the economy. Telephone penetration increased from 0.43 per 100 population in 1980 to 1.1 per 100 in 1990; in provincial and coastal areas density exceeded 5 per 100. Some 771 cities above the county level were connected to the long distance automatic network.[14] Manual exchanges were being upgraded to SPC (stored program control) exchanges, and eventually to digital switches. In the decade 1981–1991, the average annual growth of GNP was 8.8 percent; growth of imports and exports was 12.2 percent.[15] Telecommunications growth accelerated in the 1990s, with growth rates increasing from 25.7 percent in 1990 to 58.9 percent in 1993.

By 1994, teledensity in Beijing was 5.5 lines per 100 population, but the average teledensity in the rest of the country (including other cities) was only 1.0, and in rural areas was still only 0.5. There was only one pay phone per 4,300 inhabitants. Also, development was very unbalanced, with about two-thirds of total exchange volume in the eastern areas of the country. The southern province of Guangdong's teledensity was 3.71 times the national average. Some 26 percent of China's townships had no exchange at all, and 57 percent of villages had no telephone service at all.[16]

Even in cities, waiting lists remain long, with 100,000 families in Beijing waiting for telephones and 400,000 on the waiting list in Shanghai. Applicants must pay a large deposit (about $500) which remains with the local Post and Telecommunications Bureaus until installation. Despite higher levels of digitization, completion rates remain poor because of network congestion.

Wireless networks are being installed in urban areas, with tremendous growth in demand. During the Eighth Five-Year Plan, from 1991 to 1995, China spent some $7 billion on telecommunications infrastructure, especially optical fiber and digital switching. The Ministry of Posts and Telecommunications (MPT) is in the process of increasing its communication capacity by 800 percent by the year 2000. The target is 20 lines per 100 population for economically developed prefectural cities and provincial capitals, and 25 per 100 for large cities. In addition, the goal is for every village to have a telephone, with automatic service in 98 percent of county towns.[17]

Switching capacity is projected to increase by 10 million lines per year, to reach 114 million lines by the year 2000 and 140 million lines by 2004, making China one of the world's largest networks and equipment markets.[18] China is expected to purchase more than $30 billion worth of switches and transmission equipment between 1996 and the year 2000.[19] However, potential investors attracted to this massive market face daunting challenges, including a lack of foreign exchange, tough competition from other vendors eager to establish a foothold, and extensive bureaucratic red tape.

The government wants to attract more foreign investment in the telecommunications industry, and to encourage the import of advanced telecommunications technology. It also has given high priority to development of the nation's own technology and industry.[20] Indigenous manufacturing is particularly important, because much of the telecommunications revenue generated in China is in yuan, rather than the hard currency needed for imports. In order to transfer technological and manufacturing expertise, several joint ventures have been established. NEC, Fujitsu, Siemens, and Ericsson manufacture equipment locally through joint ventures incorporating technology transfer. The Shanghai Bell Telephone Equipment Manufacturing Company is a joint venture with the Alcatel group; its locally produced digital exchanges are not only installed in China, but also exported to Russia.[21] Tong Gang Nortel is a joint venture with Northern Telecom. Major U.S. suppliers in China are AT&T and Motorola.

Massive investment of more than $20 billion is necessary in the 1990s to meet the goals of the Eighth and Ninth Five-Year Plans. While the central government still retains control over investment, the scale of funding required to achieve the goals of these plans has resulted in greater flexibility in investment policy; for example, the government has permitted the MPT and the local bureaus to retain 90 percent of the profits and foreign exchange income to upgrade equipment. It also expects telephone installation charges to contribute significant revenue for reinvestment. In addition, China will use debt financing and loans from international financial organizations and private sources.

Network Facilities and Services

Cellular telephone service is available in commercial and industrial regions. Analog cellular networks have been installed in all of China's 27 provinces and three autonomous municipalities of Beijing, Shanghai, and Tianjin. Digital networks are also being installed in several cities. In 1990, there were only 18,000 mobile telephone subscribers; by 1994, there were 1.5 million cellular subscribers to about 230 cellular systems, and more than 11 million paging subscribers.[22] In Guangdong province, an extensive network supports industrial growth; 43 of 85 cities in Guangdong offer paging services. The major foreign suppliers of cellular networking equipment are Motorola, Ericsson, AT&T, Northern Telecom, Alcatel, Nokia, and Novatel.[23]

Since 1986, CHINANET, the Chinese Academic Network, has connected China to the Internet via a store-and-forward dial-up link to the University of Karlsruhe in Germany. At that time, most users were government academies. An X.25 packet-switched public data network called CHINAPAC began operation in 1989, and covers more than 400 cities with 5,500 nodes.[24] The United Nations Development Program (UNDP) provided funding of $850,000 for the years 1991–1995 to further develop data communications facilities. CHINAPAC now offers electronic mail and Internet access. A national high-speed data network called CHINADDN was inaugurated in 1994. Users include several state banks, the State Foreign Exchange Administration, and the State Administration of Taxation. Local administrators are also building high-speed data networks.[25]

The Golden Bridge project, a joint venture involving IBM China and Ji Tong Communications, an enterprise established by the Ministry of Electronics Industry (MEI), will link 500 cities to a backbone trunk network. The first commercial Asynchronous Transfer Mode (ATM) high-speed data network in Asia is being built in Guangzhou and will support pilot broadband projects such as LAN internetworking, multimedia connectivity, and video-on-demand (VOD).[26] China has waived restrictions on foreign participation in network operations for this project. The network will be operated by HuaMei Communications, a joint venture between SCM/Brooks Telecommunications of the United States and Galaxy New Technology Company, a subsidiary of China's military manufacturing and research commission.[27]

Beijing Cable TV began transmissions in 1992, using multipoint microwave transmission that can carry up to 22 channels. China's cable television subscriptions are the cheapest in Asia, at just $1 per month, resulting in a connection rate of 95 percent. The strategy appears to be to connect as many subscribers as possible and raise rates later, or perhaps to provide an attractive, but controllable, alternative to satellite television. By 1992, there were at least 700,000 television receive-only (TVRO) earth stations in China, primarily receiving StarTV on AsiaSat from Hong Kong.

In addition to installing optical fiber backbones, China is also using satellites and microwave to link its cities and towns. A large-capacity fiber network will link Beijing and six major regions as well as coastal areas. Submarine cable links to Japan and South Korea connect to other international undersea cable networks. A joint project of MPT, AT&T, and KDD laid optical fiber cable from Shanghai to Kyushu in Japan, in order to connect China, Japan, the Pacific Rim, and Transpac cables to Asia and the United States. A trans-Asia-Europe fiber network will run from Shanghai to Frankfurt, across China, Central Asia, and Eastern Europe.

Satellites are an ideal means to serve China's remote areas as well as to supplement or bypass terrestrial networks. In 1984, China launched its first experimental communications satellite, followed by a communications and broadcasting satellite in 1986 and two more satellites in 1988. There are earth stations in at least 15 cities, including the frontier cities of Lhasa, Urumqi, and Huhhot. The satellites are used for voice, facsimile, public television, and educational television programs.

China has its own launching capability, and has become a major competitor in the global satellite launching industry. The Great

Western Wall Industry Corporation was approved by the Chinese government to provide commercial launching services. Analysts estimate that China has been charging roughly $35 million per launch using its "Long March" series of rockets, as much as 50 percent below the price of a launch on a U.S. or European rocket.[28] However, several recent failures have slowed China's launching industry, including a rocket carrying an Intelsat satellite that exploded just after launch in 1996, causing extensive damage and casualties near the launch site, The United States has also raised concerns about Western satellites being launched from China because of the concerns about revealing Western technological advances to the Chinese.

China launched the AsiaSat satellite in 1990, and the Apstar satellite in 1994. China is a part owner of AsiaSat 1 through the CITIC, and leases AsiaSat transponders for domestic voice and data. The Shanghai stock exchange has a VSAT network of more than 2,000 data and audio terminals to broadcast securities information. The Xinhua News Agency also operates a VSAT network. Small earth stations provide rural telephony linking Tibetan villages with Lhasa using AsiaSat. China is also a major investor in the regional Apstar satellite. (The latter was not coordinated with other satellite systems with which it could have interfered, but was eventually moved to an unused orbital location that had been allocated to Tonga.)

Structure of the Sector

China has a dual administrative structure, operating at the national and local levels. The State Council is the highest organ of state administration to formulate regulations, issue orders, and coordinate the work of ministries and state commissions. Under the Council are the State Planning Commission (SPC), which reviews and steers industrial activities and growth, the State Science and Technology Commission, responsible for technology policy, and the Radio Management Commission, which is largely controlled by the People's Liberation Army (PLA).

The SPC is the most important and powerful agency in the central

government, responsible for national economic planning and allocation of state funds. There are many departments in the SPC responsible for review and approval of telecommunications projects. Two
ministries have responsibilities for telecommunications. The Ministry
of Posts and Telecommunications (MPT) operates national telecommunications networks, acts as national regulator, carries out technical
and economic research, and manufactures equipment through the
Posts and Telecommunications Industrial Corporation (PTIC). The
Ministry of Electronics Industry (MEI) is also a primary manufacturer
of telecommunications equipment. The Ministry of Finance (MOF)
allocates funds and monitors use of funds; MOF controls use of funds
from international agencies such as the World Bank and the Asian
Development Bank. The Ministry of Foreign Economic Relations and
Trade (MOFERT) is responsible for negotiation of contracts involving
import of equipment and projects financed with foreign loans.

At the provincial level, Posts and Telecommunications Administrations (PTAs) coordinate intraprovincial networks and the activities
of the 2,500 municipal and city Posts and Telecommunications Enterprises (PTEs). At the local level, government departments deal
more directly with foreign business customers such as the Shanghai
Posts and Telecommunications Authority (SPTA), a government department set up by the Shanghai municipal government. Telecommunication revenues are redistributed to each enterprise according to its
contribution and supplemented by cost factors such as geographical
location. The total income of PTAs consists of local revenues and redistributed revenues, with an incentive salary scheme rewarding employees for traffic growth.[29]

New enterprises started by local and provincial governments, or
state-owned enterprises associated with national ministries during
China's transition to a market economy, are still closely affiliated with
the state sector. Government agencies with private networks such as
local branches of the military, municipal government departments,
and the Public Security Bureau formed telecommunications enterprises after 1990. In Shanghai alone, there were 83 wireless networks
and 9 cable networks run by local authorities and 125 networks run
by various ministries in 1994.[30]

The authority to approve usage of radio frequencies became highly fragmented during the Cultural Revolution, and provincial and municipal authorities were allowed to assign frequencies to local users with little supervision. In addition, the People's Liberation Army (PLA) controlled many frequencies that could be assigned to local enterprises. This decentralization spawned the radio paging industry, which had low start-up costs and was affordable to large numbers of users. By 1995, there were more than 2,500 paging companies in China with more than 15 million subscribers; however, despite the number of providers, MPT affiliates hold the major market share in most cities. Thus, the entrepreneurial process was linked to political and economic competition among government agencies at both the national and local level.

In 1993, the MPT issued licensing directives for the following services that were declared "open": wireless paging, 800 MHz trunked telephone service, 450 MHz wireless mobile service, domestic VSAT, and various value-added services such as voice mailboxes and information services. By 1995, there were more than 100 value-added and trunk telephone service providers, 82 mobile telephone operators, and 15 domestic VSAT operators.[31]

Until 1994, China had only two types of national telecommunications carriers: the MPT itself and specialized ministerial carriers. An MPT unit, the Directorate General for Telecommunications (DGT) manages the public, interprovincial, and international networks. Other ministries, such as the Ministries of Electronics, Railways, Petroleum, Coal, Water, and Power, and some large industrial enterprises, the new China News Agency and the PLA, operate private communication networks. In 1994, China announced that a second network would be allowed to compete with the former monopoly, MPT. It has allowed the Ministries of Power, Electronics, and Railways to create a second operator, China United Telecommunications Corporation, known as Unicom in English and Liantong in Chinese, to challenge the MPT's monopoly.[32] Unicom has apparently signed several memoranda of understanding that involve minority stakes with foreign companies.

The government considers posts and telecommunications to be a basic industry that has significant effects on the national economy, and should therefore be run by the state, but should involve the local authorities, departments, and collectives. To date, foreign joint ventures are limited to provision of equipment, as China continues to ban any foreign equity ownership, management, or operation of its telecommunications sector. The MPT plans to create a direct investment fund designed to raise about $200 million from overseas institutional investors. However, many analysts predict that foreign investment will remain limited until the ban on foreign ownership of services is lifted, and the Chinese policy and regulatory environment become more transparent and predictable.[33]

THE FOUR TIGERS

Hong Kong, the Republic of Korea (South Korea), Singapore, and Taiwan have become known as the Four Tigers or the Four Dragons because of their dramatic economic growth and industrialization since the 1960s. Hong Kong and Singapore are economic rivals that are transforming themselves into "intelligent islands" to compete in information-intensive trade and services industries, as well as to serve as regional hubs. However, their telecommunications policies are very different, with Singapore maintaining a strong government role in operations and limited competition, and Hong Kong relying on the private sector, with extensive competition.

HONG KONG: POSITIVE NONINTERVENTION

With a per capita GDP of more than US$18,000, Hong Kong is the world's eighth largest trading economy, operating the world's largest container port and busiest airport in terms of cargo. More than 700 international businesses and organizations maintain regional headquarters in Hong Kong, using the territory as both a gateway to China and regional hub for east and Southeast Asia.

International telecommunications traffic increased more than 1,200 percent between 1984 and 1994, as a result of the dynamic growth of the Chinese economy, which funnels much of its output through Hong Kong, and growth in trade with the rest of the world. During this period, the balance in traffic has changed from a large, outgoing surplus for China, to a growing outgoing surplus for Hong Kong, as Hong Kong has moved from the role of being China's "window on the world" to being a staging point for local and foreign businesses to enter China. Hong Kong firms have moved labor-intensive production to China, increasing intra-industry trade, while China imports consumer goods. This trade, plus travel by Hong Kong residents to China, which now averages three trips per resident per year, is also reflected in the growth of international telecommunications traffic.[34]

Telecommunications regulation is the responsibility of the Office of the Telecommunications Authority (OFTA). OFTA is generally modeled on the U.K.'s Oftel and Australia's Austel; in fact, the director general was recruited from Austel. Following the Hong Kong government's approach to industrial policy of "positive nonintervention,"[35] OFTA has liberalized the sector into one of the most competitive in the world. However, recent decisions such as selections of licensees for fixed network services, have required consultation with the Chinese government. There is now competition in local fixed, mobile, and value-added services (VAS). VAS providers must be licensed by OFTA and registered under the Companies Ordinance in Hong Kong, but there are no foreign ownership restrictions. In 1995, there were 48 value-added services licensees offering data networking, Internet access, voice messaging, public facsimile, and other services.[36]

The dominant carrier is Hong Kong Telecom (HKT), with majority ownership of 57.5 percent held by Cable & Wireless. The other major investor is the Chinese International Trust and Investment Company (CITIC), an arm of the Chinese government, which owns 12.5 percent; the rest is owned by funds and private investors. HKT is the sole provider of all international service, having been granted a

25-year monopoly in 1981. The Hong Kong Telephone subsidiary HKTel had a 20-year monopoly that ran until 1995, when the government introduced local competition. HKT also operates a GSM cellular network, has interests in China, and is a partner in the second GSM cellular network in Singapore. The HKT Group is critical to the viability of Cable & Wireless, contributing 73 percent of C&W's operating profit in 1994.[37]

Hong Kong has the highest concentration of wireless communications in the world. Cellular mobile services have been available in Hong Kong since 1984. By 1994, there were 431,800 mobile subscribers, or 7.4 per 100,[38] using services provided by four cellular operators on seven networks covering five analog and digital technologies. Hong Kong has the world's highest concentration of pagers, with almost one pager for every four inhabitants. In addition to the four cellular operators, there are also four Telepoint cordless telephone (CT2) licensees (service started in 1992), and 36 paging service operators.[39] Digital mobile service is offered, combining two systems, one based on the European GSM standard and one based on AMPS (Digital Advanced Mobile Phone Service), the U.S. standard. Up to six operators are being licensed to provide Personal Communications Services (PCS), with very small handheld handsets accessing microcells.

Three operators have been licensed to compete with HKT in the provision of fixed network service. New T&T is owned by conglomerate Wharf Holdings, which also has a cable television franchise and extensive real estate holdings in Hong Kong. New T&T was able to use optical fiber strung through the subway system for cable television service. Its prime clients include tenants of Wharf's office buildings, much of the financial services industry. Hutchison Communications, the second licensee, is owned by Hutchison Whampoa, whose majority owner is Li Ka-Shing, one of Hong Kong's wealthiest entrepreneurs, with extensive holdings in many sectors overseas as well as in Hong Kong. Hutchison is also a cellular operator. The third licensee is New World Telephone, majority-owned by the New World Group, with hotel properties in Hong Kong, major cities in China, and Southeast Asia.

Hong Kong–China traffic accounts for nearly 50 percent of Hong Kong's outbound calls, and over 30 percent of international revenues. While the settlement rate on revenues between HKT and the Chinese MPT is 50:50 on calls to nearby Shenzen, it is 33.3:66.7 with the rest of China, providing an important source of revenue for China's network expansion.[40] However, although HKT International has a legal monopoly until the year 2006, the new carriers are effectively introducing international competition to other parts of the world by offering callback services. The result has been a reduction in international rates by HKT, and stiff price competition among the carriers. Callback services were ruled legal in Hong Kong in March 1995. Even government departments are encouraged to use them to save money.

In the 1980s, the Hong Kong government looked to cable television as a means of introducing a second telecommunications carrier to compete with HKT, following the British duopoly model (which the United Kingdom was about to abandon). However, HKCC, the consortium awarded the second license, collapsed in 1990, following extensive internal wrangling and the introduction of StarTV via AsiaSat. The government then separated cable television from switched telephone services, and granted a single cable TV license to Wharf Holdings (which had owned the largest share of HKCC), with a three-year exclusive license through 1996. However, HKT is also planning to introduce video-on-demand (VOD), so that there is the potential for two providers of interactive video services.

One Country, Two Systems . . .

Established as a British Crown Colony in the 19th century, Hong Kong will revert to China when the United Kingdom's lease on the neighboring New Territories expires on June 30, 1997. After that date, Hong Kong will be designated as a Chinese Special Administrative Region (SAR) under the rubric "one country, two systems." The agreement with China allows a 50-year grace period, during which Hong Kong's capitalist structure and legal system are to be maintained. While most observers believe that China will not want to kill the capitalist goose that lays the golden eggs, there is much more skepticism about the extent to which political and legal autonomy will be maintained.

The preamble of the Basic Law on the status of Hong Kong after mid-1997 states: ". . . under the principle of 'one country, two systems' the socialist system and socialist policies will not be practiced in Hong Kong."[41] What this will mean for telecommunications in Hong Kong is uncertain. At present, the two telecommunications systems would appear antithetical. The Chinese MPT is government-run on the PTT model, with a virtual monopoly in China. Hong Kong's monopoly, formerly controlled by the British company Cable & Wireless, is now one of the most competitive industries in the world (the HKT international monopoly has been effectively undermined by callback services offered through competing local carriers). Yet China is clearly hedging its bets. The authorization of the second carrier, Unicom, although now controlled by other ministries, may presage the introduction of competition and eventually private sector investment in China. China could also use its stake in HKT through CITIC to designate HKT as a "chosen instrument" to compete in China. Hong Kong investors may be considered "domestic" rather than foreigners and therefore allowed to invest in China.

The policy of "one country, two systems," however, may have a more profound effect on freedom of expression, including media content. Compared to the original Joint Declaration between the United Kingdom and China, the Basic Law strengthens the central government's control in the definition of subversion and imposition of martial law. As Milton Mueller points out, the Basic Law was issued after the uprising and massacre at Tiananmen Square, during which there were huge public demonstrations of support in Hong Kong for the Chinese students, whereas the original Joint Declaration between the United Kingdom and China was signed in 1984. In 1996, China announced that it would replace the popularly elected members of Hong Kong's Legislative Council with appointees after taking over in 1997, a policy that is widely believed to presage other interventions.

SINGAPORE: THE INTELLIGENT ISLAND

Singapore is a city-state with a population of 2.8 million. With only 620 kilometers of land and no natural resources, it has built its economy on trade and the skills of its people. As a British colony, Singapore was heavily dependent on entrepôt trade, primarily rubber and tin. After independence and separation from Malaysia in 1965, the government launched an intensive industrialization program. To diversify the economy, it developed other sectors, particularly manufacturing, finance, and tourism.

To support this economic development strategy, Singapore needed good telecommunications. It first emphasized basic telephone service, increasing the network from fewer than 60,000 telephone lines in 1960 to 270,000 in 1974. The introduction of satellite communications in 1971 marked the beginning of rapid telecommunications growth; international telephone calls grew tenfold and telex calls 15-fold from 1966 to 1974.[42] Singapore also became a maritime communications center, and now has the world's most efficient port, relying on computerized management of everything from logistics to customs clearances.

The Progressive People's Party (PPP), led until 1995 by Lee Kwan Yew (still acknowledged as the de facto national leader), has carefully engineered Singapore's economic growth and political climate since Singapore's separation from the Malaysian Federation in 1965. Its strategic economic plan is to become a "total business hub," with emphasis on high value-added, knowledge-based industries and internationalization of local companies. From 1979, the economy was restructured to shift from labor-intensive, low-skill, and low value-added activities to capital-intensive, high-technology, and high value-added activities.

Key to this strategy has been the growth of communications and information infrastructure to make Singapore an "intelligent island." Singapore Telecom (ST) geared its corporate strategy to meet the increasingly sophisticated needs of customers. Three broad areas of development were planned and undertaken: infrastructure development, international networks, and modern services for the Information Age. In 1985, Singapore Telecom collaborated with the Economic Development Board, the National Computer Board, and the National University of Singapore to draft a National Information Technology Plan (known as IT2000) to exploit information technology as a tool for improvement in productivity and business competitiveness, and as a growth industry.

Singapore Telecom and several national planning agencies have been involved in implementing this plan:

- ST has introduced islandwide ISDN, installed optical fiber networks, and introduced videotex, electronic mail, Internet access, and other information services.
- The National Computer Board has established computer networks as part of its IT2000 initiative.
- The National Science and Technology Board has funded science and technology parks with broadband networks installed by ST.
- The Economic Development Board has promoted the telecommunications industry as a leading development sector.[43]

Telecommunications Facilities and Services

Telephone density in 1994 was 47.3 per 100 population, with more than 1.3 million lines for a population of 2.8 million people, and a compound annual growth rate of 5.8 percent from 1984 to 1994.[44] The average time for installation is six working days.[45] Growth in revenue has been strong, with a 26 percent increase in net surplus in 1991, and a further 26 percent increase in 1992.[46] Singapore was the first country to be 100 percent push-button: it replaced half a million rotary dial phones with push-button pulse phones in 1979. Soon, banks started to use them for home banking and other consumer services. The network is now completely digital.

Singapore Telecom introduced an analog AMPS 800 MHz cellular network in 1988, a second analog 900 MHz ETACS system in 1991, and a digital GSM network in 1994. It also introduced a CT2 wireless service in 1993. Roaming has been available since 1989 so that Singaporeans have been able to use their mobile phones in Hong Kong. There are 8.4 cellular subscribers per 100 inhabitants; Singapore also has the world's highest pager density, with almost one person in three owning a pager.

Singapore Telecom also operates Teleview, an interactive videotex system. However, videotex has never been very popular, and is being eclipsed by Internet services. There are several Internet providers, including Singapore Telecom's Singnet, and Internet cafes have sprung up to provide community access.

ST has invested $1 billion in facilities in the last five years, and plans to spend an additional $1.5 billion in the next five years. It has installed 140,000 kilometers of optical fiber; all 28 telephone exchanges are interconnected by 140 Mbps fiber-optic cables. ST plans to introduce broadband ISDN and fiber to the home by the year 2000.[47] Meanwhile, the cable TV network uses a fiber ring installed by ST; it could emerge as a competitor to ST in providing video-on-demand (VOD).

Singapore is linked by satellite and cable to the rest of the world. International satellite service is via Intelsat Indian Ocean and Pacific Ocean satellites. Seven submarine cables link Singapore directly with 26 countries. It is a partner in the world's first integrated optic fiber submarine cable system, linking five southeast Asian countries. In addition to regional links to Indonesia and other ASEAN nations, Singapore is a node for the SEA-ME-WE cable to Europe and the Asia Pacific Cable Network (APCN), linking Singapore with other Southeast Asian countries, as well as Hong Kong, Taiwan, Japan, and the Republic of Korea.

In 1993, more than 3,000 international companies hubbed their telecommunications through Singapore, but fewer companies were regionally based in Singapore than in Hong Kong.[48] However, Singapore has been aggressively marketing itself as an alternative to Hong Kong because of the uncertainty of the post-1997 political environment in Hong Kong. Singapore has attracted several regional satellite uplink operators including HBO, ESPN, and ABN (Asian Business News), among others. Ironically, these operators can transmit from

Singapore, but their programs cannot be received by Singaporeans, who are not permitted to own satellite antennas.

Structure of the Sector

The Telecommunication Authority of Singapore (TAS) is responsible for regulation and operation of telecommunications and postal services. The Telecommunications Authority Act of 1992 confirmed the separation of posts and telecommunications, provided for eventual flotation of shares, and reconstituted the TAS into a Statutory Board. TAS, however, is not an arm's length regulator, but still interweaves regulation, operation, and policy making. Its function is not only to regulate, but also to develop and promote the telecommunications industry in support of the IT2000 Plan to make Singapore an "intelligent island." TAS is mainly self-financing. It receives about 60 percent of Singapore Telecom's surplus, and may receive revenue from licenses and administrative fees, and may raise capital through stocks and bonds.[49]

In April 1992, Singapore Telecom was corporatized, taking over the former commercial functions of TAS. The company was given a 15-year monopoly by TAS in domestic and international telephone services and leased circuits. It was also given exclusive rights to provide cellular mobile service for five years, until 1997.

The next step in restructuring was privatization. A cabinet-level decision in 1990 was made to privatize Singapore Telecom not earlier than 1993. The goals Singapore wanted to achieve in privatizing Singapore Telecom were to provide better management incentives, remove day-to-day operations from the national policy process, and accelerate innovation and capital investment, thus allowing Singapore Telecom to expand internationally.[50] The first stage of privatization began in 1993; Singapore Telecom is now the largest company listed on the Stock Exchange of Singapore. The government of Singapore will continue to hold, at least initially, approximately 80 percent of the privatized company, and will likely hold a "golden share" indefinitely.

Pricing is based on a managed "price comparison" approach, with a benchmark of being "among the lowest three" in price comparison to services offered in other countries, and taking into account the changes in cost of living of the average Singaporean. In 1990, Singapore Telecom introduced local time-based charging, partly because of the growth in fax and data traffic, to make pricing more equitable. The monthly bills for nearly 75 percent of Singaporeans decreased, while bills for the higher usage customers increased.[51]

Recognizing that the globalization of Singapore's economy would create future growth opportunities in international communications, in 1988, Singapore Telecom set up a subsidiary—Singapore Telecom International (STI)—to further international expansion through marketing, strategic partnerships, and consultancy services. STI has invested in projects and joint ventures in wireless cellular and paging networks, satellite communications, data communications, cable television, and multimedia services in Asia and Europe, including Australia, China, Hong Kong, Indonesia, Malaysia, Mauritius, Norway, Pakistan, the Philippines, Sri Lanka, Thailand, the United Kingdom, and Vietnam.[52] ST was also a founding member of the WorldPartners consortium initiated by AT&T.

In 1989, ST restructured to make the organization more market-driven, by establishing three business units in Business Communications, Residential Communications, and Mobile Communications, and spinning off a wholly owned subsidiary, Telecom Equipment Private Ltd., to compete in terminal equipment, which had been liberalized. ST also established Mobilelink to run its cellular services and Pagelink to run its CT2, paging, and mobile data services.

TAS will introduce some competition by issuing a second license for GSM service, starting in 1997. The franchise will be for 20 years, and overseas partners will be allowed to bid. No further cellular licenses will be issued for three years. TAS will also issue three additional ten-year paging licenses.

Controlled Intelligence?

The Singaporean model is unusual both in its industrial policy of investing in telecommunications as a means of supporting its economic development strategy, and in its continued reliance primarily on a managed monopoly structure. As Mark Hukill points out:

> The close link between government, regulator and operator in Singapore must be seen not in terms of conflict of interest as might be the case in other countries, but as a tripartite strategy for development opportunity. . . . The question remains as to how far this managed approach of encouraging technology development and then ensuring market success can go, especially in the absence of any real market competition outside of the CPE market.[53]

Yet, Singapore seeks to control other aspects of its information sector besides the provision of telecommunications services. Consumers are not allowed to install satellite terminals to receive television from AsiaSat, Malaysia's new Measat, or other satellites with footprints covering Singapore—even though some of the channels are uplinked from Singapore. Instead, Singapore has authorized a cable television network (partly government-owned) to provide multiple channels of television, but under government control. Singapore is also anxious to promote Internet access as part of its "intelligent island" strategy, but wants to know who is using the Internet and what information is being accessed. Customers must provide their ID number or passport number to get an Internet account; the government has blocked access to some Internet sites.

Such policies would seem paradoxical for a country that has staked its economic future on becoming an information-based economy. Singapore wants to be among the leaders in Internet access, yet seeks to control what its citizens access. It wants to attract satellite uplink services, but prohibits its own people from watching their channels. It also seeks to ensure that the government has a stake in all media and information technology ventures, including Internet providers and the cable network. Such policies would not be surprising, perhaps, among countries with less developed economies or less entrepreneurial citizens. Of course, Singapore has found that it is virtually impossible to keep out information, whether by banning certain magazines and newspapers or banning access to Internet sites and satellite channels. But, its commitment to a policy of control seems to contradict an economic development policy that is built on access to information and a highly educated workforce.

REPUBLIC OF KOREA: TOWARD AN INFORMATION SOCIETY

Known as one of the Asian Tigers, the Republic of Korea is one of the fastest growing economies in the world, with a 9 percent growth rate in 1990. Korea has made huge investments in telecommunications in the past 15 years. In 1981, there were 3.5 million telephone lines, with penetration less than 10 percent. Waiting times were six months to one year. By 1994, the telephone density was 39.7 lines per 100 population, with an increase of more than 450 percent in telephone lines, and compound annual growth rate of 12.2 percent from 1984 to 1994. With a population of 44.5 million, South Korea now ranks ninth in the world in terms of telephone lines installed. Teledensity was projected to reach 55 per 100 by 1995. In 1994, there were 960,300 cellular subscribers and 3.8 million paging subscribers.[54]

Like Singapore, the Korean industrial strategy is a "government driven scenario," with government leadership in Information/Communications (I/C) and related fields. I/C is seen as infrastructure, industry, and a network that can give competitive advantage. The I/C field is classified into information/telecommunications service, I/C manufacturing, and information providing areas. In the 1980s, Korea began a massive investment in building I/C infrastructure through the government and the Korean Telecommunications Authority (KTA). Several areas were improved through cooperation based on alliances among government, academic, and business communities. By

1987, 1.2 percent of GDP was invested in I/C fields, 30 times the amount invested in 1971.

Korea's goal is to build its own infrastructure using its own technology. The liberalization of CPE in 1980 turned out to be a huge boost to the telecommunications equipment manufacturing industry. Korea emerged as a major exporter of telephone terminals, and soon fax machines, personal computers, cellular phone sets, and PBXs, in parallel with the advent and expansion of the semiconductor industry. In 1984, Korea developed a digital electronic switching system known as the TDX-10, becoming the tenth nation to develop a switching system. About 35 percent of the total digital network uses this TDX switch, which is also exported.

In 1994, there were 2.6 million cable subscribers, while some 500,000 households had home satellite antennas.[55] The lure of foreign satellite television, primarily from Japan, was one of the factors that contributed to the Korean decision to develop its own national satellite system. Korea's first domestic satellite, Koreasat 1, was launched in 1995, and is used for direct broadcasting and fixed satellite services. Koreasat is being implemented by a partnership of GoldStar Information and Communications and Korean Air, with General Electric Astro-Space and Matra Marconi.[56] Korea plans to be able to manufacture its own domestic satellite by 2005, to replace its first generation. Korea is also involved in several submarine cable projects, including the R-J-K project linking Russia, Japan, and Korea; the Asia-Pacific Cable Network linking ten Asian countries; and the China-Korea submarine cable CKC that links China through Shangdong province to Korea.[57]

Under the Seventh Five-Year Plan (1992–97), the goals are to upgrade the network with early introduction of the intelligent network, construct a personal portable communications network, and upgrade corporate communications through the introduction of ISDN and informatization of industrial sector. Other goals include promoting information awareness through education, training, and public awareness campaigns, and advancing information/communications technology through expanding investment in R&D. These activities are to be financed by the private sector.[58]

To conduct research and development needed for its I/C strategy, the Korean government supports two telecommunications-related research institutes. The Korea Information Society Development Institute (KISDI) is a nonprofit research institute founded and funded by the Korean government under the auspices of the MOC. It conducts research on a broad range of information and telecommunications issues such as telecommunications policy studies, and applications and impacts of new technologies. The Electronics and Telecommunications Research Institute (ETRI) is a nonprofit R&D foundation established and funded by the Korean government under the auspices of the Ministry of Science and Technology (MOST). It conducts research on advanced information technology integrating telecommunications, computers, semiconductors, and automation.

The National Computerization Agency is responsible under MOST for the promotion of National Basic Information Systems (NBIS), a networking project including links of Korea's research institutes, access to databases, networks for finance, banking, and trade, and computerization of the society. The goal is to have one personal computer per household by the year 2000, 3 million to be supplied by the public sector, and 7 million by the private sector.[59] Korea plans nationwide ISDN and extension of optical fiber into high-rise buildings and eventually to the home.

Structure of the Sector

Until the 1980s, telecommunications in South Korea was a monopoly run by the Korean Telecommunications Authority (KTA), as part of the Ministry of Information and Communications, formerly the Ministry of Communications. As part of its economic development strategy, the government mobilized capital resources to provide infrastructure for export-driven economic growth. However, by the 1980s, four problems convinced the government of the need for economic reform: a waiting list of half a million for basic telephone service; call completion rates below 50 percent; outdated rural exchanges; and lack of modern data communications.[60]

Korea made three major industrial policy decisions in the 1980s: separation of business from policy, deregulation, and internationalization. The first step, in 1979, was the Interim Law for Expansion of Public Telecommunications Facilities, which increased tariffs and introduced subscriber bonds to raise funds for network expansion, along with loans from Belgium, Sweden, and the United States. With the start of the Fifth Five-Year Economic Plan, the government gave priority to telecommunications infrastructure, increasing investment from under 3 percent of the government's overall fixed investment in the 1970s to 5.7 percent in the 1980s.[61]

The first steps to liberalization were taken in 1980, when subscribers could buy their own telephone sets. When KTA was separated from the Ministry of Communications in 1982, it had unmet demand for 5 million lines, as development in the 1970s had concentrated on manufacturing and heavy machinery and chemicals. KTA got rid of the backlog, and met goals of one telephone per household and a nationwide automatic telephone system in the 1980s.

In 1982, Data Communications Corporation of Korea (DACOM) was established by a consortium of public and private sector interests to develop data communication services by leasing lines from KTA. DACOM offers data communications services ranging from leased data circuits to value-added services (VAS). Other companies were established as common carriers to provide various specialized and diversified services, including Korea Mobile Telecommunications Corporation (KMTC), established in 1988 to provide cellular and paging services. The Korean Port Telephone Company was set up to provide seaport and airport communications, while the Korea Travel Information Services Corporation (KOTIS) served the travel industry.

The government-run KTA previously had a monopoly on domestic voice communications, and DACOM had a monopoly on data communications. However, in July 1990, the Ministry of Communications (MOC) announced a Structural Reform Plan under which a duopoly system for data, mobile, and international voice was to be established. KTA was incorporated as Korea Telecom (KT), as a step toward privatization, and was divested of DACOM, which became the second international carrier. DACOM was authorized to become a full-fledged common carrier in both voice and data on a global scale, and was authorized to enter the international telephone market in 1990, with a goal of 30 percent of the domestic and international telephone service market. KT was allowed to enter the data communications business through its subsidiary, Korea PC Communications.

DACOM attributes its success to date to several factors. Equal access is important. Customers can access DACOM by dialing 002; access to KTA is 001, so people can choose a service every time they make a call. DACOM had a 5 percent price advantage when it introduced service. DACOM had also been established as a carrier in data communications, and had good contact with corporate users and a good public image. The disadvantages are that DACOM must rent domestic circuits from KTA, and there are no structural separations and transparency. Also, DACOM users get separate bills, while KTA customers have domestic and international calls on one bill.[62]

U.S. pressure to open telecommunications markets was a factor leading to reform, beginning with a bilateral agreement to partially open markets. Legislation in 1989 and 1991 amended the 1983 Telecommunications Basic Law and Public Telecommunications Business Law. The Korean Communications Commission (KCC) was established, but has not had sufficient independence or resources to be effective.

Domestic voice services are still a monopoly, but various telecommunications services have been liberalized. The VAN service market has been gradually deregulated since 1985, and in 1994, the government permitted VANs to be operated by fully foreign-owned service providers. The data transmission sector was also opened to foreign companies already registered as VAN service providers. New service categories were introduced: Network Service Providers (NSPs), which were divided into General Service Providers (GSPs) and Specific Service Providers (SSPs); and Value-Added Service Providers (VSPs).[63] Foreign direct investment in NSPs is limited to 33.3 percent, while there are no restrictions on VSPs. The GSPs, KTA, and DACOM are allowed to offer nationwide network service. SSPs require approval

TABLE 11.2
STRUCTURE OF THE KOREAN TELECOMMUNICATIONS INDUSTRY[64]

Category	Licensing	Existing Carriers
General Service Provider	By designation	KT, DACOM
Special Service Provider	By licensing	KMTC, Shinsegi
VAS Provider	By registration	34 providers

from the MIC and may offer only specialized services such as mobile cellular, paging, and port communications (see Table 11.2).

KT is diversifying into data communications, corporate communications, satellite communications, and cable television. The government plans to lower its stake from 80 percent to 66 percent by the end of 1996, and to fully privatize KT by the end of the decade.

In wireless communications, Sunkyong, Korea's fifth largest "chaebol" (conglomerate), won a controlling interest in KMTC in 1993. Its analog network covers 80 percent of the population. The second mobile license was awarded to Shinsegi Mobile Telecom in 1994, after an intense struggle by two of Korea's leading chaebols, POSCO and KOLON. The Federation of Korean Industry gave a controlling 15 percent share of the consortium to Shinsegi Mobile, a POSCO subsidiary, and a 14 percent share to Second Mobile, a KOLON subsidiary. Shinsegi divided over 22 percent of the project among AirTouch, Southwestern Bell, Qualcomm, and GTE (the latter later withdrew). The government announced in 1993 that CDMA would be the digital standard for Korea; it has required KMTC to migrate to CDMA technology, which has been under development for several years by ETRI in cooperation with Qualcomm.[65] Paging is a regional duopoly, with licenses issued in 1992.

Informatization as Industrial Policy

South Korea's strategy to reach an information society has involved government goals accomplished through strategies of developing domestic research and manufacturing, stimulating investment from public and private sources, and fostering competitive domestic and international carriers. This strategy has involved:

- Promoting informaticization of society;
- Industrialization of information;
- Liberalization of telecommunications as a platform for computer-based information networks.[66]

Although the government's approach in the 1980s favored strong domestic monopolies, it has responded to international pressure to open much of the industry, with the result that Korean suppliers and carriers have expanded both their domestic and international markets.

Teledensity targets in the 1980s were achieved; Korea has now set ambitious targets for personal computer penetration by the year 2000. Although information technologies have become a major Korean export, and Korean conglomerates are using these technologies to improve their global competitiveness, South Korea still seems less secure in the path it is on toward a national information society. As telecommunications and information technologies become widely available throughout the society, Koreans will have to be able to share and access information from sources beyond government control. However, like Singapore, the Korean government appears wary of relinquishing control over information, despite a deliberate economic development strategy based on information technology.

TAIWAN

Taiwan (the Republic of China) is known as one of the Asian Tigers because of its strong economic growth. With a per capita GDP of $10,280, it has a telephone density of 40 lines per 100 population, about the same as South Korea. In the 1970s and 1980s Taiwan invested in telecommunications to support its growing and diversifying economy. The information industry sector was seen as strategic to the island's economic security. The Tenth Four-Year Plan (1990–1993) completed a phone in every village project; the Eleventh Four-Year Plan calls for upgrading the network to become fully digital as well as installation of fiber and ISDN trials as part of a strategy for making Taiwan an information hub for the region.

A nationwide paging system and cellular network was installed by Ericsson for the Directorate General of Telecommunications (DGT). In 1994, it had more than 500,000 cellular subscribers, and 1.7 million pagers, or about 8 per 100 population. The network was to be fully digital by 1995. To help reduce the huge trade imbalance with the United States, the Taiwanese government directed the ministry to select three U.S. digital exchange suppliers for its domestic program. As a result, AT&T, ITT, and GTE now locally manufacture digital exchanges for the domestic market.[67]

Taiwan produces low-cost terminal equipment, but wants to expand its networking and transmission equipment industries. The Integrated Communications project (ICOM) was a $39 million, four-year project begun in 1989, designed to develop ISDN technology and to help local companies produce ISDN equipment. However, the budget was cut due to government financial restrictions. The DGT has since announced the construction of a broadband backbone as part of a national information infrastructure (NII) initiative.[68] Initial purchases of optical fiber were from foreign suppliers, but now local companies manufacture and install the cable.

Structure of the Sector

Since the Kuomintang retreated from mainland China to Taiwan in 1949, the island's communications facilities have been under government control. For 40 years, there was no public communication with the mainland. Since 1989, channels of communication have opened, but telephone traffic is still routed through a third country such as Hong Kong, Japan, or even the United States.

Under the 1950 Telecommunications Act, the Ministry of Transport and Communications (MOTC) issues licenses, while its Directorate General of Telecommunications operates and regulates the public switched network. All telecommunications services except VANs are provided by the government-run Directorate General of Telecoms (DGT). The DGT also acts as an arm of government industrial policy, promoting local equipment manufacture and development of information infrastructure.

With its booming economy and highly developed electronics industry, it may appear strange that Taiwan has moved so slowly to reform its telecommunications sector. However, control of communications has long been considered necessary by the Taiwanese government because of Taiwan's history and its proximity to mainland China. Proposals to split regulatory functions of the DGT from operations were stalled not only because of the government's security concerns, but also apparently because of the profitability of telecommunications. In 1993, DGT earned a surplus of NT$31 billion (about US$1.13 billion), of which NT$25 billion (about US$910 million) was taken by the Treasury.[69]

In 1986, the MOTC established a steering committee and three working groups to revise the Telecommunications Act. After many delays and refinements, the result was three bills that propose to restructure the telecommunications sector. The legislation would restructure the DGT as the regulator, and split off the operator to become a new company, Chunghwa Telecommunications Company (CTC), which is to be partially privatized, with strict guidelines on share ownership to ensure that Taiwanese nationals retain control of CTC.[70] The DGT is to remain an arm of government industrial policy, but its R&D functions are to be transferred to CTC, which will also take on an industrial policy role as part of the government's "three C's science and technology strategy": communications, computers, and consumer electronics.[71]

The law would permit foreign ownership of VAS providers (up to 33 percent) and would divide carrier services into two categories following the Japanese model: Type 1) the network and basic facilities, which would remain a CTC monopoly, and Type 2) VAS using these facilities, which could be relatively competitive. Cellular mobile telephony has been classed as a Type 1 service, and therefore part of the DGT monopoly; however, CT2 cordless telephone services as well as value-added services have been opened to domestic and international operators.

Meanwhile, customer premises equipment was opened to competition in 1988, and the VAN market was liberalized in 1989, but there are some restrictions. All circuits must be leased from the DGT, and resale is prohibited. Foreign VAS providers must operate through a ROC operator or distributor, unless the service is operated as a partnership in which an ROC citizen is the majority owner. Since 1990, more than 30 private companies have been licensed to offer services including EDI, e-mail, voice mail, database access and retrieval, remote transactions, and Internet access.[72]

As John Ure notes: "Central planning in Taiwan, as perhaps in most countries, is more indicative than literal, but in Taiwan's case it hinders the pace of telecommunications development because resources are diverted."[73] Paradoxically, the government that maintained a strong role to build a globally competitive technology sector appears forced to concede its role in order to further internationalize Taiwan's economy and respond to competition in telecommunications equipment and component manufacturing from mainland China.

TOWARD INFORMATION SOCIETIES: ACCESS VS. CONTROL

The Asian countries reviewed in this chapter have introduced policies to increase access to information and telecommunications services throughout their countries and to strengthen their own telecommunications and information technology sectors. They have invoked industrial policies that set specific goals and involve their governments directly through research, funding, and ownership of operators. Privatization and liberalization are being introduced often as a result of external pressures, but are now generally seen as elements in the overall strategy of moving toward information societies.

These countries are still reluctant to recognize that the inevitable result of their industrial policies is increasing access to information throughout their societies. Indeed, as we have seen in earlier chapters, it is access to and utilization of information, not the mere existence of networks, that contribute to socioeconomic development. But efforts to control access to information through blocking Internet sites, censoring television content, and banning satellite earth stations are disturbing evidence that the governments of Asia's booming economies fear the power of information access that their policies are designed to foster.*

---

* Of course, Hong Kong is an exception, having taken a market-oriented approach to opening its telecommunications sector, and generally following British norms of freedom of expression. Yet the Chinese shadow looms over Hong Kong, as Hong Kong reverts to China in 1997. As Mueller remarks: "The rulers of the PRC favor economic liberalization if it leads to development, but not political liberalization."[74]

CHAPTER 12

*Telecommunications in
Asia's Emerging and
Industrializing Economies*

*"In much of Asia, it's not so much a question of 'Sorry, wrong number,' as
'Sorry, no number.' "*

—*The Wall Street Journal*[1]

INTRODUCTION

This chapter examines the changing telecommunications sectors in
emerging and industrializing Asian economies. Included are India, a
major emerging market, the industrializing economies of Malaysia
and Thailand, and the emerging economies of Indonesia, the Philippines, and Vietnam. These countries are adopting a variety of innovative strategies to attract financing for telecommunications
infrastructure.

INDIA

With more than 900 million inhabitants, India is the second most
populous country in the world. Although its GDP is only $257 per

person, making it one of the world's poorest countries, its industrial base places it in the ranks of the top ten industrial nations. Also, despite great poverty among the majority of the population, India now has a middle class estimated at 300 million, or more than 40 million households."[2] Teledensity is only 1.07 lines per 100 population.[3] Of India's 570,000 villages, only 13 percent have telephone links.

Telecommunications in India dates back to 1853, when the British built a 33-kilometer telegraph line between Calcutta, the imperial capital, and Diamond Harbour, the anchorage of the East India Company,[4] an indication of the strong commercial and political links between the first multinationals, the overseas trading companies, and the British Empire. Telephone service began in 1882, when exchanges were opened in Calcutta, Bombay, and Madras. By the time of independence in 1947, India had some 300 exchanges, but only 100,000 lines for its several hundred million inhabitants.

The gap between demand for telephones and the capacity to install lines at present is 300,000 to 400,000 per year and widening. In an attempt to close the gap, the Eighth Five-Year Plan (1992–1997) targets 20 million installed telephone lines by the year 2000, and service to every village, for an average teledensity of 6 per 100.[5] The government estimates a resource gap of $2.3 billion to achieve these initial targets, although it has revised targets because of the accelerating demand for services.

Where telephone service does it exist, it is not affordable for most people: the per telephone revenue is 1.4 times the per capita income, compared to about 0.05 in industrialized countries; that is, telephone service is 28 times less affordable in India.[6] One of the problems has been operating inefficiency. The staffing ratio is about 24 main lines per employee, compared to a global average of 119.[7] The Department of Telecommunications (DOT) has relied heavily on internal generation of funds to build out the network. Without competition there has been no incentive to cut prices.[8]

There are more than five times as many television sets in India as telephone lines, an indicator not only of the availability of disposable income to spend on communications, but of the bottleneck in supplying telephone service. Small entrepreneurs have built cable television networks, originally to distribute Indian films and offer numerous channels delivered by foreign satellites. Yet people wait years for telephone service, even if they can now buy their own telephone sets. The government has encouraged some entrepreneurial activity without changing the public monopoly by authorizing privately owned public call offices. The local operator receives a commission on each call, creating an incentive for rural small businesses to install public payphones. In cities, privately run "Telecommunications Centres" offer telephone, fax, and telex services.

The Telecommunications Equipment Industry

From the 1950s until the 1990s the government restricted telecommunications manufacturing to public sector undertakings as part of its industrial policy to foster self-sufficiency. Eventually, some state government enterprises began to compete with the national government-owned manufacturers. However, the government's strategy of restricting equipment manufacturing to Indian public sector firms resulted in inefficient production of high-cost, obsolete equipment.

In 1991, the Indian government introduced sweeping economic reforms aimed at stimulating long-term growth, in response to a fiscal crisis marked by an acute shortage of foreign exchange, excess of imports over exports, and soaring government budget deficit. Strategies included regulation of domestic private industry, privatization of some government-owned enterprises, liberalization of some former controlled markets, and globalization through the opening of what was one of the world's most closed economies to foreign trade and investment.[9]

Under the new industrial policy, the government has now opened manufacturing to the private sector, with foreign equity up to 51 percent allowed. Companies including Alcatel, AT&T, Ericsson, Fujitsu, GPT, and Siemens are manufacturing switching and transmission equipment in India, in collaboration with Indian firms such as Tata, Birla, Thapar, Modi, and a Punjab state government enterprise.[10] The results have been dramatic. After introduction of competition in manufacturing, prices fell by 25 percent to 75 percent, and $350 million of private sector investment went into manufacturing within eight years, compared to $140 million in the previous 47 years.[11]

An innovative strategy to develop indigenous technology suitable for rural areas was the creation of the Centre for Development of Telematics (C-DOT), established by Sam Pitroda, an Indian entrepreneur who returned from the United States when Rajiv Gandhi was prime minister. Funded by the Department of Telecommunications and Electronics, C-DOT has developed a small modular switch of 10,000 lines for businesses and rural exchanges, and intends to manufacture a full range of digital switches from small rural to urban systems. By the end of 1992, about 7,000 C-DOT rural automatic exchanges (RAX) were installed, accounting for about 500,000 lines, and bringing direct dial services to rural areas.[12]

Customer premises equipment, ranging from telephone sets to PBXs, is also now liberalized, so that customers may purchase and install their own equipment. Value-added services were opened to private investment in 1992. Competition is allowed in services such as electronic mail, voice mail, data services, audio and videotex, and videoconferencing. A duopoly model with ten-year franchises has been adopted for cellular services. In 1994, eight licenses for GSM cellular service were awarded for four cities: New Delhi, Bombay, Calcutta, and Madras.[13] Foreign partners include France Telecom, SRF (France), Hutchison (Hong Kong), Telekom Malaysia, Telstra (Australia), Millicom and Vodafone of the United Kingdom, and Cellular Communications International and BellSouth from the United States.[14]

Structural Reforms

Telecommunications in India is still under the jurisdiction of the Telegraph Act of 1885, which has been modified and reviewed over the years, but not replaced. In 1985, the government separated posts and telecommunications, creating the Department of Telecommunications (DOT) with its own Telecommunications Board. In 1986, it converted the Bombay and Delhi Telephone Administration into Mahanagar Telephone Nigam Limited (MTNL), a public sector under-taking, and the Overseas Communications Service into Videsh Sanchar Nigam Limited (VSNL). MTNL is 80 percent government-owned, with 20 percent held by institutional investors, while VSNL is 85 percent government-owned and 15 percent privately held. The Department of Telecommunications (DOT) provides local and long distance services everywhere but in Bombay and New Delhi, where MTNL operates the local network. VSNL has a monopoly on interna-tional service. Although service has traditionally been poor, digital switching and optical fiber are helping to increase the number of completed calls. The call completion rate in Bombay and New Delhi increased from 30 percent to 80 percent in the last five years. How-ever, MTNL must still adhere to government hiring and compensa-tion policies, so it has not had a free hand in taking steps to improve the efficiency of the network and productivity of the staff.

Since 1985, the government of India has been preparing a revised telecommunications policy to strengthen the sector. The Seventh Five-Year Plan (1988–92) identified telecommunications as one of the top five development priorities. A Telecommunications Commission was established in 1989, with responsibility for policy formulation, implementation, and budgeting. Goals set in 1991 included services for all Panchyats and 150,000 villages, public telephones for every 100 households in urban areas, and, remarkably, a waiting period for individual urban service not to exceed two years. However, the liber-alized economic policies adopted by the government in 1991 and ex-ploding demand for telephone service called for drastic revisions in targets and policies to achieve them. A new Telecommunications Pol-icy was announced in 1994, envisaging that by the end of the Eighth Plan (1997), telephone service would be provided to all villages and there would be one public call office per 500 population in cities, and widespread availability of value-added services. These targets would require provision of about 10 million telephone connections during the Eighth Plan period, at an estimated cost of $3.9 billion.[15]

An intense national debate preceded the government announce-ment of the new national telecommunications policy in 1994. The government had to act in such a way that the reform process would be politically acceptable, as Indian telecommunications is a highly visible sector, with 450,000 unionized workers. Forced corporatization of DOT could lead to industrial strife because of union concerns about losing civil service job security and benefits; thus, the government apparently concluded that the only feasible way to bring massive investment funds into the telecommunications sector was to extend the demonopolization from manufacturing to network provision and services. Rejecting privatization as a viable policy option, the government chose to attract private sector investment by allowing private companies to provide cellular and local telephone services.

Thus, in 1994, the Ministry of Communications announced the following criteria for private sector participation in basic telecommunications services:

- foreign investment is limited to 49 percent in basic telephone service companies, and 51 percent in value-added services;
- international and long distance networks are to remain public monopolies, and private sector companies are limited to serving local service areas called "circles";
- in addition to the DOT, only one private operator will be licensed per circle.

The guidelines are to be reviewed after five years.

Potential bidders were required to have a track record of operating a system of at least 500,000 lines for five years, thus requiring Indian companies to form partnerships with foreign operators. At least 10 percent of connections must be in rural areas; all districts must be covered in 24 months; and prices must not exceed those charged by DOT. By liberalizing local rather than long distance services, the government disarmed potential criticism that it was selling out profitable services to private investors; however, by leaving open the prospect that local companies could enter the interstate market after five years, it held out the promise of significant long-term profits.[16] Other potential competitors for long distance service (and potential partners of the private companies) include the National Power Grid Corporation, Indian Railways, and state electricity boards, which have their own networks.

However, the strategy may have other shortcomings. The fact that the interstate market will not be opened to competition for at least five years raises the specter of enormous bottlenecks in the toll network, once new local services are installed. Also, licenses are being awarded based on the highest bid, without consideration of the proposed pricing for a basket of services (installation, monthly charges, local calling rates, etc.). The danger is that the cost of the license will be passed through to the customers in the form of high rates. While private companies are not allowed to charge more than DOT, they could try to persuade the government to allow DOT to raise its prices in order to raise their own ceiling, as happened when domestic air services were opened to competition with government-owned Indian Airlines.[17]

Foreign companies bidding with Indian partners include: NYNEX with Reliance, Bell Atlantic with Essar, Korea Telecom with Himachai Telematics, Deutsche Telekom with Videocon, and Singapore Telecom with HCL and Ashok Leyland. There have been lengthy delays compounded by the 1996 elections which defeated the ruling Congress Party and allegations of government corruption. Some circles, such as Jammu and Kashmir, received no bids, and five received only one bid. One consortium emerged as the highest bidder for nine circles; the government then set a cap of three service areas for any one bidder.

In 1995, the Department of Telecommunications (DOT) was restructured into the Telecom Commission, responsible for policy and planning, and India Telecom, the operating body. A new regulatory body called the Telecommunications Regulatory Authority of India (TRAI) was established in 1996. Although the government pledged that the TRAI would act "independently, impartially, in an objective manner to protect the interests of the private operators as also of the subscribers,"[18] it remains unclear how independent it will be from the telecommunications bureaucracy, and how much authority it will have to take on challenges from the unions and the private companies.

Data Networks and Economic Development

There has been a surge in computer networking in the last five years, with the growth of electronic mail and online services. Costs are high: a typical modem costs what an average programmer would earn in three or four months, and modem users have to pay a license fee; getting a telephone line costs almost five months salary.[19]

India's first packet-switched, satellite-based data network, known as NICNET, was installed by the National Informatics Centre (NIC), and now has 700 VSATs linking 500 centers. In 1994, the NIC added high-speed earth stations linking 14 cities as part of an information highway. Another satellite data network is operated by the National Stock Exchange (NSE), one of 23 stock exchanges operating in cities throughout the country. The NSE, which began trading in 1994, has installed a VSAT network to enable members to trade from their own offices. In mid-1995, more than 225 members had been linked to the NSE's mainframe computer, enabling them to view online market information, place orders, and execute trades directly from their offices.

The NSE founders anticipated that 750 VSATs would be required, but the network was expected to grow to about 2,000 VSATs by the end of 1996. Ravi Narain, deputy managing director of the NSE, states: "The extent of the success of this information technology-based securities exchange model can be judged from the fact that the equity volumes on the Exchange grew 800 percent in the first four months. . . . It has led to greater liquidity as the whole market is able to see pending orders and it has led to lower jobbing spreads due to the ability to negotiate prices on the screen as finely as two people would do face to face." The extent of the adoption of the system by the brokers "suggests that if their business requirements are met and can be conducted painlessly, the community takes to it with ease. In fact, many members had never touched a PC before. Within months, those same brokers were clamoring for improvements on the screen, for back office systems that interface with their front office trading application, for multiple terminals connected in a LAN environment, and so on."[20]

While international public telecommunications remains a monopoly, software technology parks are now authorized to bypass the terrestrial network using Intelsat Business Service earth stations. Such direct links have contributed to the growth of the Indian software industry. Programmers in Bangalore can now compete with those in industrialized countries by transmitting computer code directly to companies in North America and Europe.

The Video Revolution

India's Insat series of domestic satellites provides television distribution and voice and data communications via dedicated VSAT networks. Recent satellites have been designed and built by the Indian Space Research Organization (ISRO). In 1994, India successfully launched a low earth-orbiting (LEO) satellite, becoming the sixth country in the world (joining the United States, China, Russia, France, and Japan) with the capability of launching LEOs. It is also developing a rocket for launching geosynchronous satellites.

An estimated 10 million Indian families are cable subscribers and 1.5 million own satellite receivers.[21] However, Insat was not the catalyst of the video revolution. Cable television began in India in the late 1980s, when small entrepreneurs began to string coaxial cable through their neighborhoods to transmit films from their VCRs. Until 1991, there was only one state-owned television channel; in 1996 there were more than 45 cable-delivered satellite television channels in India, the majority privately owned as a result of the growth of the satellite industry in Asia. The boom started with Hong Kong-based StarTV on AsiaSat in 1991, and accelerated when Russian transponders became available after the breakup of the Soviet Union.

The first Indian commercial satellite channel, established in 1992, was Zee TV, which is uplinked from Hong Kong and carried on AsiaSat. Zee TV estimates its viewers at 150 million, 90 million of whom are in India. The company has attracted Indian advertisers and now operates additional channels: El TV, an entertainment channel for the "young-at-heart"; Zee Cinema, India's first encrypted Hindi movie channel; and SitiCable, providing an integrated cable network facility to cable operators who can become franchisees and gain a local channel that can be used as a city TV station.[22]

Doordarshan, the national TV network, was confronted with competition and forced by the government to cover its own costs from advertising rather than government subsidy. It began to offer more films and soap operas, cutting educational and development programs. It has now introduced a channel aimed at wooing the Indian elite back from foreign channels. Insat has also tried to regain ground, as other regional satellites such as Apstar and PanAmSat are available to deliver video signals. The government has allowed CNN to lease a transponder on INSAT-2C to make the satellite more attractive; however, private Indian broadcasters still cannot uplink channels to Indian satellites.[23]

Challenges of Change

The Indian model of restructuring by introducing local duopolies attempts to provide incentives to build out the network without changing the structure and much of the revenue base of the DOT. But will this approach lead to huge bottlenecks in interstate services, as new customers generate more traffic? And how will the government attract investment to the least attractive circles? Will duopoly actually result in lower prices, or a tacitly agreed ceiling set by DOT?

It is unusual to create a duopoly with one government-owned provider (another example is Australia, which did not privatize Telstra before licensing Optus as a competitor). DOT is not taxed on profits, and did not have to pay a license fee for its franchise areas, thereby giving it a potential financial advantage. Also, if the license fees or other fees paid by the private companies are allocated to DOT, it will in fact be subsidized by its competitors.

It is also unusual to introduce local competition before long distance competition. Another example is Hong Kong, where OFTA was blocked from introducing international competition by a previous government decision to grant Hong Kong Telecom a 25-year monopoly. One result of local competition in Hong Kong has been the growth of callback services, as competitors attract customers by promising much lower international rates through callback operators. Will Indian operators also find creative ways to bypass technical bottlenecks and high prices for interstate and international calls? If they do, the DOT and VSNL will be forced to find other funding sources to upgrade and extend their networks.

To date, the DOT has acted as policy maker, licensing agency, regulator, and operator. The model of the independent regulator designated as TRAI has yet to be implemented, but its success will be critical in resolving issues of interconnection, allocation of revenues, and standards, as well as ensuring consumer protection.

INDONESIA

Indonesia is the world's fourth largest country in population, with nearly 190 million people consisting of 300 ethnic groups living on 17,000 islands strung over 5,000 kilometers along the equator and stretching 2,000 kilometers north to south. Approximately 60 percent of Indonesians live on the island of Java.

As the only OPEC member in the region, Indonesia was the beneficiary of oil crises in the 1970s, which provided capital for economic development. However, the country faced difficulties when oil prices collapsed during the years 1985 and 1986; the value of exports declined, and the value of its debt increased, as most of the debt was in Japanese yen, which appreciated in value against the U.S. dollar. The devaluation of the Indonesian rupiah, plus implementation of market- and export-oriented policies, helped move the growth rate back to 7 percent by the end of the decade.

Facilities and Services

Indonesia has only 2.5 million telephone lines, or 1.33 lines per 100 population. In Jakarta, there are 7.7 lines per 100, with a waiting list of about 80 percent of existing lines. Only 50 percent of 4,600 sub-district capitals and 25 percent of 65,000 villages have access to a telephone. At least 7 million lines are needed just to meet current demand.[24] There are about 77,000 pay phones, or one pay phone per about 2,500 people.

Telecommunications is included in the National Five-Year Plan (Repelita) of the National Planning and Development Agency (BAPPENAS). Indonesia's goals are to increase access in rural areas and improve telephone density in high-demand areas; priorities also include increased provision for tourism-related activities, banking, financial centers, and industrial plants, all of which are highly dependent on telecommunications. Indonesia's Fifth Five-Year Plan aimed to add 2.1 million lines by 1994, of which 100,000 were to be built in cooperation with the private sector. In Repelita 6 (1994–98), 5 to 6 million lines are to be added, 80 percent of which are to be funded entirely by the private sector. The goal is to increase telephone density to 3.2 per 100 by 1999, and eventually to have telephone access in every village.[25]

Restructuring the Sector to Encourage Investment

The Directorate General of Posts and Telecommunications (DGPT), under the Ministry of Tourism, Posts, and Telecommunications (Parpostel), has both policy and regulatory functions. A Telecommunications Advisory Board, chaired by the Minister of Parpostel, was established in 1989 as a forum to frame government policies and strategies. Its members include experts in telecommunications and related scientific fields.

Provision of basic telecommunications services is by state-owned monopolies: PT Telkomunikasi Indonesia (PT Telkom) for domestic services and PT Indosat for international. PT Indosat was established in 1981; Perumtel became PT Telkom in 1991 when it changed its status from a private company to a limited company. PT Indosat was 32 percent privatized in 1994, raising $1.1 billion.[26]

To accomplish its goals of extending and upgrading facilities, Indonesia needs financial resources: $3 to $4 billion for the fifth plan; and $8 to $10 billion for the sixth plan. In the past, funding came either from the government budget or from multilateral and bilateral public borrowing. This approach has changed, as economic reality forced new strategies such as public/private partnerships.[27] Indonesia has resorted to innovative approaches to raise capital and provide incentives for investment. To expand infrastructure rapidly, infrastructure policy was redefined, so that private investors can not only build, but take equity stake in projects, in order to speed up expansion without additionally burdening public debt.

In 1989, the Indonesian government passed a law allowing private participation in the telecommunications development program. The Law on Telecommunications of 1989 supersedes the 1964 law, which allowed only Perumtel and PT Indosat to provide telecommunications services in Indonesia unless provision was made with their consent or their participation. The new law opens nonbasic telecommunication services to the private sector. It does not change the basic structure, but allows for foreign investment through joint ventures. The law preserves the state monopoly; the government owns and operates facilities and services, but there is some liberalization. There are no restrictions on entry for private entities in providing of nonbasic or value-added services (VAS). In provision of basic services, private entities must cooperate with the national operators, for example, through investments in telecommunications facilities and joint operation of services.[28]

The 1989 law allowed private companies to run basic telecommunications services in cooperation with the state company operators under a Build-Operate-Transfer (BOT) scheme, where the private sector finances and constructs the network and operates it for a specified period to recover its investment, after which it is transferred to PT Telkom. This scheme was implemented with several projects from

1990 to 1992, and incorporated in the Sixth Development Plan as the Joint Operating Scheme (JOS). Another version is the Build-Operate-Own (BOO) scheme, whereby national private sector companies are allowed to enter into joint ventures with Telkom to develop and operate basic telephone services. An example of BOO is Ratelindo, which is 45 percent owned by PT Telkom and 55 percent by Bakrie Telecommunications Company.[29] This model can also be used for satellite networks. For example, in 1989, a private company signed a contract with Perumtel (now PT Telkom) to establish a VSAT network for data communications. The company was required to invest in the project and set up a network in ten months, and to market, install, and maintain the VSATs. PT Telkom operates the hub. After ten years of operation, the network will be turned over to PT Telkom. PT Telkom also collects the transponder lease fees and 25 percent of the company's profits.

Indonesian investors are also participating in revenue-sharing ventures. For example, a project was awarded to Telekomindo (owned by the pension fund of PT Telkom) and Astra, an Indonesian conglomerate with principal operations in automotive products, motorcycles, heavy industry, agribusiness, timber, chemicals, electronics, financial services, and telecommunications.[30] The lack of an operating phase when the investor can operate the network has made the scheme unattractive to foreign operators, but the relative safety and predictable returns on investment have persuaded the Indonesian financial institutions that it is a sound investment.[31]

In 1993, the government allowed private operation of value-added services such as electronic mail, data networking, online services, and paging. A duopoly model has emerged for analog mobile telephony, with PT Telkom operating one network and a joint venture including PT Telkom operating the second, typically with private sector investment under BOT revenue-sharing schemes for a maximum of ten years.[32] A duopoly model has also been adopted for digital cellular service. The first digital GSM network is operated by Telkom, with infrastructure by Ericsson and Siemens. In 1993, PT Satelindo was awarded nationwide rights to operate a second GSM network, which is being built by Alcatel. In radio paging, there are more than 20 companies with approximately 80,000 subscribers.

The telecommunications targets set by the Sixth Plan require a compound growth rate in installed lines of 26 percent as well as enormous capital investment. As the government takes about 60 percent of PT Telkom's pretax profits, other funding sources are required. Thus, in 1994, the government announced a deregulation package known as PP20, which permits up to 95 percent foreign direct investment in the industry, overturning a previous requirement that Indonesian companies hold majority stakes. PT Telkom's operations were divided into 12 regions, with each operating company free to enter a Joint Operating Scheme (JOS) with private companies contracted to build turnkey networks. The private JOS partners will share 70 percent of installation fees and fixed monthly rentals and 70 percent of outgoing and international call revenues generated by their networks.[33]

Manufacturing

Indonesia is continuing the policy to develop and manufacture telecommunications equipment within the country. Foreign companies may participate in manufacturing equipment by forming a joint venture with a foreign investor and local company. PT INT is a government-owned manufacturer of telecommunications equipment that has a joint venture with Siemens to produce digital exchanges, and with Bell Manufacturing of Belgium to produce PBXs and pay phones. PT INT developed a small, digital rural exchange that is now being installed in subdistrict capitals and remote areas.

Private Indonesian equipment manufacturers include Elektrindo Nusantara (EN), RF COM, Centronix, and Citra Telekomunikasi Indonesia (CTI). In November 1990, a $330 million contract for digital switching was awarded to two consortia, one involving AT&T and Siemens, the other involving NEC and Sumitomo. CTI manufactures AT&T switches; EN manufactures NEC switches. PT INT and RF COM manufacture small earth stations and radio equipment.

The Palapa Satellite System

Indonesia was the first developing country to launch its own satellite. The Palapa system was inaugurated in 1976, as a fulfillment of the Palapa oath made by Gjah Mada of the ancient Majpahit Kingdom, who vowed that he would continue to work until all the islands in the archipelago were united.[34] Domestic applications include TV distribution, data communications using VSAT networks, and telephone links to remote islands and rural areas. There are approximately 485 satellite antennas for telecommunications[35] and an estimated 1 million antennas for television reception.[36] However, despite the fact that Indonesia has operated its own domestic satellites for nearly two decades, the "last mile problem" of providing terrestrial links from satellite stations to customers with terrestrial links, combined with inefficient management, has prevented the satellite system from achieving its full potential.

Transponders on Palapa have been leased to other countries in the region, including Malaysia, Thailand, the Philippines, and Papua New Guinea. The Palapa B1 satellite was withdrawn from domestic service in 1992 and put into an inclined orbit to offer service in the Pacific region. Indonesia's third generation Palapa C series, built by Hughes, is being launched by Arianespace. Previous satellites and launches were supported by the U.S. Export-Import Bank, but the company's new partially privatized status has made it ineligible for guarantees of such financing. Another concern is whether private VSAT networks should be allowed to compete with state-run PT Telkom to serve customers such as airlines, banks, and hotels. In addition to replacing its existing fixed service satellites with the new Palapa C series, Indonesia plans to launch a direct broadcasting satellite system called Indostar.

Indonesia partially privatized its satellite operations in 1993; the satellite system is now operated by PT Satellite Palapa Indonesia, known as Satelindo. A private consortium led by Indonesian President Suharto's second son, Bambang Trihatmodjo, controls a 60 percent share of Satelindo. PT Telkom holds 30 percent, and PT Indosat, the state's long distance company, controls 10 percent.[37]

Rural Services

Indonesia is also using innovative strategies to extend telecommunications services in rural areas. These include:

- use of exchange capacity of the nearest town plus digital radio line concentrators, with a capacity of up to 3,000 subscribers;
- long distance subscriber connections using locally manufactured radios and remote line concentrators;
- automation of subdistrict capital telephone services and provision of services to surrounding villages using small earth stations or spur route radio systems (there are 3,644 subdistrict capitals of which about 25 percent have been provided with automatic telephone service);
- implementation of wireless local loop and interconnection through the cellular system;
- establishment of *"wartels,"* communications services shops that provide postal and telecommunications services and are managed by private operators on a revenue-sharing arrangement with the local telecommunications company.[38] Wartels are 50 percent privately operated; they keep 20 percent of domestic revenue and 6 percent of international revenues.[39]

The verdict is still out on how successful Indonesia's strategies to encourage private investment will be in bringing telecommunications to its largely rural population. Potential dangers include interconnection disputes, pricing disparities, and incompatible equipment from numerous vendors that will be expensive to maintain and upgrade in years to come. But its incentive strategies, from BOT and BTO to entrepreneurial telecommunications shops, or wartels, offer promising models for developing countries and industrializing economies throughout the world.

MALAYSIA

Malaysia has one of the highest economic growth rates in the region, and is the most developed ASEAN country after Singapore, with a per capita GDP of $3,392. Telephone density is 14.7 lines per 100 population. The government has placed great emphasis on developing telecommunications infrastructure and improving service, and considers the partially privatized Syarikat Telekom Malaysia Berhad (STMB) to be an arm of its industrial policy. In 1986, Malaysia launched an Industrial Master Plan that included the electronics and computer technology sectors. As a result, Telekom Malaysia was urged to upgrade its networks for data communications. STM plans to spend $3.9 billion on infrastructure development during the next five years.[40] Malaysia's Vision 2020 aims to bring the country to developed country status by that year; there is a major emphasis on information technology as the key element in the economy. The telecommunications goal is 45 lines per 100 population by the year 2005, an increase of more than 9 million lines.

The Telecommunications Sector

The Ministry of Energy, Telecommunications, and Post was set up in 1978, and was responsible for operations and policy through the Telecommunications Department of Malaysia, Jabatan Telekom Malaysia (JTM) until 1987. In 1984, in line with the government's privatization policy, a public limited company, Syarikat Telekom Malaysia Berhad (STMB), was incorporated, and JTM's operations were transferred to it. JTM became the regulatory body. Corporatization led to rapid improvements in service access and quality, with growth in subscriber lines accelerating, and service quality indicators improving substantially. Labor productivity also improved about 13 percent per annum.[41] Telekom Malaysia was partially privatized in 1990, with the Ministry of Finance holding 76 percent, local shareholders 16 percent, and foreign investors 8 percent of the shares.[42]

There are no clearly defined foreign ownership restrictions except that foreign ownership in Telekom Malaysia is limited to 33 percent. Telekom Malaysia is listed on the Kuala Lumpur stock exchange; by 1993, shares were trading at three times the initial offering price.

Four private companies have been given international gateway licenses. In mobile communications, there is limited competition by licensing, with four cellular license holders and 36 radio paging licenses.[43] All of the new entrants are local companies, and many have strong ties to Prime Minister Datuk Seri Mahathir Mohamad, or other members of the ruling party UNMO (United Malays National Organization).

Until the licensing of Time Telecommunications in 1993, Telekom Malaysia was the only supplier of fixed wireline services, although there was no monopoly specified in the telecommunications act. Time has laid an optical fiber network along the federal highways; its subsidiary has a license to offer value-added data services. Cellular operator Mobikom is a consortium of four state-owned enterprises: Telekom Malaysia, Edaran Otomobil, Permodalan Nasional, and equipment supplier Sapura Sdn Bhd. Technology Resource Industries Sdn Bhd (TRI) owns Celcom (Cellular Communications Network Malaysia), which operates a cellular network and international gateway. TRI has joined with Teleglobe of Canada to purchase 50 percent equity in Orbcomm, a U.S.-based, Low Earth-Orbiting satellite project. Malaysian companies are also entering other Southeast Asian markets such as Vietnam and the Philippines.

Telekom Malaysia operates the Maypac packet-switched network, and offers other value-added services such as electronic mail, videotex, and videoconferencing. Telekom Malaysia also has a universal service obligation, including service to rural areas, and is implementing wireless networks in some areas. In some major cities, it also operates a cordless telephone (CT) network called "Smartfon," but subscriber growth has been very slow. One of the two other licensees returned its license, while the other, Sapura, was allowed to convert to a PCN system license. There are two operators of public pay phones: Telekom Malaysia in rural areas, and Uniphone Sdn Bhd, a subsidiary of Sapura Holdings, which has a license to provide urban pay phones in a revenue-sharing arrangement with Telekom Malaysia.[44]

The major equipment vendors are Ericsson and NEC, which have set up joint ventures. There are also several local equipment companies. Malaysia has a well-developed electronics assembly industry that is centered in the northern city of Penang. American- and Japanese-owned companies assemble consumer electronics, computers, and telecommunications equipment primarily for export.

Malaysia leases capacity on Indonesia's Palapa satellite for television distribution, and is also covered by the AsiaSat footprint. Licenses are required for all earth stations; however, the law has not been enforced. Officially, there are approximately 200 satellite earth stations in the country, but unofficial estimates exceed 40,000. In the early 1990s, the Malaysian government decided that the country should have its own domestic satellite system. One of the international gateway and cellular network operators, Binariang Sdn Bhd, was authorized to build and operate a domestic satellite system: its geostationary satellite, Measat, was launched in late 1995. Measat's footprint covers not only Malaysia, but Southeast Asia and much of eastern Asia. Measat is a very "hot bird"; its high-powered Ku-band signals can be received by antennas as small as 60 centimeters (about two feet) in diameter.

Policy Issues

Although Malaysia has separated its regulatory agency, JTM, from its now partially privatized operator, JTM is really not independent of either the government or STM. Its employees formerly worked for the government-run operator. JTM only recommends decisions; it can be overruled by the minister or by the cabinet. Also, the government still sees Telekom as an important element of its industrial policy, and, as the majority shareholder, wants it to be commercially successful. Other potential issues that JTM may find difficult to resolve include network quality, network standards and compatibility, revenue sharing, and interconnection agreements.

The licensing process is opaque, with no obvious criteria, and evidence of political cronyism in selection of successful licensees. The terms may also not be fixed; in 1996, the government decided that too many licenses had been awarded during the previous administration and election period, and indicated that it would like some licensees to buy out others, despite the fact that all had made investment plans and were building networks based on their license terms.

The reliance on Telekom Malaysia as "carrier of last resort" is also problematic. Telekom is required to provide service in less profitable rural areas, while competing with carriers that apparently can cream-skim the most lucrative business and urban customers. Yet reserving a rural monopoly for Telekom Malaysia provides no incentive for the company to reduce costs for rural services. A test may come with the availability of the Measat satellite, which would be very suitable for providing telephone service in isolated areas of Sarawak and Sabah. However, Telekom has indicated that it may build its own terrestrial networks rather than leasing capacity from Measat.

JTM has not authorized resale. Without the option of obtaining capacity from Telekom Malaysia at wholesale prices, new carriers are choosing to build their own backbone networks, which are highly capital-intensive. Meanwhile, the government is urging Malaysian industry to reduce imports in order to cut its foreign exchange deficits. Resale could help to reduce this deficit by creating an incentive for new carriers to use surplus fiber and satellite capacity rather than building their own facilities.

THE PHILIPPINES

The Philippines consists of more than 7,000 islands with a total population of 66.2 million. The largest concentration is on the island of Luzon. The economy of the Philippines languished for two decades during the Marcos regime, but under President Fidel Ramos, the Philippines improved its economic growth rate from 2 percent per year in 1991 to 4.5 percent in 1995. Another legacy of the Marcos era is very poor telephone infrastructure, with a telephone density of only 1.6 telephones per 100 population. Of the total 920,000 telephones, 80 percent are in Manila.[45] In Manila there is one line for ten people; in other towns, one line for 50; and in rural areas, one line for 1,000 people.[46] Many rural communities and remote islands have no telephone service. The goal in the next five years is to increase the teledensity to 4.5 per 100 Filipinos, requiring an investment of about $13 billion.[47]

The Telecommunications Sector

The policy-making body of the Philippines is the Department of Transportation and Communications (DOTC). The regulatory body is the National Telecommunications Commission (NTC), which is largely modeled on the U.S. Federal Communications Commission, and is under the administrative supervision and control of DOTC. Established in 1979, the NTC took over the regulatory and quasi-judicial functions of the Board of Communications and Telecommunications Control Bureau that it replaced. In 1987, it was attached to the Department of Transportation and Communications. Its decisions are appealable only and directly to the Supreme Court. The NTC has contributed substantially to the preparation of the National Telecommunications Development Plan (NTDP), which is the strategic plan to guide the development of telecommunications for the next 20 years.

The Philippines is perhaps the most open of the ASEAN markets, with more than two dozen companies offering telecommunications services. Most are controlled by a small circle of ruling families who form alliances with foreign companies for financial backing and technical expertise; the alliances are volatile, with considerable turnover. The Constitution of 1987 limits foreign investment in public utilities to 40 percent, but foreign companies can also own up to 40 percent of the local joint venture partner, or potentially, 64 percent of the enterprise.[48] The Build-Operate-Transfer concept for public sector infrastructure projects was extended in 1994 to include information technology networks.

The Philippine Long Distance Telephone Company (PLDT) is the dominant operator, with an 86 percent share of the market: the remaining 14 percent is the responsibility of some 50 other franchises. PLDT has been a fully private operator since 1928; its franchise extends until 2028. During the Marcos era, there were numerous reports that the president and his cronies appropriated PLDT revenues, resulting in a growing telephone shortage and higher prices.[49] Now, PLDT is also adding lines in Manila and the provinces, and installing digital switches. PLDT's Zero Backlog Program, launched in 1993, aims to have a total of 654,000 lines installed in Manila, and 302,795 lines installed in the provinces by the end of 1996.

Several private carriers have entered the market to provide local, long distance, and value-added services. In 1995, seven record carriers provided domestic telex, facsimile, and leased-line services; four provided international services. There are also several private carriers for paging and data communications. Five companies have been granted provisional authority to establish national cellular networks: Piltel, Express Telecommunications (Extelcom), Smart Communications, Globe Telecom, and Isla Communications. Seven major families dominate many of these new companies; PLDT is also an investor in several of them. For example, Piltel, a domestic PSTN franchise holder, is jointly owned by PLDT and the Cojuangco family, which also has close connections to Philcom, an international gateway franchise holder, and Extelcom. Foreign partners include First Pacific of Hong Kong, Singapore Telecom International, Shinawatra and TelecomAsia of Thailand, Cable & Wireless, Telecom New Zealand, Bell Atlantic, NYNEX, and US WEST.[50]

Rural Projects

The government's attempts to encourage investment in rural telecommunications are rooted in a conviction of the importance of telecommunications to the country's rural development. In addition to contributing to economic development, telecommunications in the Philippines has proved to be critical in coping with natural disasters. For example, cellular phones were used to coordinate relief after a major earthquake, and VHF radios distributed by the NTC were used to manage evacuation logistics during the volcanic eruption of Mount Pinitubo.[51]

More than 80 percent of telephone lines are in Metro Manila; outside Manila, small companies provide services often confined to one locality, and contribute the bulk of their revenues to PLDT for carrying long distance traffic. Local governments can also set up and operate public call offices (PCOs). Some companies are using cellular for fixed service as a way to bring telephone service to rural areas. The National Telephone Program, launched in 1988, planned 600,000 new lines in five years. The project was funded by Japanese, French, and Italian concessionary loans and grants. The Municipal Telephone Act of 1990 mandates installation, operation, and maintenance of a public telephone in every municipality. The Municipal Telephone Project, which aims to establish PCOs in more than 1,000 unserved municipalities, is a joint government and private project, with the government serving the financially unattractive areas. Funding for this project came from Canada, France, and Germany.

The government's Telecommunications Office (Telof), formerly the Bureau of Telecommunications, provides service in some rural areas, operating 27 exchanges plus the telegraph service. Telof's primary role is to develop and recommend plans for communication services, especially in areas not covered by services of a private operator. In 1990, the DOTC mandated the eventual privat-ization of all government-owned telecommunications facilities.

The government has adopted an innovative strategy to create incentives to install telecommunications networks in unserved areas. Licenses for international gateways and domestic services now require that operators also undertake the installation of several hundred thousand lines in an unserved region. Executive Order 109, issued by President Ramos in 1993, requires a total of 5 million landlines from gateway and cellular telephone operators. Each cellular mobile telephone service (CMTS) operator is required to install a minimum of

400,000 local exchange lines. Similarly, each international gateway facility (IGF) operator is required to install a minimum of 300 local exchange lines per international switch termination and a minimum of 300,000 local exchange lines, within three years from the date of authority to operate and maintain local exchange carrier service.[52]

Executive Order 59, also issued in 1993, requires compulsory interconnection of authorized public telecommunications carriers to create a nationalized, integrated network and encourage greater private sector investment. This decision paved the way for other authorized carriers, including small "mom-and-pop" operators, to interconnect with the national backbone of large carriers. The Philippines was also divided into ten local exchange carrier (LEC) service areas. The NTC has left PLDT and the local exchange carriers to negotiate among themselves, intervening only if invited or if the parties cannot agree within 90 days. Since PLDT's network includes 90 percent of all installed lines, the other carriers have felt pressure to agree to PLDT's terms. For its part, PLDT had not designed its network for interconnection with numerous other carriers and had not budgeted for the transition. Also, PLDT acts as a bottleneck in high-demand areas where the new carriers depend on the PLDT network. In mid-1996, the new carriers cited unfilled backorders for hundreds of trunks in Manila from PLDT.[53] Other issues remain unresolved, including billing protocols and cost allocations among the carriers.

Satellite Policy

The government's satellite policy is intended to promote use of satellites for telecommunications, broadcasting, education, the public sector, and other services. In 1993, the government lifted virtually all controls in the operation of domestic satellite communications facilities and services, allowing for the entry of more satellite operators and installation of private networks. Today, five private companies provide domestic satellite services. They include CRS (Clavecilla Radio Systems), ICC (International Communications Company) for interactive data communications, Liberty Broadcasting for data broadcasting,

Capitol Wireless, and Globe Mackay Radio and Communications Corporation.[54] These companies primarily use Indonesia's Palapa satellite, although other satellites with coverage of the Philippines, such as PanAmSat's Pacific Ocean satellite, could be used in the future.

However, the Philippines has also announced plans for two domestic satellite systems. Mabuhay Philippine Satellite Corporation (MPSC) and Philippine Agila Satellite (PASI) are racing to be the first with a satellite in orbit. PLDT, with a 51 percent share in Mabuhay, has committed to leasing 14 transponders, while Piltel, one of the cellular operators, plans to lease six.[55] Mabuhay is also designed for regional traffic, with coverage of Southeast Asia, the eastern Pacific Rim, and Hawaii. The government and 17 operators have signed a memorandum of understanding to launch Agila (Eagle), with a target market made up primarily of Philippine carriers.

Bold Experiments

The Philippine policy requiring licensees of international gateways and cellular systems to build networks in unserved parts of the country is a very innovative way to create incentives for investment in unserved areas. Although some observers have questioned the wisdom of this approach because it may perpetuate internal cross-subsidies, it looks promising as a model that could be emulated in other countries with large unserved territories. Some new operators apparently see the policy as a burden, while others welcome the opportunity, and anticipate profitable operations in their new franchise areas. Perhaps in a few years it will be possible to make a market in these rural franchise areas, so that rights to build and operate may be traded, with operators interested in expanding their franchise areas buying the obligations from those who find them a burden.

Regulating the multiplicity of new operators is a formidable challenge, especially given the continued dominance of PLDT. Enforcing interconnection agreements and determining how to allocate costs among the carriers are among the daunting issues facing the NTC.

However, the policies being developed by the NTC could become models for other countries in the region.

The Philippine plans for two domestic satellite systems appear to be a more dubious experiment. Some officials state that the satellites would help reduce the national debt at a faster pace, since scarce foreign exchange would no longer be needed for foreign transponder rental. However, these systems themselves represent an enormous investment of hard currency, as domestic satellite systems (including the satellite, launch, insurance, and master control stations) typically cost about $200 million.

THAILAND

Thailand, with a population of 57.8 million, is classified as a lower-middle-income country by the World Bank. However, its economy is growing rapidly, with an increase in per capita GDP from $800 in 1985 to $1,543 in 1994, mostly from tourism and export-oriented industries. Thailand is also the world's largest rice exporter.

The development of infrastructure cannot keep pace with Thailand's economic growth. In 1994, there were about 2.8 million lines, or 4.7 per 100 population, slightly more than half the average rate of all lower-middle-income countries. The current waiting list is about 1.6 million. During the 1980s, there were delays in implementing development projects because of a shortage of funds.

Telecommunications services in Thailand are provided by two state enterprises: the Telecommunications Organization of Thailand (TOT), and Communications Authority of Thailand (CAT). Telecommunications development was one of the highest priorities of the Seventh National Education and Social Development Five-Year Plan (1992–1996). The plan's guidelines for development emphasize:

- use of advanced technology;
- need to accelerate expansion of service and improvement of service quality throughout the country;

- expansion of telecommunications services in provincial areas to span the Bangkok-provincial gap;
- introduction of new services to meet demand from business, and to be internationally competitive.[56]

Thailand's goals include eliminating the waiting list, increasing the telephone density to 10 lines per 100 population, and fully digitizing the network. The Rural Long Distance Public Telephone Project (1992–1996) aims to link 4,500 locations to the national network: 36,000 villages are to be linked by the year 2001. The government plans to expand the basic telephone network to 6 million lines by the end of 1996 and 12 million lines by 2001. The target is a teledensity of 10 per 100 by 2001 and 20 per 100 by 2006, requiring a total of 24 million lines, 18 million of which must be installed within ten years.[57]

The Ministry of Transport and Communications (MOTC), established in 1912, is responsible for setting policy, in cooperation with the National Economic and Social Development Board (NESDB) for all national transportation and communication planning. The Telephone Division of the Post and Telegraph Department (PTD) was separated from the MOTC in 1954. It was then established as the Telecommunications Organization of Thailand (TOT), a state enterprise under MOTC that provides for domestic services and connections to neighboring countries.

In 1977, the Communications Authority of Thailand (CAT) was established to separate completely the operations of post and telecommunications services from the former parent PTD, which now has only administrative functions such as international and regional cooperation, and frequency management and monitoring. Originally, TOT was responsible for domestic public switched telephone services as well as trunk calls to neighboring Malaysia, Cambodia, and Laos, while CAT was responsible for international services. In practice, they are emerging as competitors in four services: cellular mobile, paging, data networks, and leased circuits.[58]

Incentives for Investment

In 1991, there were 900,000 requests for telephone service, and TOT could provide between 150,000 and 250,000 lines per year. The government decided to look for private investment to overcome the lack of funds and dramatically increase the number of lines. It introduced a policy known as Build-Transfer-Operate (BTO), which allows private companies to build out the networks, transfer ownership, and operate the service through revenue-sharing arrangements with CAT and TOT in order to recover their investments. All new telecommunications services are given to private companies to operate in the form of concessions, including cellular, paging, data communications, card phones, and satellite services.[59] The advantage of BTO, which transfers ownership before operation, is that it allows circumvention of telecommunications laws requiring basic services to remain under state control. However, this massive involvement of the private sector could be viewed as de facto privatization of over 50 percent of the telephone network.

The government then invited the private sector to submit proposals to build 3 million lines as a TOT concession: 2 million in Bangkok and 1 million in the provinces. The licensees are responsible for investment, financial arrangements, procurement and installation of equipment and materials, and operation and maintenance of the 2 million-line network during the 25-year term of the concession agreement. The investors are fully protected from competition for the first five years. Additional terms include:

- the network must be installed within five years;
- 50 percent of cable and switching is to be provided from Thailand to promote local industry;
- TOT tariffs must be applied: any changes need government approval;
- concessions must use TOT manpower where possible, to ensure continuing involvement of TOT staff;
- customer service must remain with TOT;

- assets must be transferred to TOT before the network is operated;
- revenue is shared between TOT and the contractor: there is also a formula to share high profits.[60]

The Telecommunications Industry

The major Thai telecommunications companies began as suppliers of government, military, and corporate communications equipment. A major player is Shinawatra, founded by Dr. Thanksin Shinawatra (an entrepreneur who was appointed foreign minister in 1994). Shinawatra has cellular, paging, broadcasting licenses, and the right to operate the domestic satellite system, as well as contracts in Cambodia, Laos, Vietnam, and the Philippines. TelecomAsia, a joint venture between Charoen Pokphand (Thailand's leading agribusiness conglomerate) and NYNEX, also has contracts in Cambodia, China, and Vietnam.

While TOT and CAT retain their legal monopolies, the BTO policies have stimulated the growth of the domestic telecommunications industry. Two consortia were licensed under the TOT concessions to install 3 million lines. TelecomAsia has the Bangkok franchise. The 1-million-line provincial telephone project concession was awarded to Thai Telephone and Telecommunication (TT&T), a consortium composed of Jasmine International, Loxely, Ital-Thai, and Thanakit.

There are several other BTO franchises. Shinawatra and United Communications (UCOM) have BTOs with TOT and CAT for cellular mobile services, in addition to TOT's and CAT's own networks. Total Access Communications, a cellular operator licensed by CAT, also launched the first PCN system in Asia in 1994. There are four BTOs in paging, two each with CAT and TOT. Five data and data/voice operators serve banks, finance and security houses, manufacturers, airlines, and hotels. The local Comlind group has a BTO to install fiber-optic cables along railway rights-of-way in a cooperative effort between TOT and the State Railway of Thailand. Jasmine and Northern Telecom have a BTO with TOT to lay a submarine optical fiber across the Gulf of Thailand, linking coastal towns and cities.

Thailand's first satellite, Thaicom 1, was launched in 1993, followed by Thaicom 2 in 1994. The satellites are hybrids for fixed services at C band and television broadcasting at Ku band. Thaicom 2 has a larger footprint, extending to the Philippines and Malaysia and west to part of Europe. Shinawatra Satellite, a subsidiary of the Shinawatra Computer and Communications Group (SC&C), was awarded a 30-year concession, with eight years of exclusivity to operate the Thaicom system. The Jasmine International Group operates VSAT voice and data corporate networks, and plans to install 500 VSATs for rural communications, competing with TOT.

The government has plans for gradual liberalization between 1995 and 2001, including possible privatization of TOT and CAT. It may be difficult to encourage additional private sector participation because of the long concession period which usually includes periods of exclusivity. However, introduction of digital mobile services, a teleport, and data processing zones may provide opportunities.[61]

VIETNAM

One of the poorest countries in Asia, Vietnam is struggling to expand and diversify its economy after the ravages of the Vietnam War. Vietnam's per capita GDP in 1994 was only $181, but the economy is growing rapidly. There were only 0.61 telephone lines per 100 people,[62] and most were concentrated in Ho Chi Minh City (Saigon) and Hanoi. Telephone lines had been installed in 350 of the 527 districts, most serving government offices, the military, and businesses.[63]

Vietnam has invited foreign investors to help modernize and extend its telecommunications networks, but retains a PTT structure for the sector. The lifting of the U.S. trade embargo with Vietnam in 1993 opened the doors to U.S. investors and to multilateral agencies including the World Bank, International Monetary Fund, and the Asian Development Bank. Vietnam also uses vendor credits from French, German, Swedish, Italian, and Korean companies.[64] However, reliance on so many equipment sources raises the danger of incompatible transmission and switching equipment and difficulty in obtaining spare parts.

Telecommunications policies are set by the Directorate General of Posts and Telecommunications (DGPT), which was separated from the Ministry of Communications and Transport in 1992. A new entity, Vietnam National Posts and Telecommunications (VNPT), was established to operate the national network under the regulation of the DGPT. The VNPT established several subsidiaries, including Vietnam Mobile Services (VMS), Vietnam Data Corporation (VDC), Vietnam Telecoms National (VTN), Vietnam Telecoms International (VTI), and the Vietnam Postal Service (VPS). VNPT is also responsible for transmission of radio and television.

The new Foreign Investment Law, part of the constitutional reform of 1992 that adopted the guiding principle of "doi moi" or economic innovation, permits joint ventures and 100 percent ownership of assets. However, telecommunications is still regarded as important to national security; to retain government control, the DGPT has authorized only one form of foreign participation, the Business Cooperation Contract (BCC), which is an agreement between a foreign and Vietnamese partner for "the mutual allocation of responsibilities and sharing of product, production or losses, without creating a joint venture enterprise or any other legal entity."[65]

Vietnam's first BCC in telecommunications was with Telstra of Australia (formerly OTC) in 1988 to install Intelsat earth stations in Hanoi, Ho Chi Minh City, and Danang for international communications. International revenues are shared between Telstra and the DGPT, which uses part of these revenues to expand the domestic network. Telstra also financed the DGPT's share of an optical fiber submarine cable linking Vietnam to Thailand and Hong Kong.[66] Vietnam also has two Intersputnik earth stations, and uses the Intersputnik system to transmit television nationwide for six hours per day.

Operating companies from Singapore, Malaysia, Thailand, and Hong Kong are also participating in ventures in services such as cellular, paging, and pay phones. Mobile cellular and paging services have been introduced in Hanoi and Ho Chi Minh City; a packet-switched data service is available in these cities and Danang. The Ho

Chi Minh City P&T entered into a BCC with Singapore Telecom International to operate a cellular network known as Saigon Mobile Telephone Company (SMTC). In 1993, the DGPT awarded a contract to Ericsson to install a national GSM system. Another cellular operator, to be formed by the army, plans to enter the market. The DGPT has plans to authorize a total of four cellular networks to compete eventually. Two public telephone operators, Malaysia's Sapura and Schlumberger from France, have installed pay phones in Hanoi and Ho Chi Minh City. These phones operate only with telephone cards, because Vietnamese currency has no coins.

The Prime Minister's Office has also authorized the Ministry of Defense to set up an Army Telecommunications Company (ATC) to operate a second network.[67] This approach, which appears similar to the model in China, provides an additional revenue base for the government while keeping the networks under state ownership. The prime minister also suggested that Ho Chi Minh City might be allowed to set up a separate postal and telecommunications service. As a result, the People's Committee of the First District in Ho Chi Minh City announced that it would seek approval to set up the Saigon Post and Telecommunications Service Corporation.[68]

In 1996, Korea Telecom (KT) won a $40 million BCC to install and operate networks in three northern provinces. However, contracts with Telstra, Cable & Wireless, NTT, and France Telecom to install and operate networks in Hanoi and Ho Chi Minh City were delayed, apparently because of the reluctance of the VNPT to allow so much foreign participation, as well as disagreements over revenue forecasts.[69]

Recognizing the need to develop its national infrastructure, the government issued an information technology master plan called ITP-2000 in 1993. It may be possible for separate, government-sponsored IT networks to become alternatives to the VNPT networks. However, the DGPT has retained control to date, despite pressure from some research institutes and government departments to obtain full Internet access, which was not available through the VNPT.

INNOVATIONS AND CHALLENGES

The countries reviewed in this chapter have developed innovative strategies to encourage foreign and domestic private sector investment. Yet they have all had to contend with tension between the need to attract foreign investment and resistance to foreign ownership of a sector considered politically sensitive or important to national security. The various incentive strategies of BOT and BTO in several countries, the JOS in India, and BCC in Vietnam are all variations on a theme of providing incentives for private investors without giving up control over the sector. While it is too early to judge their degree of success, they appear to offer promising models for other developing regions.

But there are still many potential pitfalls. The regulators in most countries are not operating at arm's length from either vested interests in the government or a publicly controlled dominant carrier. Decisions made now in such matters as durations of licenses and exclusivity periods, may have repercussions only in future decades. In the near term, there are likely to be problems in setting terms for interconnection agreements and revenue settlements.

The "national flag carrier" syndrome that appears to require every country to have its own airline seems to have mutated to satellite systems. While satellites offer many advantages to link scattered islands and remote villages, demand in each country does not appear to justify investment in its own domestic satellite system. To obtain more revenue, the regional model pioneered by Indonesia's Palapa has now been adopted by Malaysia, Thailand, and the Philippines for their satellite systems. If these new entrants are not successful in attracting enough business, the benefits of their satellites in serving their commercial and rural customers may be overshadowed by hard currency debts to pay for their systems.

13

Telecommunications in Latin America and the Caribbean

"A credible and stable legal and economic environment is key to the ultimate success of the privatization process."

—ELIANA CARDOSO, The World Bank[1]

OVERVIEW

In the early 20th century, Latin American countries such as Mexico and Chile attracted U.S. multinationals that established profitable and powerful telecommunications monopolies. The remaining countries adopted the European PTT model. In the 1950s and 1960s, Latin American governments nationalized many foreign-owned enterprises to break their hold on local economies. As a result, by the 1980s, almost all telecommunications services in Latin America were operated as government monopolies; in general, they were poorly managed, undercapitalized enterprises that could not meet the demand for basic service, let alone new enhanced services.

Average teledensity in Latin America is only 8.4 lines per 100,[2] with many people in rural areas and inner cities having no access to a telephone. In 1993, there were a total of 8 million people registered on waiting lists, with an average waiting time of more than three years, and waiting periods in some countries of up to 21 years.[3] Of course, the numbers underrepresent actual demand, because in many developing countries people simply don't bother to put their names on a list if they face a wait of several years—or decades. Latin American governments have, however, pledged to improve and extend telecommunications networks as part of their economic development strategies. The whole region adopted the goals of the International Telecommunication Union's 1992 Acapulco Declaration to at least double the number of lines in each country, with a view to reaching a teledensity of 20 lines per 100 population by the beginning of the 21st century.[4] Latin American countries spent an average of about $725 per subscriber on telecommunications facilities in 1994. Brazil, Jamaica, and Honduras invested more than 60 percent of their revenue in upgrading and extending their networks; Ecuador and Paraguay spent more than 50 percent, and Argentina and Chile more than 40 percent.

There are major disparities in access to telecommunications throughout the Americas. For example, teledensity is only 2.4 lines per 100 population in Guatemala and Honduras, and just over 3 lines per 100 in Bolivia, Cuba, and Peru. There are more than four times as many telephone lines per 100 in Argentina, Belize, Costa Rica, and Trinidad and Tobago (see Table 13.1). Generally, television sets are much more accessible in Latin America, particularly in poorer countries such as Bolivia, Cuba, and El Salvador, which have about five times as many television sets as telephone lines. The relatively greater accessibility to television demonstrates that even in poor countries, many families have enough disposable income to buy a television set, whereas telephone service is still not an option.

European equipment suppliers have been active in the region for many years. For example, Ericsson has subsidiaries in Brazil and Mexico, and supplies equipment to many other countries in the region. Siemens has local manufacturing ventures in several countries. Japanese suppliers, led by NEC, are also very active in Latin America. Japan provides more tied aid than anyone else in the region, offering 3.75 percent loans over 40 years with a ten-year interest holiday.[5] The

TABLE 13.1
TELEPHONE LINES AND TV SETS IN
SELECTED LATIN AMERICAN AND CARIBBEAN COUNTRIES

Country	Tel Lines per 100	TV Sets per 100	Ratio of TV Sets to Telephone Lines
Argentina	14.1	32.2	2.3
Barbados	33.4	26.3	0.8
Belize	13.4	17.0	1.3
Bolivia	3.0	14.3	4.8
Brazil	7.4	24.5	3.3
Chile	11.0	24.9	2.3
Colombia	9.7	22.0	2.3
Costa Rica	13.0	21.6	1.7
Cuba	3.2	18.9	5.9
Dominican Republic	7.9	8.9	1.1
Ecuador	5.9	13.4	2.3
El Salvador	4.2	22.5	5.4
Guatemala	2.4	5.2	2.2
Honduras	2.4	7.9	3.3
Jamaica	10.3	29.6	2.9
Mexico	9.3	19.0	2.0
Panama	11.1	16.6	1.5
Peru	3.3	9.8	3.0
Trinidad and Tobago	15.8	31.4	2.0
Uruguay	18.4	51.5	2.8
Venezuela	10.9	18.2	1.7

Derived from ITU, *World Telecommunication Development Report,* 1995.

United States does not provide tied aid, a factor that apparently contributed to a 15 percent drop in U.S. telecommunications exports to Latin America in the 1980s.

As a result of an industrial policy that has fostered indigenous technology development and production, Brazil manufactures equipment for its own markets and exports to other countries in the region. Mexico is also an equipment exporter; in fact, Mexico actually generates a net surplus in telecommunications trade with Canada. Mexican exports are primarily low-cost consumer goods such as handsets, television sets, and computers, most of which are produced in the tax-free zones along the United States border, while Canada exports high value-added equipment such as central office switches, PBXs, digital microwave equipment, and multiplexers.[6]

SATELLITE COMMUNICATIONS: THE LURE OF BYPASS

Several Latin American countries have looked to satellite technology as particularly appropriate for serving customers spread over vast territories, especially where jungle or mountains hinder installation of terrestrial networks. Brazil, Mexico, and Argentina have their own domestic satellite systems. Intelsat is used for domestic service by Argentina, Bolivia, Chile, Colombia, Peru, and Venezuela. Demand for private network services is also high. Since the launch of PanAmSat in 1988 (see chapter 16), Latin America has had the option to use services from competitive carriers, if the governments will allow these carriers to compete. PanAmSat operates a shared hub in Florida; it provides domestic and regional VSAT services, and transmits Mexico's Televisa programming to the Caribbean, and Central and South America.

Private satellite networks offer VSAT services to bypass the public networks in Brazil, Venezuela, Mexico, and Argentina. There are about 8,000 VSATs installed in Latin America, and the regional market for VSAT equipment is estimated at about $110 million per year. Major customers include banks such as Banco do Brasil and Banco Itau, which use Brasilsat to connect their branches throughout the country, manufacturing operations such as General Motors do Brasil, and mining and petroleum companies that use VSATs to link remote operations to regional headquarters.

Bypass is a sensitive issue in countries that have not liberalized telecommunications services. It is perceived as siphoning traffic from the public network and setting a precedent for authorizing alternative networks to the public monopoly. The compromise is generally to allow private VSAT networks, but not to authorize interconnection with the public switched network, although satellite-based rural telephony has been introduced in some countries. The major VSAT operators are Impsat of Argentina, which has established subsidiaries in Colombia, Mexico, Venezuela, Ecuador, Chile, and the United States; Comsat, which has a presence in virtually every Latin American and Caribbean market; and Arnet, jointly owned by Entel Chile and systems integrator Coasin.[7]

Some administrations impose severe restrictions or surcharges on VSAT networks and equipment. Brazil is the most protected market; the VSAT market is liberalized in Mexico, Argentina, Colombia, Venezuela, and the Andean countries, although some countries impose terms on network operations. American companies dominate the VSAT equipment market. Hughes Network Systems has about 65 percent of the market; in Brazil, it has established a successful partnership with Promon, a Brazilian company licensed to manufacture VSATs. Other major VSAT suppliers in Latin America include AT&T Tridom, Scientific Atlanta, and GTE Spacenet.[8]

Bypass of national broadcasting networks is also being introduced in the form of direct-to-home (DTH) satellite television. The first DTH satellite, a venture of Hughes-controlled DirecTV, was launched in late 1995. Other media companies, including Rupert Murdoch's News Corporation, the cable operator TCI (Telecommunications Inc.), Globo of Brazil, and Televisa of Mexico, are planning DTH services. Like AsiaSat's StarTV, these systems will deliver foreign entertainment programs that are likely to lure viewers away from national television channels. However, regional commercial production is much more advanced in parts of Latin America than in Asia, particularly in Mexico and Brazil. As a result, a major impact of DTH systems in Latin America may be to extend the dominance of Televisa and Globo, as distributors of their own productions and of foreign commercial programming.

THE SHIFT TOWARD PRIVATIZATION

Although many governments had nationalized their key industries and were discouraging foreign investment by the late 1970s, most are now actively seeking foreign investment to obtain the benefits of modern technology and to expand their economies. Among the industries that are receiving attention for foreign investment are telecommunications, computer equipment and services, banking, petroleum, and travel and tourism. Yet the region's reputation for government intervention and inefficiency, as well as political instability, had to be overcome to attract investment: "With Latin America, it's not just a matter of the state of the various national economies but a function of the global perception of those economies. Latin America is now perceived to be strong enough to justify long-term investment."[9]

The poorer Latin American countries suffer from a crippling debt and lack of foreign exchange; therefore, the public sector still faces credit rationing with virtually no access to world capital markets and high borrowing costs.[10] As a result, maintenance and expansion of capacity in the public sector have not taken place. Lack of investment leads to a lack of infrastructure, which impairs profitability and competitiveness of private investment. Privatization relaxes the credit constraints faced by governments. The private sector and foreign investors do not face credit rationing, and therefore are better placed to conduct operations requiring major capital investments, such as telecommunications.

While Eastern Europe has also attracted significant foreign investment, privatization will likely be easier in Latin America than in parts of Eastern Europe: "In principle, privatization is justified on the grounds that private ownership is more efficient in terms of resource allocation than public ownership because of different incentive structures. But privatization is unlikely to generate major gains in efficiency unless accompanied by other reforms."[11] Such reforms are much more common in Latin America. The public sector in Eastern Europe represents 80 to 90 percent of GDP, whereas in Mexico, where it was dominant until the mid-1980s, the public sector now represents only 17 percent.[12]

However, privatization is not without detractors in Latin America. Labor has resisted privatization because of fears of job losses, and has blocked privatization in Brazil. One strategy to overcome worker opposition, which was adopted in Mexico, is to allow employees to buy shares at discounted prices. Worker participation in ownership may also help to bring about productivity increases.

The typical privatization model in Latin America is large, long-term investment, with the government imposing performance obligations. In return, investors receive an exclusivity period that can range from five years for Peru to ten years for Argentina and 25 years for Jamaica. Typically, the government allows one or two investors in the incumbent telecommunications operator, rather than authorizing formation of a new company on which it can impose fresh obligations, as is more common in Asia.

Other than Colombia and Brazil, all South American countries have effectively completed privatization of their fixed and mobile operators. While the privatizations generally have similar goals of attracting investment to modernize and extend national networks, the governments have adopted different models and terms. For example, in Mexico, the government permitted 49 percent foreign investment, but reserved portions of the shares for Telmex employees, and stipulated that 40 percent of the shares must always be held by a Mexican majority. The Venezuelan government sold 51 percent of CANTV to a consortium led by GTE, while Chile sold 44 percent of CTT to the Australian Alan Bond, who subsequently sold out to Telefonica de Espana.

The following sections examine the status of the telecommunications sector and recent developments in Argentina, Brazil, Chile, Mexico, and Venezuela. There are also short analyses of other countries in South America, Central America, and the Caribbean.

ARGENTINA

President Carlos Menem's primary goal since taking office in 1989 has been to stabilize Argentina's collapsing economy. His strategies include reducing the federal deficit to stop inflation and privatization of government-owned industries. One of the privatized industries is telecommunications, which was previously operated by the notoriously inefficient monopoly, Empresa Nacional de Telecommunicaciones (Entel). Equipment was antiquated, switches were only 5 percent digital, and there was a lack of redundancy, except in Buenos Aires's frail network. Also, Entel had been required to buy from suppliers with factories in Argentina (set up by Siemens and NEC) and paid twice the world market price for equipment. To obtain service, subscribers had to pay outright for equipment and the cost of installation, approximately $500 in a country where the average monthly wage rarely exceeds $80. Some residents had waited 14 years for a telephone. As a result, there was anarchy in Buenos Aires, with clandestine telephone wires strung everywhere, and piracy worth $30 million per year.

The privatization of Entel was South America's largest ever debt equity transfer; it enabled Argentina to clear $5 billion in foreign debt.[13] Entel was 60 percent privatized by the end of 1990, in an unusual restructuring that split the country in two through the middle of Buenos Aires, and awarded franchises to two private monopolies. The remaining 40 percent of the equity of Entel was to be shared between employees and local cooperatives (15 percent), with 25 percent floated on the Buenos Aires stock exchange.

The consortia were led by Bell Atlantic and Telefonica, with several banks holding major interests. However, Bell Atlantic's financing fell through, and in November 1990, France Telecom and the Italian operator STET led a consortium, including J. P. Morgan and the Argentine group Perez Companc, that successfully bid for the northern half of Entel. France Telecom and its partners paid $100 million in cash and $2.3 billion in a debt equity swap. The southern half was acquired by Citibank (holding 57 percent) and its partners Telefonica de Espana (33 percent) and the Italian company Techint (10 percent).[14]

The two operators, known as Telecom and Telefonica, hold regional monopoly concessions to provide voice services until 1997, when they will be allowed to enter each other's territory, and new operators will be allowed to compete with them. To make the franchises equally attractive, the city of Buenos Aires was split between the two operators. The dual monopolies are supposed to create some incentives to match the rival's standards, but there have been noticeable disparities. While the network division should be transparent to users, they found that billing practices differed and calling cards for one carrier would not operate from the other's pay telephones.

The National Telecommunications Commission (CNT), a quasi-independent agency under the Undersecretariat of Communications, was established in 1990 to regulate the new carriers. A new regulatory framework was prepared with the help of U.S. consultants. However, users have already complained that the carriers have increased their profit margins above the allowed 16 percent by overcompensating for inflation through higher rates. It is unclear how CNT will enforce the other terms of the agreement, which state that both operators must expand their networks to 1.5 million lines, make repairs within one day and installations within 15 days of service request, and provide free emergency calls, directory information services, and directories. There are also indications that CNT may not allow competition in 1997, but will extend the regional monopolies until the turn of the century.

In 1989, the government took a step toward privatization of telecommunications operations by awarding the cellular service franchise in Buenos Aires to a consortium called Movicom, which included BellSouth, Motorola, Citicorp, and the Argentinean Macri Group. A second carrier owned by Telecom and Telefonica will also offer cellular service in this market. Telecom and Telefonica now jointly operate StarTel, a value-added services company that took over Entel's public data network, Arpac. International service is provided by a private monopoly jointly owned by Telecom and Telefonica, known as Telintar (the Argentina International Service Company). The companies may choose to relinquish their international monopoly after 1997 if they are allowed to get into nonvoice or imaging services.[15]

In addition to the terrestrial network, Entel operated a domestic satellite network using Intelsat to reach the isolated western and southern regions (Patagonia). However, in 1993, Argentina pur-

chased two satellites from Canada, Anik C1 and C2, which are operated by Paracom, a company established by Deutsche Aerospace,
Aerospatiale, Alcatel, Alenia Spazio, and Embratel of Brazil. Paracom
operated the two satellites on a temporary basis until the launching of
the first custom-built Argentine satellite, Nahuelsat, in 1996. Based
on the Eutelsat 2 platform, Nahuelsat covers Brazil, Uruguay, Chile,
and Paraguay, as well as Argentina.[16]

Argentina is becoming a regional satellite hub, allowing private
satellite networks and direct access to the Intelsat system. The Argentine network operator Impsat has established VSAT networks
throughout Latin America. Impsat has 3,000 earth stations in Argentina, Colombia, Venezuela, Ecuador, Chile, and Mexico, and also
owns a fiber-optic ring in Buenos Aires linking its customers to its
satellite gateway.

BRAZIL

With nearly 160 million inhabitants, Brazil is the fifth most populous country in the world; it is also the fifth largest in area, occupying
48 percent of the South American land mass. Some 77 percent of the
population now lives in cities, but the rest is scattered from agricultural areas in the southeast to remote jungle regions of the northwest.
Telephone line penetration is 7.4 per 100 population, but vast regions
remain without service, and there are severe problems of poor reliability and congestion of networks in some urban areas.

Where service exists, Brazilians have to wait nearly one year for a
telephone; offices and dwellings with telephone service command a
significant premium in price. More than 510,000 people are on the
official waiting lists for telephone service installation, a procedure
that can itself take up to a year to complete. Telephone service is sold
under a "self-financing" plan in which the future user pays part of the
investment in advance and gets shares in the national company, Telebras, in return. To cope with the chronic shortfall of telephone lines,
the Brazilian government is setting new targets for network develop-

ment. To meet the targets, investments in the telecommunications sector are expected to rise to more than $7.2 billion in 1996 and should reach $69 billion in the year 2003.[17]

Structure of the Telecommunications Sector

Until the 1960s, Brazilian telecommunications was administered by hundreds of companies, most at the city level. Tariffs were not coordinated, and service was poor and inadequate. In 1962, telecommunications services were put under a national monopoly. A public company, Embratel, was created to operate interstate and international services. A 30 percent surcharge was levied on all telecommunications services to finance expansion. There was a massive investment in telecommunications in the 1960s and early 1970s, followed by a period of underinvestment, resulting in deterioration in facilities and services. In the economic crises of the 1970s, funds were diverted to other parts of the economy so that the network became unreliable and congested.

Brazil has reorganized its telecommunications structure to include a Ministry of Infrastructure, which is responsible for transportation, mining, energy, and communications. The National Secretariat of Communications (SNC) includes the postal service, telecommunications, and spectrum management. Telebras, the national telecommunications holding company, is under the secretariat, and is in charge of policies, standards specifications, and equipment approval.

More than 95 percent of Brazil's local telecommunications is provided by Telebras affiliates, which include 28 state telephone operating companies, one research and development unit, and Embratel. Embratel provides long distance services for interstate and international communications, provides data and telex services, and operates Brasilsat, the domestic satellite system. Embratel offers public nonswitched data networks and packet networks, as well as international public data networks. International private data networks include Airdata for airlines and Interbank, which connects banks to the SWIFT network.

The government is beginning to open the market to foreign

companies. Some competition and new entry are allowed in cellular networks, paging, leased lines, private data services, satellite data networks, and customer premises equipment (CPE). Private companies may also participate in private build-out projects such as new residential developments and condominiums.[18] Typically, foreign companies form partnerships with Brazilian companies to enter these markets.

Privatization would require amendment of the 1988 constitution, which mandated government monopolies for basic telephone and telegraph services, electric energy production, and petroleum extraction and refining. However, the minister of communications has stated that the entire telecommunications system, with the possible exception of long distance telephony, will be transferred to the private sector by the end of 1998.[19]

Embratel's monopoly on satellite service ends in January 1998.[20] Another part or the reform involves the breakup of Telebras, which has revenues of $7.5 billion per year and controls the world's eleventh largest network. Currently, the state holds 22 percent of the total capital of Telebras (51 percent of the voting stock); private investors hold 78 percent of the capital.[21] The government proposes to restructure the Telebras subsidiaries into five or six regional companies before privatization, which may not occur until 1999. Instead of being enshrined in law, limits on foreign ownership will be considered on a case-by-case basis.

Cellular and Satellite Systems

To stimulate competition, the market for cellular mobile services has been divided between the state-owned operating companies (Band A), with provision of equipment through tender, and the private sector (Band B), with a single provider selected through a bidding process. Cellular mobile service was introduced in 1991 in Rio de Janeiro and Brasilia; cellular service is also available in the states of Parana and Minas Gerais. The cellular equipment contracts for Rio went to Ericsson and NEC via local subsidiaries, and for Brasilia, to Northern

Telecom (Nortel) with NovAtel and Elebra SA, a local equipment manufacturer.

Full privatization of the state-controlled cellular system is planned for 1997. New digital licenses are also to be awarded. In 1996, the government lifted the 49 percent ceiling on private ownership of cellular and satellite services, which can be privatized without changing the constitution. About ten groups have expressed interest in cellular franchises, mainly Brazilian banks and international carriers. The cellular networks are likely to add 3 to 5 million cellular subscribers in the next three years.[22]

To help relieve backlogged demand, private networks may now be constructed with connections to the public telecommunications system. Foreign companies bidding for these networks must have a Brazilian partner, and several of the state telephone companies have indicated interest in participating, despite the fact that it is underinvestment in their own networks that is creating the demand for private networks. Telebras is also implementing a rural telephony program called Ruracell (Rural Telephone Advanced Public Service), which will be implemented in all operating companies supported by a cellular network. The project will serve 50,000 users in its first phase, and plans to add another 100,000 users in its second phase.

In northeastern Brazil, Telebahia is using an innovative approach to bridge the gap between supply and demand for telecommunications by offering "virtual telephone service." Small businesses that do not yet have individual lines can lease a voice mailbox, which can be accessed from a public pay phone. Businesses are provided with a telephone number where customers and suppliers can leave detailed messages.

Brazil's first domestic satellites, Brasilsat 1 and 2, were launched in 1985 and 1986. The second generation was launched in 1994. Brasilsat is used for data networking, rural telephony, and television distribution. There are several private VSAT networks. Banco Bradesco, the largest bank, has a 700 VSAT network with equipment from a local manufacturer, Digilab, under an agreement with Contel (now GTE). Banco Itau, Brazil's second largest private commercial bank, operates a network of some 600 VSATs, provided by Hughes and the

Brazilian company Promon. Banco Bamerindus and Banco Real also link their branches using VSAT networks.

Achieving Brazil's Telecommunications Potential

For many years, Brazil followed a strict industrial policy which favored industries based in Brazil and Brazilian technical development, banning imports of foreign technology in order to stimulate the development of its own industries. SNC supported this policy by using Telebras's purchasing power to promote domestic industry.[23] For example, Brazil spent $240 million on research and development in a recent five-year period; new products included digital exchanges, optical transmission systems, and packet-switching exchanges. However, this policy was controversial, not only among equipment suppliers who could not penetrate the market, but among Brazilians who felt their economy was suffering because they could not get access to modern technology.

Since 1990, most of Brazil's nontariff barriers to trade, such as import quotas, import prohibitions, restrictive import licensing, and local content requirements, have been lifted. Yet the government must still purchase from local industry unless the product is not in the local market, a policy designed to justify the investment made by large telecommunications companies in Brazil. Some import duties remain high in comparison with other countries, and other taxes and fees can significantly increase the cost of imports. The Department of Informatics and Automation Policy (DEPIN) has the authority to approve import licenses for data communications equipment.

Brazil's political and economic climate remains unstable. The country has been plagued by inflation, which successive governments have failed to control. Former President Fernando Collor de Mello, the first elected president since 1961, resigned under threat of impeachment over corruption charges in 1992. However, many nontariff trade barriers were lifted during his administration. His successor, former Vice President Itamar Franco, was unable to control inflation, which grew to an annual rate of 1,500 percent, nor to implement long promised economic reforms.

Still, Brazil has the potential both to expand its exports in telecommunications equipment and to harness these technologies for its own development. The commercial television networks now reach viewers throughout Brazil via satellite. To date, the potential of Brasilsat to bring educational opportunities to Brazilian villages, homes, and workplaces has not been achieved. The Brasilsat system could also be used more extensively for rural telephony if there were more incentive for Telebras companies to use the satellite instead of investing in their own terrestrial networks. Restructuring could provide the financing and incentives to accelerate provision of telecommunications services to both rural and urban households and businesses.

CHILE

Following a decade of liberalization, Chile has one of the most deregulated telecommunications markets in the world. Unlike other Latin American countries, Chile does not restrict foreign ownership of telecommunications. The telecommunications sector is one of the most highly developed in Latin America, due to the influx of foreign investment. Under General Augusto Pinochet, Chile privatized more extensively than any other Latin American country. From 1975 to 1982, 135 companies and 16 banks were sold off.[24] The network operated by the dominant carrier, Compania de Telefonos de Chile (CTC), is 100 percent digital. Teledensity in 1994 was 11 lines per 100 population. All services are liberalized.

The two main telecommunications operators in Chile, CTC, which is dominant in the local services market, and Entel, the dominant long distance carrier, were approved for privatization in 1987 and 1988 respectively. In 1987, Australian financier Alan Bond purchased 52 percent of the CTC monopoly, outbidding Spain's Telefonica. However, there was a change of fortunes in Bond's financial empire, and Telefonica bought Bond's controlling interest for $393 million in 1990. CTC's owners are now Telefonica Internacional de Chile (44 percent), Bank of New York (16.7 percent), Chilean pension funds and insurance companies (23 percent), and private investors 16.3 percent.[25] Telefonica also purchased 20 percent of Entel of Chile, the long distance company. There are small local carriers in the southern part of the country, and two small private companies have concessions overlapping parts of CTC's service areas.

In 1990, Entel announced it would compete for domestic service with CTC, and began to install phones and a digital network in Santiago. However, when Telefonica bought out Bond, it was ordered by the Chilean antimonopoly commission to sell its stake in Entel; Telefonica then appealed to the Comision Resolutiva. In April 1993, an antitrust tribunal decided that Chile's telecommunications market should not be segmented, and that CTC and Entel should be allowed to enter each other's markets, which are also open to other carriers. A 1993 Supreme Court decision upheld the antimonopoly commission's position that Telefonica must divest itself of its holding in either CTC or Entel.[26] Telefonica retained its interest in CTC; Entel is now a publicly traded company.

The local network has seven major operators, of which CTC is the largest. There are eight long distance carriers for a population of just 14 million, and some analysts expect a shakeout during the next few years.[27] There are several international services license holders in addition to Entel.[28] CTC still has 95 percent of the local market, but has captured only 35 percent of the long distance market and 20 percent of the international market since 1995.[29] Foreign operators include BellSouth, SBC, and Bell Atlantic. BellSouth is licensed to operate long distance, international, and cellular networks.[30]

There are four cellular concessions, two of which are held by a consortium including NovAtel of Canada and Cidcom SA, a Santiago-based joint venture with Pacific Telecom of the United States. They serve Santiago and Valparaiso and the corridor in between. Entel leases capacity from Intelsat to serve remote regions and islands. CTC operates a domestic VSAT data network with capacity leased from PanAmSat.

MEXICO

Mexico is the eleventh largest economy in the world, with a population of nearly 92 million. Teledensity is 9.5 lines per 100 population.[31] However, in 1989, it was 83rd in installed lines per capita, with a telephone density of only about 5 lines per 100 population. The growth rate was only 6 percent per year, with 1.5 million on the waiting list. There were also 10,000 communities of more than 500 people with no phone service. Service was very poor; Telmex revealed in 1989 that it received an estimated 12,000 complaints in a single day, and 20 percent of the phone lines were out of service at any one time.[32]

Telephone service was introduced in Mexico in 1882 by a subsidiary of ITT. Ericsson began operating in Mexico in 1907 and joined forces with ITT in 1941. A joint company, Telefonos de Mexico SA (Telmex), was incorporated in 1947. A group of Mexican entrepreneurs acquired ITT's and Ericsson's interests in 1958. In the early 1970s, President Luis Echeverria expanded the state's role in the economy, and the Mexican government acquired 51 percent of the company in 1972. The government helped to finance expansion, but in the 1980s it faced pressures of financial crisis and slow economic growth. President José Lopez Portillo's interventionist strategy culminated with the expropriation of banks in 1982. Finally, President Miguel de la Madrid began to initiate economic reforms in the mid-1980s. His successor, President Carlos Salinas de Gortari, vowed to accelerate these reforms.

Restructuring the Telecommunications Sector

Until 1989, telecommunications services in Mexico were provided by two state-owned bodies, the SCT (Secretariat of Communications and Transportation) through the General Directorate for Telecommunications, and Telefonos de Mexico (Telmex). Telmex was 51 percent state-owned, with over 120 federally licensed and franchised subsidiaries; its board chairman was also secretary of SCT, with responsibility for planning, regulation, and tariffs.

In 1989, President Salinas de Gortari began a six-year term in office by dismantling the traditional corporatist structure of the state, which had been held together for 70 years by his own party, the PRI (Partido Revolucionario Instituticional). Determined to integrate Mexico into the world economy, Salinas pushed ahead with the North American Free Trade Agreement (NAFTA). Mexico had already established "maquiladoras," industries along the United States border that could import raw materials and components duty free and reexport their finished products duty free into the United States.

Salinas also believed that development of telecommunications was central to plans for expansion of the economy. However, the investment required for telecommunications modernization was more than $14 billion. Salinas saw privatization of the remaining 49 percent of Telmex as the only solution both to raise the necessary capital for infrastructure and to help reduce the foreign debt, which had reached $105 billion.

During 1989–1990, the Ministry of Communications and Transportation (SCT) decided that the Ministry would concentrate on regulatory and development functions and would not operate networks; it retained control over satellite and telegraphy services which are constitutionally reserved to the state; and it established a new carrier, Telecommunicaciones de Mexico (Telecom), by merging DGT and Telegrafos Nacionales (Telenales), a public company created to carry out services that belong to the state exclusively. Telecom is to work alongside Telmex to provide local and long distance services. The carrier operates the packet data network and is modernizing its telegraph offices to provide telegrams, money orders, faxes, telexes, electronic mail, and phone mailboxes.[33] The creation of Telecom finalized the institutional changes required for privatization of Telmex.

In preparation for privatization, the finance ministry overhauled Telmex's tariffs, so that its profits increased 75 percent in 1989. The government decided to sell 51 percent of its shares, first changing the share structure so that 40 percent became voting and 60 percent non-voting L shares; it also required that the purchaser have Mexican majority control. The winning consortium consisted of Grupo Carso (run by Mexican financier Carlos Slim), which put up $859 million, and partners Southwestern Bell ($485 million), and France Telecom ($412 million) for a 51 percent stake of the full voting shares. This price was 20 percent higher than the commercial value and seven times greater than its 1988 value. In May 1991, the finance ministry placed 14 percent of the company's capital in over 20 countries (L shares) at a price 12 times greater than the 1988 value, raising a further $2.271 billion.[34]

The investors pledged to invest $14 billion over the following five years. The goal was to install a million lines per year—to double the network in four years and triple it in ten years. Telmex aims for a penetration of 30 lines per 100 population by the year 2000. Public call boxes were to be increased by 40,000 per year, the total number available in 1988. All unserved communities with at least 500 population were to be covered in five years, using terrestrial wireless and satellite links. Other goals were to improve service quality, including reliability and repair times. More than 80 percent of main lines were digital in 1994, and there were 217,000 pay phones. However, despite substantial progress, only 25.3 percent of Mexican households had telephones in 1994, while 77.4 percent had television sets.[35]

The tariffs for the new license are based on a price cap system to reward efficiency. The government has also eliminated a tax that was one of the highest in the world.[36] Local service is still subsidized; it now costs about $3 per month with 100 free local calls. International rates were lowered 40 percent. Yet income per line increased from $425 to over $700, so that 70 percent of investment could be financed from Telmex's own resources.

Telmex's long distance monopoly ends on December 31, 1996, and the government has decided to waive entrance fees for long distance competitors. New foreign operators have been making strategic alliances to enter the market. Among the eight new challengers will be a joint venture between Alestra (owned by AT&T and Grupo Industrial Alfa) and Unicom (owned by GTE and a Mexican-Spanish consortium).[37]

Concession titles will allow new competitors to develop their networks in stages. First they will be able to serve the lucrative Mexican Triangle (Mexico City, Puebla, and Guadalajara and Monterrey) in 1997, before having to cover the 50 to 60 cities they must serve by 2001. However, there have been disagreements over interconnection fees, with Telmex saying it wants at least $3 billion in fee income to make up for losses on local service.[38] Telmex fears it will lose much of its business in the triangle (which represents 70 percent of its calling volume) to competitors who will not have to serve less profitable areas in the short term.

Cellular and Satellite Services

SCT has also opened cellular services to competition. In 1990, the government awarded eight cellular licenses worth $1 billion in license fees and construction costs to consortia partially owned by foreign interests including BellSouth, McCaw Cellular Communications, Racal Telecoms, Millicom, Contel Cellular, and Bell Canada International. It was the biggest cellular program outside Europe.[39] Now, nine regional operators compete with Telmex, which has the only national cellular license and 55 percent of the market. Telmex operates a cellular subsidiary of Telcel in Mexico City, Tijuana, Guadalajara, and Monterrey. Cellular penetration is only 0.7 percent, with scope for growth to 3 percent by the end of the century.

Cellular technology may help to solve some of the problems associated with extending services. For example, cellular can be used for some fixed rural telephony such as community pay phones. Some cellular capacity is also used for bypass and urban primary service, as many subscribers opt for more expensive cellular service, rather than waiting for wireline service. Business customers can use the cellular systems installed by Ericsson in Tijuana to reach San Diego and Mexico City. Ericsson is also building an overlay digital network for high-speed data communications.

Satellite communications is also another important means of extending services to unserved communities and offering value-added services, such as data networks, to user groups including banks, hotels, and other industries. Telecom operates the domestic satellite system which was launched in 1985; the original Morelos satellites have been replaced by the second generation, known as Solidaridad, which has twice the capacity. Each Solidaridad satellite has 18 C-band transponders and 16 Ku-band transponders and mobile services on L band. Solidaridad covers the U.S. border and areas of heavy traffic within the United States so that it can link companies with operations in both countries. It also has beams for Central and South American coverage.

Growth in satellite use was slow until 1989, when the government introduced permits for private satellite networks and teleports. In two years, the occupancy of Morelos I increased from 30 percent to 100 percent, and on Morelos II, to 70 percent. Satellite communications is used for telephone service to some remote communities. There are also more than 140 private satellite data networks, including SARH (Mexican Ministry of Natural Resources) for agriculture and weather information; Servicios Industriales Penoles SA, the central mining organization, which has a VSAT network for voice, data, and video; Petroles Mexicano (Pemex) and Industries Resistol SA, with networks for voice and data; and Banco Serfin, with a network linking its headquarters in Mexico City with four regional offices and 172 branches throughout the country.[40] Satellite bypass is also provided by the Houston-based Stars network for use in hotels and resorts.[41]

VENEZUELA

Venezuela had one of the highest economic growth rates in the region (9.2 percent in 1992), but has been plagued by a stagnant economy and inflation in recent years. While still heavily dependent on oil exports, Venezuela is attempting to diversify its economy by increasing exports of other natural resources such as minerals and forest products, as well as manufactured goods. The country has also embarked on a program of economic restructuring, opening the economy and eliminating market distortions. Inflation remains a stubborn problem, as does the nation's external debt.

The administration of President Carlos Perez announced its privatization plans in early 1989, and the first state assets were sold by the FIV (Venezuelan Investment Fund) in late 1990. During 1991, FIV sold off 60 percent of its shares in Viasa, the largest state airline, and auctioned off two commercial banks. The sale of the state-owned telephone company, CANTV (Compania Anonima Nacional Telefonos de Venezuela), was one of the biggest privatizations to date in Latin America. VenWorld, a GTE-led consortium, bought 40 percent of CANTV for about $1.9 billion. Other foreign investors include Telefonica de Espana and AT&T. The government retained 49 percent for later sale, and placed 11 percent in trust for CANTV employees.[42] Further moves to privatization have been stalled in the face of strong political resistance and unfavorable economic conditions. However, an IMF adjustment program introduced to stem inflation has opened the way for the government to sell its remaining 49 percent interest.

Telecommunications is the most dynamic sector in the Venezuelan economy. In 1992, $1 billion was invested for service expansion, new services, and technology development. Teledensity increased to 10.9 lines per 100 population in 1994. Investments in the sector are expected to total $10 billion by the end of the decade. The newly privatized CANTV is to install 4.5 million new lines by the year 2000, increase digitization to 80 percent of exchanges, and add 100,000 pay phones. Service quality is also to improve, with the target for waiting times to be reduced to one week. Previously, Venezuelans had to wait up to eight years for telephone service.[43]

Venezuelan telecommunications are regulated by the Consejo Nacional de Telecommunicaciones (Conatel), an autonomous organization within the Ministry of Transportation and Communications. CANTV has a 35-year concession, renewable for another 20 years, and a monopoly over domestic local and long distance service until the year 2000.

Mobile and paging services have been liberalized. Currently, there are two mobile operators: Movinet, a subsidiary of CANTV, and Telcel, which is owned by a consortium headed by BellSouth. The cellular market is booming, with some 450,000 cellular customers in the country.[44] Conatel plans to license a third operator.

Private and public satellite services, VSAT data networks, value-added services, private line, and certain other mobile services have also been liberalized.[45] Nineteen licenses have been granted for operators of private networks; they can offer data, text, and video, but not voice during the restricted period.[46] Foreign providers face strict currency controls, but there are no limits on foreign investment.

OTHER COUNTRIES IN LATIN AMERICA AND THE CARIBBEAN

South America

With a per capita GDP of $714, **Bolivia** is one of the poorest countries in South America. Its teledensity is only three lines per 100 population. The country's goal is to increase teledensity to seven lines per 100 by the turn of the century. Bolivia's economic reform program, introduced in 1995, has brought partial privatization to six leading state-controlled enterprises, including telecommunications. STET, the Italian state holding company, purchased 50 percent of Entel Bolivia (Empresa Nacional de Telecomunicaciones) for $610 million in 1995. The partially privatized carrier has a six-year monopoly and a 40-year concession for long distance, paging, and cellular services.[47]

Seventeen Bolivian cooperatives provide local telephone service. Establishing interconnection agreements has been a stumbling block for Entel. Cellular service is provided by Entel's subsidiary Telecel. The government plans to license a second cellular operator in 1997.

The major carrier in **Colombia** is the government-owned Empresa Nacional de Telecomunicaciones (Telecom). The government plans to introduce competition by auctioning two licenses for long distance and international services. Licensees will have to pay 2 percent of revenues to the government to subsidize rural telephony, which is provided by municipal or regional monopoly operators. The government has no plans at present to privatize Telecom, which will compete with these new entrants.

Colombia's cellular network is one of the most competitive and highly developed in Latin America. In 1994, Colombia was divided into three regions, with two cellular operators in each region. One of the operators in each region is completely privately owned; the others are mixed companies comprised of local telephone companies and private investors.

Access to telecommunications in **Ecuador** is severely limited, with only five lines per 100 population. Telecommunications service is provided by state-owned Emetel Ecuador. However, the government has announced plans to privatize Emetel. Under the terms of a 1995 privatization bill, Emetel will separate into two operating companies, one for the northern highlands region, and a second for the southern coastal region. A 35 percent stake of each company is likely to be sold off to private investors, with another 10 percent reserved for employees. The successful bidders will operate the new companies under 15-year concessions, with exclusivity for the first five years. Cellular and paging services are liberalized, with cellular services provided by two carriers, the government-owned Otecel, and Conecel, which is a consortium of Mexico's Iuscell, Telecel of Colombia, Bell Canada, and local companies.[48]

Peru has a per capita GDP of almost $2,000, but a teledensity of only 3.3 lines per 100 people. In 1994, Telefonica of Spain bought controlling interests in two formerly state-owned telephone operators, which were merged to form Telefonica del Peru (TdP). The new operator has pledged to install 1.9 million telephone lines by 1998, raising teledensity to about 7.2 lines per 100. There is already some competition in the local network. Resetel has obtained a local carrier concession to compete with TdP. InterAmericas Communications Corporation of the United States plans to purchase Resetel and to build and operate a fiber-optic and wireless network throughout Lima. The cellular market is partially liberalized, with Tele 2000, a private operator, providing cellular service in Lima and Callao.

Central America

Costa Rica has the highest teledensity in Central America, with 13 lines per 100 population. Service is provided by the government-owned Instituto Costarricense de Electricidad (ICE), which has a reputation for being one of the best run operators in the region. ICE has placed a high priority on providing telecommunications services to rural communities. In 1994, ICE launched a $50 million Euro-bond issue to finance network expansion and installation of a cellular system for urban and rural areas.[49] Cellular and paging services are liberalized. Privatization of ICE remains controversial, with opposition from employees and much of the electorate.

With 10.6 million inhabitants, **Guatemala** has the largest population in Central America, but the least developed telephone network, only 2.4 lines per 100 population. With the exception of paging, all services remain monopolies. The national operator is government-owned Guatel (Empresa Guatemalteca de Telecomunicaciones). In 1994, Millicom International Cellular was authorized to participate in the development of a cellular network. It is expected that a second cellular license will be issued in the near future.

Honduras is one of the poorest countries in the hemisphere, with a per capita GDP of only $600 and teledensity of only 2.4 lines per 100 population. As part of its economic restructuring, the government plans to partially privatize the state-owned operator, Hondutel (Empresa Hondurena de Telecomunicaciones). Earlier proposals to sell off 100 percent of Hondutel were withdrawn because of the company's outdated infrastructure and huge debts incurred to install 220,000 new lines. Scaled-down plans call for privatization of only the more efficient parts of Hondutel, leaving the remainder under government ownership. Cellular service is also a monopoly, operated by a consortium of Millicom International Cellular, Motorola, and Inversiones Rocafuerte, a local partner. The group paid $5.1 million in 1996 for a ten-year license.

Although the land area of Honduras is small, satellite communications have proved to be a cost-effective means of providing basic communications in rural areas. Hondutel uses PanAmSat and earth station equipment from Vitacom of California to provide rural telephony in isolated areas.

Panama's national telecommunications operator is government-owned Intel (Instituto Nacional de Telecomunicaciones), which has a monopoly on local, long distance, and international services. The Panamanian government plans to sell 49 percent of Intel; it will retain 49 percent and place 2 percent in a trust fund for employees. The legislative assembly has passed Law 73, which creates and defines the role of the new regulatory agency, ENTE Regulador; this agency will oversee public services once they are privatized.

Cellular services have been liberalized. In 1996, BSC de Panama, a consortium headed by BellSouth, won a 20-year concession to build and operate Panama's first cellular network. Intel also plans to operate a competing cellular network.

The Caribbean Region

The conditions of telecommunications in the Caribbean indicate the problems of small states, which are also shared by the island nations of the South Pacific. They are characterized by small markets in terms of population; however, high-quality and affordably priced telecommunications is important not only for domestic links, but to support the international business on which their economies depend; tourism in particular relies on good communications, as do resource-based industries such as sugar, fisheries, and mining. These small countries typically have few telecommunications experts, and are at a disadvantage in negotiating with large foreign carriers and equipment suppliers.

Two strategies have proved beneficial for these small countries. First, they can aggregate demand to become a more attractive market, and to increase their bargaining power with equipment vendors. The Caribbean Association of National Telecommunications Organizations (CANTO) has helped Caribbean nations to collaborate on planning and upgrading their networks. A second strategy is to obtain outside advice when evaluating options for privatization or restructuring. Such expertise is provided by the International Telecommunication Union (ITU), and may also be offered by regional telecommunications organizations. Administrations may also hire consulting firms with experience in valuing telecommunications assets and negotiating with carriers. Without expert assistance, they risk selling their assets too cheaply and locking themselves into long-term concessions for the new owners. While private advisors may seem expensive for small countries, the cost of their services can be included in the valuation of the corporation that is to be privatized.

Many of the smaller islands have found a willing buyer for their telecommunications operations. Cable & Wireless (C&W) is thoroughly entrenched in the Commonwealth Caribbean (as well as the Commonwealth South Pacific), where it formerly operated telecommunications facilities in the British colonies. C&W owns 100 percent of the telecommunications operators in Anguilla, Bermuda, the British Virgin Islands, the Cayman Islands, Dominica, Montserrat, St. Lucia, St. Vincent, and the Turks and Caicos. C&W also owns 79 percent of the carrier in Jamaica, 70 percent in Grenada, 65 percent in St. Kitts and Nevis, and 49 percent in Trinidad and Tobago.[50]

Although located in Central America, **Belize** has more affinity with the Commonwealth Caribbean than with its Latin neighbors. In Belize, the government has a joint venture with British Telecom (BT) to operate the network. Belize Telecommunications is 37 percent owned by the government and 25 percent by BT, with the remainder owned by local institutions.[51] Previously, the government had operated the domestic network, and Cable & Wireless had operated the very profitable international network. The arrangement with BT results from a Belize government decision to take over international telecommunications from C&W, in order to use the revenue from international services to expand and upgrade the domestic network. The government decided to go into a joint venture with BT to operate both the domestic and international networks.

Cuba's telecommunications network languished for decades, without funds to modernize its antiquated equipment or extend its network. However, in 1994, the government agreed to sell 49 percent of

the state telecom operator, Emetel Cuba (Empresa de Telecomunicaciones de Cuba), to a joint venture between Mexico's Grupo Domos
and STET of Italy. The new joint venture plans to invest about $1.4
billion to modernize the network. Wireless local loop is being used to
speed expansion of the local network. In 1993, Emetel entered a joint
agreement with Telecomunicaciones Internacionales de Mexico
(TIMSA) to launch a cellular network called Cubacel; Ericsson is supplying the equipment. All services remain monopolies.

International traffic is to be switched by Telmex. Cuba apparently
expects its international services to be highly profitable. However, the
U.S. Department of State blocked applications filed by five U.S. companies, including MCI and Sprint, for direct telephone services between the United States and Cuba because the collect call surcharge
to be imposed by Cuba was considered too high.[52]

In 1995, an act of parliament created a single company to provide
Jamaica's telecommunications services, replacing the former international and domestic monopolies, the Jamaica Telephone Company
and Jamaica International Communications, which were jointly
owned by the government and Cable & Wireless. The act established
Telecommunications of Jamaica (TOJ), which is 79 percent owned by
C&W, and 21 percent by others, including employees. Cable & Wireless has exacted very favorable terms for TOJ from the government,
including a 25-year monopoly on public domestic and international
services.

TOJ has been upgrading and expanding services as part of an ambitious five-year program that will double line capacity. It also has a
$38 million contract with Northern Telecom (Nortel) for digital
switches. Other ongoing projects include expansion of the cellular
network, and a fiber-optic link to the Cayman Islands. Cable & Wireless is also the major shareholder in Jamaica Digiport International
(of which AT&T owns 35 percent), a provider of international private
line services for Jamaica's free trade zones.[53]

Puerto Rico is a Commonwealth of the United States, with fewer
rights than a state, but its own telecommunications system. In 1974,
the government purchased the Puerto Rico Telephone Company
(PRTC) from ITT for about $200 millon, in effect, nationalizing

telecommunications. In 1990, the governor proposed to reprivatize PRTC by selling it for at least $2 billion. The proceeds from the sale were to be used to improve education, roads, and other public facilities, but the privatization did not go through. The long distance carrier, Telefonica Larga Distancia de Puerto Rico (TLD), was privatized in December 1992, when the FCC approved the sale of 79 percent of TLD to Telefonica de Espana through a holding company, LD Acquisition Corporation (LD). The government retains 19 percent through the Puerto Rican Telephone Authority (PRTA), and 2 percent was placed in an employee stock option plan. Domestic services continue to be provided by PRTC, which is wholly owned by PRTA. Both domestic and long distance services remain monopolies.[54]

LESSONS FROM LATIN AMERICA

Global trends toward privatization and liberalization can work to the advantage of developing countries because many operators are now looking for attractive overseas investment opportunities. Even the smallest country may be able to take advantage of this new investment climate to attract competitive bids. Among the foreign telephone companies that have invested in Latin American telecommunications are BellSouth, France Telecom, GTE, SBC (Southwestern Bell), and Spain's Telefonica. Cable & Wireless has increased its investment in the Caribbean, and British Telecom has invested in Belize. The unmet demand has also attracted investment from commercial banks.

Typically, governments require significant investment and specific targets to be achieved in installation of lines and improved service quality as part of the terms of privatization. However, the mechanism for enforcement of these terms is often weak, and may not even be formalized at the time of the sale. Since Latin American and Caribbean countries typically adopted the European PTT model, they have had no experience with independent regulation. If they do not establish a regulatory structure with the legal authority to enforce privatization terms and introduce new regulations as necessary, these countries may soon find themselves without any leverage over a private monopoly; ironically, it was this lack of control over foreign companies that turned many Latin American countries against foreign investment in the 1960s and 1970s.

There is a marked difference between the level of investment in richer urban areas and in rural areas across the region. Most Latin American governments are not requiring obligations as significant as those imposed on investors in Asian countries, such as Indonesia and the Philippines, to build out networks in rural areas. This missed opportunity to provide strong incentives for extending rural access could pose a potential problem in the long term. As one analyst points out: "If you get a lot of competition without that sort of obligation, then you can get to the point where it's never done because there's not enough payback to do it."[55]

Danger also lies in the granting of exclusive authority to provide services for an extended period of time. Typically, foreign investors demand a period of guaranteed monopoly to reduce their investment risks. However, an extended period of exclusivity may seriously reduce incentives to improve services or reduce prices. Outside advice from nonpartisan experts and comparison with privatization strategies in other parts of the world may help countries in Latin America and other developing regions to avoid such pitfalls.

International Players and Policies

CHAPTER *14*

International Satellite
Communications

*"In short, satellites not only enabled us to watch the revolution, they
helped cause it."*

— STEVEN D. DORFMAN, President, Hughes
Telecommunications and Space [1]

INTRODUCTION

During the past decade, satellites have played a major role in transforming the world into what Marshall McLuhan heralded in the
1960s as the coming "global village." We have watched live transmissions via satellite of events unfolding in China, Eastern Europe, and
the former Soviet Union. Satellites have also brought reliable two-
way communication to many isolated and developing regions. Telephone service to Pacific islands, remote parts of China, and isolated
communities of Asia and Africa is provided by satellite; satellite circuits also link Moscow and Eastern European cities with the West.
Businesses are linking management, research, production, and merchandising units via dedicated satellite networks.

The Intelsat system was established in 1964 to provide global
communications via satellite. Today, it links more than 180 nations

and territories, but faces challenges from new international and regional satellite systems. The Intersputnik system was the Soviet and Eastern bloc answer to Intelsat, with links to Eastern Europe and socialist states around the world. The Inmarsat system provides communications for ships and more recently for airplanes, and intends to offer global mobile services. It will compete with several other commercial mobile satellite systems designed to provide communications "anytime, anywhere."

This chapter examines these organizations, the services they provide, and their global and regional impacts on the telecommunications industry and users. The following chapter examines international cable services and the emergence of competitive regional and global satellite systems.

INTELSAT

Intelsat's Origin and Structure

Intelsat is a cost-sharing cooperative that owns and operates satellites used by most countries for international communications. Intelsat carries more than half of all intercontinental telecommunications traffic. In 1995, it had 21 operational satellites providing global connectivity to more than 180 user countries and territories.[2]

The origins of Intelsat go back to the "space race" of the 1960s. In 1957, the Soviet Union launched Sputnik, a little satellite that beeped as it crossed the sky. Its beep was a wake-up call to the Western world, particularly the United States, where politicians and educators urged Americans to "catch up with the Russians." In his State of the Union message in 1961, President Kennedy invited "all nations—including the Soviet Union—to join with us in developing . . . a new communications satellite program. . . ."[3] Kennedy's special 1961 presidential message, "On Urgent National Needs," contained not only the much-publicized goal of landing a man on the moon and returning him safely to earth by the end of the decade, but also asked Congress to "make the most of our present leadership" through accelerating development of space satellites for worldwide communications. Responding to the challenge, Congress passed the Communications Satellite Act of 1962, which established the Communications Satellite Corporation (Comsat) as a "chosen instrument" to "establish as expeditiously as practicable a commercial communications satellite system as part of an improved global communications network" and to "direct care and attention toward providing . . . services to economically less developed countries and areas as well as those more highly developed."[4]

Intelsat was founded in 1964, when the United States joined with 18 other nations in a cooperative to operate the first commercial satellite. The United States took a leading role in the formation and early operation of Intelsat, promoting the organization as a means of sharing the benefits of satellite technology with the rest of the world. The United States was not only the largest user, but also provided the technical and managerial expertise for the organization through its signatory, Comsat.

From the original signatories, Intelsat's membership has grown to 139 nations, more than 80 percent of which are developing countries. Its 21 operational satellites are used by more than 180 countries, territories, and dependencies. During the 1970s and 1980s, new members were primarily from the developing regions of Latin America, Africa, and Asia. In the 1990s newly independent republics that were formerly part of the Soviet Union, such as Armenia, Azerbaijan, Kazakhstan, and the Kyrghyz Republic, have also joined Intelsat to establish direct links with the outside world.

Intelsat's traffic growth and capacity have far outpaced its increase in membership. Early Bird, the first Intelsat satellite launched in 1965, had 240 telephone circuits; its Intelsat VII series has the capacity for 90,000 voice circuits per satellite, using digital circuit multiplication equipment (DCME). The Intelsat VIII series will have 44 transponders and up to 112,500 circuits per satellite.[5]

As established in the Definitive Agreements, Intelsat has a four-tier structure consisting of the Assembly of Parties, the Meeting of Signatories, the Board of Governors, and the Executive Organ. The Assembly of Parties is composed of representatives of governments of

Intelsat member countries (Parties to the Agreement) and normally meets every two years to consider resolutions and recommendations on general policy and long-term objectives. The Meeting of Signatories consists of representatives of Signatories to the Operating Agreement (member governments or their designated telecommunications authorities), and generally meets once a year to consider issues related to the financial, technical, and operational aspects of the system.

The real power of the organization is vested in the Board of Governors, composed of representatives of Signatories whose investment shares either individually or as a group meet or exceed the minimum share for membership on the Board. Two or more members may combine investment shares to meet the minimum for membership and be represented by one Governor on the Board. In addition, groups of five or more Signatories within an ITU region may be represented by one Governor, provided that there are not more than two groups from any one region or more than five governors for all of the ITU regions. In 1995, the Board comprised 27 Governors representing 117 Signatories.[6] The Board normally meets four times per year.

The Executive Organ, which is responsible for management and operations, is headed by the Director General, who is the chief executive and reports directly to the Board of Governors. The current Director General is from the United States. Intelsat staff is drawn from more than 50 countries.

Intelsat operates as a financial cooperative, with ownership shared among the members in proportion to their use of the system (adjusted on an annual basis). The investment share determines each Signatory's percentage of the total contribution required to finance capital expenditures, and the portion of net revenues it receives. After deduction of operating costs, revenues are distributed to the Signatories in proportion to their investment share as repayment of capital and compensation for use of capital.

The Role of Developing Countries in Intelsat

At first glance, Intelsat's developing country members would appear to be in a position to dictate the organization's goals and priorities if they chose to do so, since they comprise more than 80 percent of the members. Indeed, in the 1980s Intelsat emphasized the importance of its third world members in its efforts to stave off competition: ". . . it is Intelsat's special duty and obligation to promote the cause of communication development that distinguishes Intelsat in a substantive and important way from U.S. entrepreneurs who are promising private transoceanic satellite systems."[7]

However, several other factors must be taken into consideration. First, the real power of Intelsat lies in the Board of Governors, where voting is weighted by usage. The OECD countries together control 61.56 percent of the shares, with the five largest users (the United States, the United Kingdom, Japan, France, and Germany) together controlling 41.12 percent (see Table 14.1). Thus, any change in policy requires the endorsement of several of these countries. Developing countries' shares range from 2.31 percent for China and 1.95 percent for India to 0.05 percent (the smallest unit allocated) for the smallest users such as Afghanistan, Croatia, Honduras, and Zimbabwe.[8] The developing countries themselves should not necessarily be considered a unified bloc. Some, including Indonesia, Brazil, India, Mexico, and the Arab States, now have their own satellites, and may therefore not share the concerns of those who rely on Intelsat for domestic as well as international communication. In addition, third world countries are at widely varying stages of development; the needs of Thailand differ greatly from Tanzania, and Sri Lanka from the Sudan.

Intelsat Technology and Services

While Intelsat continues to provide global coverage, recent satellites have been modified to provide higher power or more flexibility, to allow use of smaller earth stations and provision of more domestic services for developing countries.

In the early years, only the Standard A series (30 meters in diameter and costing several millions of dollars) could be used with Intelsat. The B series (10 to 13 meters in diameter) was introduced in the 1970s for smaller developing countries and others with more modest traffic requirements. The C series was introduced for international

TABLE 14.1
PERCENTAGE OF INTELSAT INVESTMENT SHARES HELD BY
OECD COUNTRIES 1995

Country	Investment Shares
Australia	2.82%
Austria	0.41
Belgium	0.34
Canada	1.81
Czech Republic	0.08
Denmark	0.43
Finland	0.09
France	3.66
Germany	3.47
Greece	0.37
Hungary	0.09
Iceland	0.15
Ireland	0.08
Italy	4.87
Japan	4.48
Luxembourg	0.05
Mexico	0.42
Netherlands	1.41
New Zealand	0.59
Norway	2.45
Portugal	0.09
Spain	1.58
Sweden	0.55
Switzerland	0.84
Turkey	0.50
United Kingdom	9.19
United States	19.14
Total	59.93%
Total of 5 top countries	39.93%

Derived from Intelsat, *Annual Report 1995–1996.*

services using Ku band. Intelsat has in recent years introduced several new earth station standards to facilitate provision of new services. They include the D series for Intelnet services, the E and F series for International Business Services (Ku and C band), the G series for international lease services, and the Z series for domestic lease services.[9]

International public switched services still account for more than 50 percent of Intelsat's operating revenues. However, television distribution is its fastest growing service; Intelsat's television transmission facilities are used for program distribution, news coverage, sporting events, entertainment programs, videoconferencing, and direct broadcast services. International television services are provided through full-time leased transponders and occasional use leases. Among programs transmitted globally through occasional use are special sporting events such as the Olympics, Wimbledon tennis, and World Cup soccer; and coverage of major news events such as the ouster of Ferdinand Marcos in the Philippines, the fall of the Berlin Wall, and student protests in Beijing's Tiananmen Square. International teleconferences, or "space bridges," link groups in several countries to increase international understanding. Multisatellite hookups have been used to raise funds to fight famine and disease through the Live Aid concerts and other globally shared events.

Several television networks lease transponders for daily transmissions. CNN now transmits its news channel to Europe and Japan; U.S. networks feed programming to Australia; Japanese television is available on cable systems in New York; and U.S. networks use regular feeds from Europe. The BBC distributes its BBC World Service Television via Intelsat; the U.S. Armed Forces distributes its TV network service to U.S. bases around the world. The Intelsat VII series of satellites includes higher powered transponders that can be used for direct-to-home (DTH) television distribution.

Intelsat Business Service (IBS) is an integrated digital system designed to carry a full range of services including voice, data, and video. IBS enables antennas to be located on or near end user premises and in major cities, providing direct satellite access and minimizing dependence on terrestrial switched networks. IBS offers global connectivity for applications including private line networks, interconnection of mainframes and local area networks (LANs), electronic mail, electronic data interchange (EDI), and videoconferencing. IBS users are generally large multinational corporations. For example, Ford Motor Corporation communicates directly between its headquarters near Detroit and its installations in the United Kingdom. Hewlett Packard uses IBS between Palo Alto, California, and operations in Europe and Asia.

The Intelnet service was designed for low-volume data transmission to VSATs operating through a central hub terminal. The system uses spread spectrum techniques that eliminate interference even in urban areas with severe microwave congestion. Applications include dissemination of wire service copy, stock market data, and weather reports to receivers scattered over an area as large as one-third of the earth's surface. Reuters uses Intelsat to broadcast its news services to more than 1,800 subscribers throughout Asia, Africa, the Middle East and Latin America.

The Vista service provides telecommunications services such as voice and low-speed data to rural and remote communities, and can be used for both domestic and international links. Small earth stations with five-meter antennas or less are designed to provide thin route voice and data service in developing regions.

Vista service has been very slow to take off, despite the evidence of enormous demand for basic communications in developing countries. Its problems reflect the many difficulties of increasing investment in telecommunications in developing countries. One problem is convincing manufacturers that there is enough demand to justify building units in sufficient quantities to drive down the price to an acceptable level. A second problem is that many developing country officials give higher priority to improving the more profitable interurban services and to meeting a backlog of urban demand, rather than extending basic services to rural areas.

In addition, Intelsat cannot market directly to end users in developing countries, but must work through its Signatories, which may have other priorities. Thus, businesses and rural customers may not be aware that affordable telecommunications facilities could be made available, and may not request service from their administrations.

Doing Business via Intelsat

The People's Republic of China leases capacity on Intelsat for its Xinhua News Agency's 150-site VSAT network, using the Intelnet data broadcasting service. Applications include data broadcasting of news and financial information and remote printing. Intelsat's IBS on-premises earth stations link businesses in Beijing and Shanghai with Japan and the United States. High-tech companies in Malaysia's Industrial Free Trade Zone areas use Intelsat's IBS service for high-speed data, and voice and video links between manufacturing sites in Europe, Japan, and the United States. Electronics manufacturing companies, including Intel, Siemens, Hewlett Packard, Matsushita, Panasonic, and Sharp-Roxy, use IBS networks for inventory control, order processing, and shipment scheduling with real-time interactive data communications between manufacturing plants and headquarters.

Texas Instruments (TI) communicates with its locations around the globe via Intelsat. TI Singapore runs its Singapore factory 24 hours per day, seven days per week, and relies on Intelsat to transmit real-time information and production data. Samsung Corporation, South Korea's largest microchip manufacturer, uses Intelsat IBS service to link its headquarters in Seoul with branch offices in the United Kingdom and Germany. Other Korean customers include Korean Airlines for worldwide flight reservations, Hyundai Corporation for import-export trade, and foreign multinationals such as McDonnell Douglas for voice and data links between its branch offices in South Korea and headquarters in the United States.[10]

Domestic Services

Intelsat was founded to provide international satellite communication, that is, links between nations. However, it has also been used to provide domestic communications in countries where terrestrial facilities are unavailable or prohibitively expensive to construct. Intelsat began offering domestic service at the instigation of Algeria in 1976. Algeria negotiated a reduced rate for a preemptible transponder on a backup satellite, which theoretically could have been taken over by Intelsat to maintain service for full-rate customers if satellite problems caused any interruption in their service.

More than 20 developing countries in Africa, Asia, and Latin

America use Intelsat capacity for domestic services, to provide voice and data telecommunications links between major cities and to distribute television programming throughout their countries. For example, in Ghana, Intelsat earth stations in two major cities link nodes in the country's new cellular telephone network operated by Millicom.[11] Nigeria and Zaire link their provincial capitals via Intelsat. In some cases, these Intelsat networks are intended as temporary solutions until terrestrial fiber or microwave links are built, or regional satellites are available.

Planned Domestic Service (PDS) was introduced to enable member and user countries to lease or purchase transponders on a nonpreemptible basis for domestic communications. PDS, coupled with the design of recent Intelsat spacecraft to include higher power beams for domestic use, appears to be a shrewd move by Intelsat to offer domestic services to countries that could benefit from domestic satellite services but cannot afford or do not need dedicated satellites. PDS also allows Intelsat to compete with regional satellites that offer domestic services.

Other Services

Intelsat has developed a 1.8-meter "fly away" C-band earth station for provision of single-voice channel service in remote areas. The service is designed to provide temporary voice and data communications, for example, for emergency and disaster relief operations, diplomatic missions, and travel by heads of state. A digital satellite news gathering (SNG) service enables television crews to transmit coverage from the scene of news stories.

Intelsat also leases temporary satellite capacity for cable restoration after undersea cable failures, and has a special "cable in the sky" service for the restoration of optical fiber systems. However, fiber may still take a sizable share of Intelsat's trans-Atlantic and trans-Pacific business. To compete with optical fiber, Intelsat has revised its tariff structure to encourage growth in its digital services and transition from analog to digital technology. It now offers ISDN

(Integrated Services Digital Network) services that are globally interconnected.

The Future of Intelsat

Intelsat faces both commercial and technological challenges to its de
facto global satellite monopoly. Does Intelsat have the flexibility to
change with the times and operate in an environment which increasingly reflects deregulation and competition in telecommunications?
Among the major new competitors are regional systems such as
AsiaSat and Apstar, which cover most of Asia, and domestic systems
with beams covering many surrounding countries such as Indonesia's
Palapa, Malaysia's Measat, and Thailand's Thaicon in Asia, and Mexican, Brazilian, and Argentine satellites in Latin America. These regional systems are designed to provide television distribution and
direct-to-home television services to many countries from a single
satellite, and to provide domestic voice and data communications. A
more significant threat is the loss of the global monopoly. PanAmSat,
which began as the first international competitor to Intelsat, with service in Latin America and links to the United States and Europe, now
offers global connectivity with satellites positioned over the Atlantic,
Pacific, and Indian Oceans.

It should be noted that Intelsat is not regulated or overseen by any
external authority. There is no mechanism for external analysis of
costs, tariffs, or internal cross-subsidies. Critics have argued that Intelsat's costs are higher than necessary because of overcapacity and un-
realistically high projections of traffic growth. Competitors such as
PanAmSat have contended that Intelsat is cross-subsidizing its new
offerings, such as domestic sales, with revenues from other services.
Intelsat denies these allegations, but is not required to make public
any documentation of its costs and rate-setting procedures.

Intelsat's primary technological challenge comes from undersea
fiber-optic cables that are being installed to link continents and regions. Point-to-point voice, data, and video are increasingly being
carried by fiber networks that have very high capacity and eliminate

the delay caused by the distance signals must travel to and from geostationary satellites. While Intelsat now obtains 3.1 percent of its revenues from cable restoration services, it appears to have lost much
more voice and data revenue to submarine cable operators, in addition
to the loss of video transmission revenue to regional satellite operators. Proposed new mobile satellite communication networks that
plan to offer global connectivity to handheld terminals could also
threaten Intelsat's public switched services revenues by the end of the
decade.

Intelsat is innovating by introducing new services and offering
more domestic capacity. The Intelsat VI, VII, and VIII series are designed with higher power and beams that can be configured for domestic and direct-to-home service; Intelsat is also increasing its
offering of digital services for voice, data, and video. In addition it has
entered into joint ventures with India and China to obtain capacity
on Insat and Chinasat satellites for Asian customers.[12]

As a carrier's carrier, Intelsat cannot market directly to end users,
but must work through its Signatories, the telecommunications administrations. Most developing country administrations are still government-operated monopolies with pent-up demand for basic
telecommunications, leaving little incentive to introduce new technologies or price aggressively. Intelsat must rely on equipment vendors to spread the word to potential customers, and hope that the
same threat of competition that spurred its own reluctant innovations
will mobilize the PTTs as well. Thus, Intelsat's innovation in service
offerings is a necessary but far from sufficient, strategy for ensuring
future viability.

The movement toward privatization and competition in the 1990s
has also challenged the assumptions of Intelsat's Signatory relationships. When the Intelsat agreements were drafted, countries typically
had only one international carrier, which was designated as the Signatory. Now many countries are opening international services to competition, and many of these new carriers are Intelsat users. Thus, a
former monopoly carrier may not adequately represent a nation's international communications interests.

A more fundamental issue is the structure of Intelsat itself. Major industrialized countries that have embraced privatization and competition within their borders believe that Intelsat must be privatized to survive. Others, particularly developing countries that feel they have benefited from Intelsat's cooperative framework, fear that privatization would create disincentives to serve the needs of smaller and poorer countries. Recent proposals favor setting up a privatized subsidiary, rather than dismantling the cooperative structure. Despite the weighted voting described above, it is likely that the major users will urge a consensus-building approach, rather than trying to force a restructuring decision on an unwilling majority of developing country members.

Intelsat has lowered tariffs to encourage more efficient use of its system, such as reducing prices for digital services and underutilized portions of space segment, including satellites in inclined orbit and off-peak periods. It has also undertaken an aggressive marketing campaign to convert 80 percent of the public switched traffic to long-term contracts for 10- or 15-year periods, noting: "Intelsat as a business is acutely conscious of the needs of its smaller members and users and equitable tariffs are an essential ingredient of Intelsat's long-term marketing strategy."[13] But Intelsat is now forced to confront more radical responses, including restructuring with at least partial privatization, if it is to survive the commercial and technological challenges of the next decade.

INTERSPUTNIK: THE SOVIET RESPONSE

The launch of Sputnik by the Soviet Union in 1957 was the impetus for the commercialization of space communications and the formation of Intelsat. Concern about Soviet advances in space resulted in U.S. commitment to a civilian space program, and the commercialization of satellite communications. In 1961, the president of the Soviet Academy of Sciences stated that a high priority was being given to space communication and that "the use of communications, and satellites, and of satellites for relay services would revolutionize communications and TV services."[14] However, the Soviet Union initially showed little interest in Intelsat and then expressed strong opposition to the U.S.-initiated international cooperative. Eventually, the Soviet Union, along with communist and socialist states, responded by establishing its own international satellite system, known as Intersputnik. In 1968, the U.S.S.R., Bulgaria, Cuba, Czechoslovakia, Hungary, Mongolia, Poland, and Romania submitted a draft agreement for Intersputnik to the United Nations. Intersputnik began operations in 1974.

In the 1980s, prior to the dissolution of the Soviet Union and the reunification of Germany, Intersputnik members included Afghanistan, Bulgaria, Cuba, Czechoslovakia, the German Democratic Republic, Hungary, North Korea, Laos, Poland, Romania, Vietnam, the People's Democratic Republic of Yemen, and the U.S.S.R. Other countries that used Intersputnik were Algeria, Angola, Ethiopia, Iraq, Kampuchea, Libya, Mozambique, Nicaragua, and Sri Lanka.[15]

In principle, Intersputnik was financed from a statutory fund made up of members' contributions on a pro rata basis. The size of each contribution was based on its utilization of the space segment, with profits distributed in the same proportion. However, the U.S.S.R. paid the total costs of the space segment and bore part of the costs of the earth segment for several countries (including Cuba, Laos, Mongolia, and Vietnam).[16] The only guidelines on procurement were: "Communication satellites owned by the organization shall be launched, put into orbit, and operated by members which possess appropriate facilities for this purpose on the basis of agreement between the organization and such members."[17] In practice, the Soviet Union was the only member that possessed such "appropriate facilities."

Intersputnik has been used for voice communications and television distribution, and for exchange of programming among members (e.g., the Eastern Bloc and Cuba). Intersputnik was responsible for about 40 percent of the television transmissions by Intervision, the Eastern European television exchange system managed by the Organization of International Radio and Television (OIRT) in Prague. However, following the breakup of the Soviet Union and the disintegration of the Eastern bloc, Intersputnik's use has declined, and it now has surplus capacity which is being offered to developing countries for domestic networks. Russia is also marketing surplus satellites, and intends to become a major supplier of rocket launches, in competition with the United States, the European Space Agency, and China. Its Globostar satellite system will provide capacity not only for domestic VSAT networks, but also communications providers in the Asia-Pacific region.

INMARSAT: THE INTERNATIONAL MARITIME SATELLITE ORGANIZATION

Inmarsat is an international cooperative organization similar to Intelsat, which provides satellite communications for ships and offshore industries. Beginning in 1973, the International Maritime Organization, a specialized agency of the United Nations, convened a series of conferences to consider establishment of an international maritime satellite system. Inmarsat was established in 1979, with the signing of the Inmarsat Convention and Operating Agreement, and became operational in February 1982.[18]

The structure of Inmarsat is similar to that of Intelsat. The organization is financed with contributions from its signatories, which are to be repaid with compensation set at 14 percent.[19] Its Assembly is similar to Intelsat's Assembly of Parties, and meets every two years to review long-term policy and objectives. Each Signatory has one vote. The Council, like the Intelsat Board of Governors, holds the real power in the organization. The Council consists of the 18 Signatories with the largest investment share plus four representatives of the other signatories elected by the Assembly on the principle of geographical representation (taking into account the interests of developing countries). Members of the Council have voting power equivalent to their investment share, and like a board of directors, make the major policy decisions for the organization. Administrative functions are carried out by a London-based directorate under a Director General appointed by the Council for a six-year term.[20]

Inmarsat has 79 member countries, representing more than 75 percent of the world's population, and also serves nonmembers. Inmarsat was the first provider of mobile satellite services with its Maritime Mobile Satellite Service. It now provides communications using C-band and L-band frequencies for approximately 25,000 vessels, ranging from fishing boats and yachts to supertankers, as well as offshore drilling platforms.[21] It provides voice, facsimile, telex, and data communications at up to 1.5 megabits, and group calls (broadcast calls by shore-based users to selected ships).[22] Interactive data services include videotex, navigational/weather information, and search and rescue.[23] Inmarsat satellites can also be used for land mobile communications for emergency relief (for example, in drought, famine, or earthquakes) to reestablish communications, or to provide basic service where there is no alternative. Inmarsat may also be used for the Future Global Maritime Distress and Safety System (FGMDSS), to alert people onshore for coordination of rescue activities.[24]

Signatories operate the coast earth stations, which provide links between the satellites and the onshore networks. The ship earth stations are purchased or leased by ship owners and operators. These stations typically use 85- to 125-centimeter parabolic antennas housed in a fiberglass "radome" and mounted on a stabilized platform which enables the antennas to track the satellite despite movement of the ship. The Standard C station is a much smaller terminal, with an omnidirectional antenna and 5 KHz channel spacing to be used for low-speed data (600 bits per second) and telex. This antenna is aimed at the small boat market, including fishing boats and yachts, and can be used for tracking trucks and railroad rolling stock. The Inmarsat M terminal provides voice and data services in a briefcase-sized unit for users such as reporters covering wars and natural disasters, or remote exploration crews and mountain climbers.[25] Inmarsat now also offers aeronautical communications, and is planning to introduce a family of personal mobile satellite communications services to offer communications via handheld terminals from virtually anywhere in the world using a new generation of advanced satellites.[26]

Trends in privatization and competition confronted Inmarsat with many of the same structural issues that are facing Intelsat. As a result,

Inmarsat decided to establish ICO Global Communications Limited as an affiliate in 1995 to implement its personal mobile satellite system. More than 60 commercial carriers and industrial partners now make up the alliance. Inmarsat is to develop, manage, operate, and market the system under contract from ICO Global. Ten low earth-orbiting (LEO) satellites are planned, with service to begin in the year 2000.[27] However, by the end of the decade, Inmarsat is likely to face competition from several other "Big LEO" systems that promise global connectivity for personal mobile communications, and from other geostationary MSS systems.

What's in a Name? FSS vs. BSS vs. MSS

The International Telecommunication Union (ITU) classifies communication satellites as Fixed Service Satellites (FSS), Broadcasting Service Satellites (BSS), and Mobile Service Satellites (MSS). Fixed service satellites provide point-to-point communications for voice and data, and for distribution of radio and television signals to broadcast stations and cable headends for terrestrial retransmission to end users. Broadcast satellites transmit television signals directly to households equipped with very small antennas. Mobile satellites are used to communicate with ships, airplanes, trucks, and other vehicles, and will soon provide personal communications.

Although these distinctions may have appeared logical to engineers 20 years ago, they make little sense today. The fixed/broadcasting distinction was ignored by rural residents of the United States and Canada in the early 1970s and 1980s. They found that they could pick up satellite signals intended for redistribution to cable systems by installing their own earth stations. The antennas were fairly large—12 to 15 feet in diameter—but millions on farms and in isolated villages were willing to install them. Similar antennas have sprouted throughout many parts of Latin America, Asia, and Africa to pick up television signals from Intelsat and regional "fixed service" satellites.

Today, BSS systems operate in Europe, and have recently been introduced in North America, where they are usually referred to as direct broadcast satellites (DBS). However, new, higher-powered fixed service satellites have blurred the distinction even more. AsiaSat, for example, is classified as FSS, but millions of viewers receive its signal directly rather than over cable systems. The industry has coined a new term, "direct-to-home" (DTH), to describe direct transmission to households using FSS frequencies. Even Intelsat is including high-powered transponders for DTH on its new satellites. Of course, fixed satellites can also be used for "data broadcasting"— transmission of weather, financial, or news data directly to small rooftop antennas.

The distinction between FSS and MSS may also be blurring. Cellular telephone operators in developing countries are installing village pay phones and other "fixed" telecommunications using the same technology and frequencies as in their mobile networks. In the future, we may also see MSS used for fixed services such as village pay phones as well as mobile links to handheld telephones. Some remote users have also argued that mobile frequencies should be available for fixed use so that they can take advantage of MSS systems covering their territory.

Increasingly, we are likely to see multipurpose satellites that carry transponders designed for voice, data, and video applications, whether fixed or mobile, and whether point-to-point or point-to-multipoint. This convergence poses new challenges for spectrum planners, but offers more flexibility for satellite users and operators alike.

15

International Networks and Competition

"The problem is not demand, which is substantial for all types of specialized services, but access to demand."
—DOUGLAS GOLDSCHMIDT, Former Chief Economist,
of PanAmSat [1]

INTERNATIONAL CABLE SYSTEMS

The Growth of Undersea Fiber-Optic Cables

In 1996, investment in optical fiber submarine cables worldwide exceeded $11 billion.[2] Despite the cost of laying cable on the ocean floor and the difficulty of repairing cable damaged by fishing boats or marine animals, optical fiber cables are preferred to satellite systems for point-to-point services because they have a much smaller signal delay, are much more secure, and are not susceptible to atmospheric interference. In addition, optical fiber has much greater capacity than previous generations of coaxial submarine cable. In late 1994, 65 countries were using fiber-optic cable; the number was expected to increase to 87 by 1997.[3]

Undersea cables were introduced in the 1850s to link North America with Europe. The limited capacity of coaxial cable could not handle the growth in international traffic in the 1980s. The first fiber-optic submarine cable in the series of Transatlantic Telecommunications (TAT) cables, TAT-8, began service in 1988 to link Europe and North America. Like its predecessors, TAT-8 was a joint undertaking of North American and European telecommunications administrations; it is owned by 29 carriers, of which AT&T is the largest, with a share of 35 percent.[4] TAT-8's capacity was exhausted after just 18 months.[5]

The total bandwidth of the cables is divided into Minimum Assignable Units of Ownership (MAUOs), each equivalent to 64 kilobits per second (Kbps). MAUOs are the standard of measure for investment in and usage of the cable. TAT-8 cable has a capacity of 40,000 voice conversations using digital compression known as DCME (digital circuit multiplying equipment).[6] TAT-9 has twice the capacity of TAT-8, and branches to three European locations—Spain, France, and the United Kingdom. AT&T owns 28 percent of the link, and was responsible for approximately half of its construction. TAT-10 and TAT-11 each can handle the equivalent of 80,000 telephone calls. TAT-10 began service in 1992, linking the United States with the United Kingdom and Germany. Demand was so great that TAT-11 was brought into service in 1993 to link the United States with the United Kingdom and France.

Undersea cables also cross the Pacific and Indian Oceans. The first trans-Pacific fiber cable was TPC-3, which began service in 1989 to link Japan to Hawaii. There it interconnects with HAW-4 and other cables to the U.S. mainland (there is also a midocean spur to Guam). TPC-3 is owned by a consortium of international carriers and PTTs including AT&T, British Telecom International (BTI), KDD, France Telecom, and Deutsche Telekom. The link cost approximately $700 million, and has a capacity of 560 Mbps.[7]

A second trans-Pacific fiber, TPC-4, began operation in 1993 to link Japan with Canada and the United States, bypassing Hawaii. TPC-4 is owned by a consortium of 31 common carriers from 18 countries, with AT&T responsible for construction of half of the cable. TPC-4 has twice the capacity of TPC-3.[8] Even more capacity is available on TPC-5, designed as a self-healing loop, which has 300,000 voice channel equivalents, at a construction cost of some $1.2 billion.[9]

There are several other international fiber cables in the Pacific, with fiber now linking Japan and Korea with Hong Kong, Singapore and other Southeast Asian nations, and Australia and New Zealand with each other and with Asia and North America. Regional cables also link North and South America, Europe and Africa, Europe and the Middle East, and the islands of the Caribbean (see Figure 15.1).

The world's longest optical fiber submarine cable is Southeast Asia-Middle East-Western Europe 2 (SEA-ME-WE 2), which spans some 18,000 kilometers across the Indian Ocean and the Mediterranean to southern France. In contrast to SEA-ME-WE 1, an analog system, which supported 7,500 calls at any one time, the digital SEA-ME-WE 2 can carry up to 60,000 simultaneous voice conversations.

There are also numerous regional submarine cables in Latin America. The Americas-1 system is an 8,000 kilometer cable network linking the United States, the Virgin Islands, Trinidad, Brazil, and Venezuela. The Columbus II system connects the United States, Mexico, the Virgin Islands, Spain, Italy, and the Canary Islands. The Unisur cable joins Argentina, Brazil, and Uruguay. The Pan American Cable (PAC) will link Chile, Peru, Ecuador, Colombia, Panama, Venezuela, and the Virgin Islands, with connections to Mexico.[10]

An innovative strategy for improving telecommunications in Africa was proposed by AT&T at the ITU's Telecom Africa conference in 1994. AT&T proposed a 32,000 kilometer submarine optical fiber network, dubbed Africa One, to encircle Africa.[11] Coastal countries would be able to connect to the fiber at a relatively small incremental capital cost. The continental fiber ring may be an attractive solution for adding reliable capacity to supplement the Panaftel microwave network and Intelsat services. However, the major problem in Africa is not international communications, but the lack of domestic infrastructure. With a teledensity of one telephone line per 100 population or less, most African countries require massive investment to provide access for their citizens. In addition, landlocked countries would still have to rely on terrestrial networks through neighboring countries to reach Africa One.

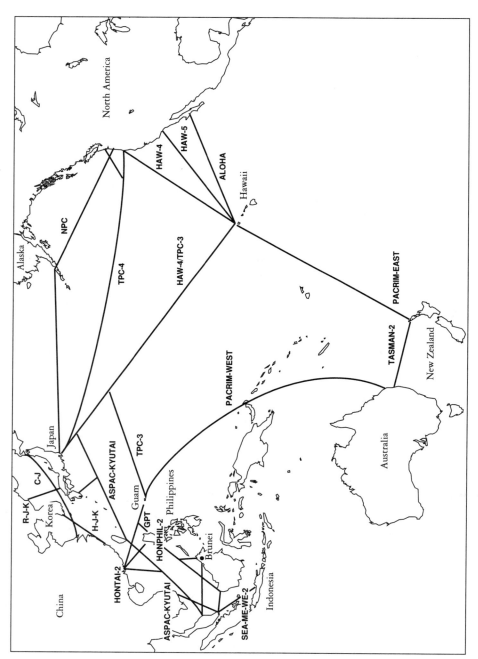

FIGURE 15.1 International Fibre Optic Submarine Cable Systems in the Pacific Region.

Private Cable Systems

Until 1989, all international undersea cables had been owned by consortia of international telecommunications carriers. The first undersea
cable owned by private business interests was PTAT, the Private
TransAtlantic Cable, which was built to exploit the potential market
for telecommunications links between the United States and the
United Kingdom. The partners included Cable & Wireless from the
United Kingdom, and a group of American investors under the name
Private TransAtlantic Telecommunications System, Inc. (PSI). In
1989, shortly before PTAT went into service, Sprint purchased the
U.S. part of the venture from PSI, putting it under a new corporate
entity, Sprint International. (NYNEX, which had an option to purchase PSI's interest, was forced to withdraw as a result of a ruling by
the Federal District Court which oversaw the divestiture of AT&T.)
    The cable runs from New Jersey to Ireland and England, with a
spur to Bermuda to serve Cable & Wireless customers there. PTAT-1
is capable of carrying 27 DS-3 circuits, or approximately 1.2 Gbps,
nearly 100,000 telephone circuits using DCME.[12] PSI was licensed by
the FCC to be a private international carrier, as opposed to a common
carrier. As a private carrier, PSI could lease bulk capacity to major
users as well as to common carriers.
    Another private cable venture in the Pacific has built the North
Pacific Cable (NPC), which links North America and Japan, with a
spur to Alaska. Partners include Pacific Telecom (which provides long
distance service in Alaska), Cable & Wireless, Pacific Telesis, and International Digital Communications (IDC), one of the new Japanese
international carriers. The cable has a capacity of 1.2 Gbps, or up to
100,000 voice circuits using DCME.[13]
    The Fiberoptic Link Around the Globe (FLAG) project is the
world's longest submarine cable, linking the United Kingdom to
Korea and Japan via Singapore, Hong Kong, and Indonesia. Parallel-
ing SEA-ME-WE 2, it will link 13 countries along its 29,000
kilometer route. FLAG's two fiber pairs will have a capacity of
120,000 digital 64 kbps circuits. FLAG will connect to the PTAT-1

transAtlantic cable, the North Pacific Cable (NPC), and Teleglobe Canada's CANTAT-3.

FLAG Limited, which is underwriting the project, is sponsored by a consortium including NYNEX Network Systems, Marubeni Corporation of Japan, the Dallah Al Baraka Group of Saudi Arabia, and U.S.-based Gulf Associates. Unlike most other cable projects that are typically funded by carriers, the project will be entirely funded by the private sector, and managed by FLAG Limited. FLAG will operate as a carrier's carrier, and has already sold capacity to more than 45 international carriers.[14]

These fiber systems not only provide capacity for large and small international users, but extend the reach of international carriers. For example, Cable & Wireless can offer private lines and switched digital services through digital links to Europe, North America, and Asia. KDD (the Japanese international carrier) can offer terrestrially based switched 56 and 64 kbps services to Europe and the United States, for applications such as ISDN, compressed video teleconferencing, and Group 4 facsimile.[15]

The proliferation of intercontinental and regional cable systems raises two questions:

- Is there sufficient demand to justify such massive investment?
- Will prices of bandwidth drop significantly with the enormous increase in capacity?

The answer to both questions at present is at best "maybe." Growing demand for point-to-point video transmissions and Internet connectivity, as well as accelerating economic growth in many of the developing countries linked by the cables, is certainly fueling demand for capacity. To date, this increase in capacity has contributed to a reduction in prices, but bandwidth across the oceans is still far from cheap. However, there may well be a glut in the trans-Pacific and trans-Indian Ocean routes, as competing fiber networks are installed.

COMPETITION IN INTERNATIONAL SATELLITE COMMUNICATIONS

The Separate Systems Decision

The 1980s brought major changes in international communications, with the introduction of fiber-optic submarine cables and the first private satellite system, as well as new institutional and regulatory structures for domestic and international communications in many countries.

In April 1983, a U.S. firm called Orion filed an application with the FCC for authorization of a separate international system, touching off extensive controversy within the federal government. By 1984, five organizations had filed applications with the FCC to construct, launch, and operate international satellite systems. A Senior Interagency Group (SIG) was established to review U.S. policies. However, government opinions on international competition were mixed. The Department of Commerce supported the administration's view that international competition was in the national interest, following the general competitive and deregulatory policies of the Reagan administration. The State Department was concerned that the United States had taken the leading role in establishing Intelsat 20 years earlier, and had been committed to a single global system. U.S. embassies in many countries had received requests from other governments (primarily developing countries) not to approve the applications, citing the founding role of the United States and the potential for significant economic harm to Intelsat. Finally, on November 28, 1984, President Reagan signed a presidential determination that alternative satellite systems were "required in the national interest" within the meaning of the Communications Satellite Act.[16]

In 1985, the FCC issued its Separate Systems decision, finding that "separate systems will provide substantial benefits to the users of international satellite communications services without causing significant economic harm to Intelsat."[17] The conditions it imposed included no interconnection with the public switched network (i.e., message telephone service could not be offered over the alternative systems), restrictions on separate systems operators from operating as common carriers, and no minimum unit of capacity.[18]

The FCC's Separate Satellite Systems order gave provisional authorization to enter the international satellite market to five companies: Cygnus, ISI, Orion, RCA, and PanAmSat. However, the new competitors faced formidable entry barriers. Each had to secure a foreign partner or correspondent. To do this, it had to convince a foreign administration, which was virtually certain to be the Intelsat signatory, to allow the new carrier to provide international services to and from its country.

All of the applicants planned to offer service between the United States and Europe; PanAmSat was the only one of the original applicants that also planned to provide service to the developing world. Entel Peru was the first entity to become a PanAmSat correspondent. Intelsat launched an aggressive lobbying campaign to discourage its members from approving PanAmSat's application for coordination and to attempt to prevent any nation from becoming a correspondent with a new entrant. Intelsat also attempted to warehouse orbital slots requested by the new entrants, and introduced a series of price cuts for domestic services, including reduced prices for domestic transponder leases, nonpreemptible leases at prices lower than preemptible leases, and sale of transponders.

While the latter strategies of introducing new services and reducing prices for existing services were predictable responses to competition, the competitors and some economists argued that Intelsat was engaging in predatory pricing. However, without a regulatory forum that could compel Intelsat to file information on costs and pricing formulas, analysis could be based only on what limited documentation was publicly available. As the U.S. signatory to Intelsat, Comsat had access to internal Intelsat documents, but refused to make them available. Comsat is now required by the FCC to provide informational access to Intelsat documents and technical proposals to other companies on an equal basis to that which it provides its subsidiaries.

In retrospect, it appears that the PTTs, and their terrestrial monopoly on communications services, as well as Intelsat, were the targets of changes in U.S. policy.[19] The telecommunications administrations were also the Intelsat signatories. Most were government-run monopolies that were slow to introduce new technologies and had no incentive to cut prices. Private international systems could bypass these administrations. In Latin America, the first customers of PanAmSat were domestic television networks frustrated by the limited capacity and high rates offered by the PTTs. Banks and other businesses with pent-up demand for private data communications were also early customers. A similar trend followed in Asia in the 1990s when AsiaSat was launched.

The New Competitors

Pan American Satellite (PanAmSat) was the first of the U.S. companies authorized to provide international satellite services separate from Intelsat to commence operations. Its first satellite, PAS-1, launched on an Ariane rocket in 1988, offered high-powered, C-band satellite services to all of Latin America, the Caribbean, and the southern United States, and Ku-band services to the United States and Europe. PAS-1 was originally to be marketed as a satellite condominium, so that customers would buy space segment that they could then use as they saw fit. However, PanAmSat found that international carriers preferred the traditional half-circuit format, and few customers had the expertise or interest in purchasing raw bandwidth, preferring instead end-to-end services.[20]

PanAmSat services include video networking for distribution to broadcasters and cable systems, video trunking for intercontinental video feeds to television networks, and private data networks for customers such as news services and banks. By avoiding telephone service and concentrating on data and video requirements, PanAmSat focused on video and data markets that had traditionally been peripheral to the PTTs' business interests, and found support from business users who were both eager for satellite services and influential with their own governments. As PanAmSat economist Douglas Goldschmidt pointed out, the problem in Latin America was "not demand, which is substantial for all types of specialized services, but *access to demand.*"[21] Regarded by many telecommunications administrations in Latin America and Europe as a threat both to Intelsat and to their own monopolies, PanAmSat used not only innovative business strategies, but economic and legal expertise, to gain a foothold in Latin American and European markets, in a style reminiscent of MCI founder Bill McGowan's savvy tenacity in challenging AT&T.

In 1994, PanAmSat launched its first Pacific satellite, with coverage of much of Asia and the South Pacific, and links to North America. Today, PanAmSat is the first global challenger to Intelsat, with global connectivity through satellites positioned over the Atlantic, Pacific, and Indian Oceans. PanAmSat will not only challenge Intelsat in global networking, but will offer a choice for domestic and regional services. Its impact on developing countries may be greatest in Africa, where it offers the only alternative to Intelsat. Future plans include a high-powered satellite for direct satellite broadcasting to Latin America.*

Orion, another of the companies that first filed with the FCC to offer international satellite services in 1985, finally launched its first satellite in 1994. Its high-powered satellite links the United States and Europe, with three spot beams and a regional beam over Europe. Orion is targeting private VSAT networks and video distribution.[22] Orion also plans to add links between the United States and Asia, with a satellite positioned over the Pacific Ocean.

Initially, the FCC did not allow so-called "international separate satellite systems" (i.e., separate from Intelsat) to interconnect to the public switched network, in order to minimize competition with Intelsat. However, in 1993, the Clinton Administration reversed this policy, notifying the FCC that interconnection of separate systems was considered consistent with U.S. telecommunications and foreign policy. Accordingly, in December 1993, the FCC allowed U.S. separate satellite system operators to provide up to 1,250 64-Kbps equiv-

---

* In late 1996, Hughes Electronics Corporation acquired PanAmSat from the heirs of the founder, Rene Anselmo.

alent circuits per satellite to interconnect with the public switched network.[23]

MOBILE SATELLITE COMMUNICATIONS

As well as providing communications between fixed points on the earth's surface, satellites may also be used for mobile communications. The first application was for ships, but new mobile satellite systems now provide communications with airplanes, trucks, and other vehicles. These systems not only transmit data back and forth between trucks or railroad cars, but also provide fleet tracking, showing the location of each vehicle in the fleet.

Inmarsat offers global mobile communications services using geostationary satellites, as described in the previous chapter. Eutelsat provides a data communications and position reporting service in Europe. Euteltracs is a joint venture of Alcatel Qualcomm and Eutelsat; it provides messaging service for companies such as long distance haulage firms that need to keep in touch in remote locations. AMSC, the American Mobile Satellite Corporation, launched a geostationary satellite in 1995 to provide a mobile service called Omnitracs in North America. An identical satellite launched by TMI of Canada provides similar services; the two systems provide mutual backup.

Satellites are also being designed to provide communications for individuals on the move. Some systems will use the geostationary (GEO) orbit, while several others plan to use nongeosynchronous low earth-orbiting (LEOs). The advantage of LEO satellites is that, being closer to earth, they require less power from the user's terminal to reach the satellite, an important consideration in reducing terminal size and price. However, LEOs' coverage of the earth is limited, so that at least a dozen satellites are required to cover the earth's surface. Further, these satellites move across the sky, rather than appearing stationary.

The most publicized of the so-called Big LEO systems is Motorola's Iridium, which plans to place 66 satellites in polar orbit to offer global personal communications. Other Big LEO systems under contruction are GlobalStar, a 48-satellite system with major financing from Space Systems Loral, and TRW's 12-satellite Odyssey. Inmarsat's spinoff, ICO Global Communications, also plans to operate a LEO system.

Meanwhile, some operators are building more modest LEOs known as "Little LEOs" for relatively low-speed data communications. The smallest is VitaSat, owned by Volunteers for Technical Assistance (VITA), which will operate two satellites to offer store-and-forward data communications. VITA pioneered store-and-forward LEO service with its first satellite, which is used by Healthnet to link African medical schools, hospitals, and field posts. Orbital Communications's Orbcomm system will also provide data communications, eventually using 26 satellites.

At the other extreme is Teledesic, a system proposed by Microsoft and McCaw, designed to provide bandwidth on demand through a system of 840 satellites. The concept is a global information skyway, so that people anywhere will be able to transmit and receive high bandwidth data and video for Internet access, videoconferencing, and file transfers without using a terrestrial network.

This investment in mobile satellites for personal communications parallels the introduction of terrestrial wireless personal communications services (PCS), also designed to provide portable communications, but through microcells connected to existing wireline networks. The satellite investors argue that there is demand for personal communications in many parts of the world where terrestrial wireless networks do not exist. Target customers appear to be international business travelers and remote enterprises such as mines and tourist lodges. Some LEO operators also point out that their systems can be used to provide basic communications in rural areas of developing countries. While this may be true, solutions to this problem already exist, such as VSATs linked to GEOs, and terrestrial wireless networks. Also, at $3 per minute, the tariff proposed by Iridium, developing country demand is likely to be very limited.

It remains to be seen whether there will be sufficient demand to ensure the viability of several Big LEO systems. While some in the industry point to ubiquitous personal communications as a "new paradigm,"[24] most seem to be betting on demand for a fairly limited amount of communication. With the exception of Teledesic, the LEOs are designed for narrowband voice and data communications, not the high bandwidth capability of "information superhighways" or even access to the Internet's World Wide Web. Also, it is not clear how inelastic demand will be for ubiquity. Demand may be limited to major corporate users, or those wanting emergency communication while traveling or working in isolated areas. Where price is a severe constraint, Vitasat's cheap store-and-forward data service may suffice.

REGIONAL SATELLITES: THE ASIAN EXPERIENCE

The largest domestic user of satellite communications is the United States, with approximately 30 satellite systems. Other industrialized countries with their own satellite systems are Russia, Canada, Australia, France, Germany, and Sweden. Developing countries with their own domestic satellites include Indonesia, India, Brazil, Mexico, Argentina, China, Korea, Thailand, and Malaysia. These satellite systems are discussed in the chapters that deal with specific regions.

Perhaps the most significant satellite trend in this decade is the growth of regional systems, through the proliferation of regional satellites and the design of domestic satellites to include extensive regional coverage. The first regional systems were the Arabsat system, serving 22 Middle Eastern and North African nations, and Eutelsat, which provides regional services in Western Europe.[25] However, the most dramatic growth has been in Asia, where the AsiaSat system covers one-quarter of the world's population. Apstar and PanAmSat also offer regional coverage, and domestic satellites from Indonesia, China, Malaysia, and Thailand have footprints covering Southeast Asia and beyond.

AsiaSat, based in Hong Kong, launched its first satellite in 1990,* with beams reaching from the Pacific Ocean to the Mediterranean, covering more than one-quarter of the world's population. Within 18 months, AsiaSat 1's capacity was fully leased, including 12 transponders used by Satellite Television Asia Region Ltd. (StarTV). AsiaSat's owners are three conglomerates: Cable & Wireless, which is also the major shareholder in Hong Kong Telecommunications; China International Trust and Investment Corporation (CITIC), China's leading state investment corporation; and Hutchison Whampoa, Hong Kong's largest multinational company.

Millions of satellite antennas sprang up across Asia and the Middle East to receive StarTV's multichannel entertainment television service. AsiaSat has had a profound effect on Asian telecommunications, as more than 53 million Asian households, or 220 million viewers in 53 countries, now watch entertainment that was not offered on their national television networks. While much of the content consists of Western television programs and movies, StarTV has also introduced Asian content in music and movie channels.

Several countries also lease capacity on AsiaSat for domestic communications. Pakistan uses AsiaSat to provide telephone services to rural areas and data communications for business and government, and to distribute Pakistan TV. Myanmar also uses AsiaSat to provide telephone service in isolated areas and to distribute television. Mongolia, with very limited infrastructure and a widely scattered population, transmits domestic radio and television via AsiaSat, and uses AsiaSat for international telecommunications circuits between Ulan Bator and Hong Kong.[26]

The AsiaSat 2 satellite will have higher-powered regional coverage, allowing the use of smaller satellite antennas, and nine Ku-band transponders optimized for China, Hong Kong, Korea, Taiwan, and Japan. StarTV plans to introduce pay television on AsiaSat 2, with

---

* AsiaSat's roots can be traced back to 1984, when two satellites launched from the U.S. space shuttle were stranded in a useless low earth orbit. Another shuttle crew rescued the spacecraft, which were then refurbished and resold. Asia Satellite Telecommunications Company Ltd. purchased one of the satellites, formerly Western Union's Westar 6, and renamed it AsiaSat 1. It was relaunched by China's Great Wall Industry Corporation.

programming in several major languages in addition to their current Mandarin, English, and Hindi services. The new services will be subscription-based, and transmitted in compressed digital format. Germany's Deutsche Welle and other international broadcasters are also expected to transmit to Asia via AsiaSat. Hong Kong Telecom also plans VSAT services to support data broadcasting, transaction processing, LAN networking, and other data networking services.[27]

However, many Asian countries will have several choices for satellite services, as additional satellites with regional coverage are launched. Apstar-1, a regional competitor to AsiaSat, was launched in 1995. Indonesia's Palapa satellites cover the ASEAN region, and its newest satellite will have broader Asian coverage. Thailand's Thaicom and Malaysia's Measat also cover the ASEAN region, as will a Philippine system. China's Chinasat series also includes much of Southeast Asia in its footprints, while India's Insat series covers the Indian subcontinent.[28] Yet another Asian entrant is Russia's Informcosmos, which plans to offer DBS services to Asia.[29]

SATELLITES AND THE GLOBALIZATION OF TELEVISION

In 1964, the transmission of television coverage of the Tokyo Olympics across the Pacific via Syncom III was considered a technological marvel: "Live television coverage of this morning's opening of the Olympic Games in Tokyo was of superlative quality, a triumph of technology that was almost breathtaking in its implications for global communications. . . . The clarity and definition of the images from Japan were . . . nothing short of extraordinary."[30] Barely eight months later, the launch of Intelsat I or Early Bird, heralded the age of television "live via satellite." Today, audiences for the Olympics and World Cup soccer exceed 1 billion people. The result of this growth in satellite television transmissions has been nothing less than the globalization of television.

Since the inception of Intelsat, Western European powers have feared that the satellite industry would be dominated by the United States. In their 1976 report to the French government, *The Informaticization of Society,* Simon Nora and Alain Minc stated:

> . . . satellites are at the heart of telematics. Eliminated from the satellite race, the European nations would lose an element of sovereignty with regard to NASA, which handles the launching, and with regard to the firms that specialize in managing them, especially IBM. By contrast, if they were capable of building them, launching them, and managing them, the same nations would be in a position of power.[31]

To develop technology and expertise, the Europeans established the European Space Agency (ESA), an experimental satellite series and the Ariane launch program. Based in French Guiana, Ariane has become a major commercial success, with most of its business now from non-European satellite systems. European telecommunications administrations established the regional Eutelsat cooperative in 1977. Modeled on Intelsat, Eutelsat's satellites provide voice, data, and video communications within Europe. Its most recent satellites, dubbed the "Hot Bird" series, are designed for direct-to-home television service.

Yet Nora and Minc failed to realize that control of content would be more critical than control of technology. For example, satellites have had a major impact on the European televison industry; the dominant factor has not been Eutelsat, but Luxembourg-based Astra. Luxembourg's tiny size and population make it among the least likely candidates for satellite use. However, its linguistic and geographic advantages in terms of its proximity to the major markets of France and Germany (as well as Belgium and the Netherlands) and the central importance of RTL (Radio-Télé-Luxembourg, the operating arm of CLT, Compagnie Luxembourgeoise de Télédiffusion) in the national economy as a program exporter and major contributor to government revenues, made Luxembourg an attractive base for a regional DBS system. After much opposition from France and other sources, a consortium called Société Européene des Satellites (SES) launched its first Astra satellite in 1989.³² By 1994, Astra had launched its fourth satellite, and was transmitting commercial television to more than 12 million households throughout Europe, including formerly communist Central and Eastern European nations.

The most successful of the European satellite programmers is Rupert Murdoch's British Sky Broadcasting (BSkyB). Sky now reaches 5.2 million subscribers in the United Kingdom and continental Europe, more than any other pay television satellite service in the world. Sky offers both distribution to cable systems and direct-to-home services, including British and international sports, news, films, and primarily American entertainment channels. Since going public in 1994, Sky's market capitalization has appreciated to more than $10 billion, making it the most valuable publicly traded television company in Britain.³³ Sky now offers 28 channels, and plans to increase its offering to 100 channels or more when it goes digital. Many of these channels will be time shifted, so that viewers will have two options as to when to view the programs.

While the number of households wired for cable in Western Europe has increased dramatically in the past decade, there has also been tremendous growth in satellite reception in Eastern Europe. The number of homes in Eastern Europe receiving satellite television increased 90 percent in 1993 to more than 5 million. Poland is now Europe's third largest satellite antenna market.³⁴

Early on, the French dubbed Astra the "Coca-Cola satellite," fearing that Luxembourg would become a beachhead for penetration of the continent by U.S. commercial programs purveying American culture.³⁵ Their fears appear to have been well-founded, as U.S. programming dominates the packages fed to cable headends and pizza-pan-sized antennas. Perhaps equally significant, Astra sounded a wake-up call to television broadcasters in Europe much as AsiaSat has in Asia. France introduced new private channels, and many other broadcasters revamped their programming to try to hold on to their viewing audience. Imitating popular imported programs, broadcasters began to stage more game shows and soap operas (modeled not only on U.S. sitcoms, but on Latin American "telenovelas"). The result has perhaps been the "dumbing down" of European television. Yet some European production houses have succeeded, like the BBC, in developing new markets for their programming, not only throughout Europe but in other regions.

Worldwide, there are now 21.7 television sets per 100 inhabitants. While density is highest in industrialized countries, growth rates are highest in developing countries, particularly in Asia, whose 13 percent compound annual growth rate is twice the global average.[36] In the past decade in Europe, and more recently in the developing world, viewers of these television sets have seen news, sports, and entertainment from around the world via satellite.

AsiaSat's StarTV offers a package of six channels including a news and information channel featuring the BBC, and sports, music, entertainment, a Chinese channel, and an associate Hindi channel known as Zee TV. The channels are advertising-supported, and distributed through direct reception, retransmission through satellite master antenna television (SMATV), cable, terrestrial rebroadcasters, and multipoint microwave distribution systems (MMDS). Star also plans to introduce subscription channels, using digital video compression. It is interesting to note that the countries with the highest penetration rates are at the extremes of the immense AsiaSat beam: in both Israel and Taiwan, StarTV reaches more than 41 percent of television households. The next highest viewership is in Hong Kong, followed by the United Arab Emirates and India.

When StarTV began broadcasting in 1991, it was not taken very seriously by national public broadcasters because it required purchase of a satellite antenna and relied upon advertising. But Star's mix of sports, music, and entertainment, free to anyone who could find money for a satellite antenna, proved enormously popular. In India and some other countries, a cottage cable industry sprang up to sell Star programming to those who could not afford their own satellite equipment. Star now estimates that it reaches 53 million households.

Technology has upstaged regulation: in many countries, the government did not enact legislation to ban TVROs before thousands were installed; such legislation would now be considered a classic example of closing the barn door after the horse has bolted. Many countries in South and Southeast Asia have turned a blind eye to the Star invasion. Singapore has banned satellite receivers, effectively keeping Star out. In Malaysia, the antennas are officially banned, but the ban is not enforced. In Hong Kong and the Philippines, reception is allowed, but redistribution is regulated.

China has seriously considered banning satellite antennas because of the potential corrupting influence of StarTV. The official policy is that no access to "foreign television signals" is allowed unless permission is granted. These policies are particularly paradoxical because China is a major shareholder of AsiaSat, programs produced in China are regularly transmitted over its Mandarin language channel, and the sale of satellite antennas has become a lucrative business for the Army General Staff Department and the Ministry of Radio, Film, and Television.[37]

In India, Star is associated with a Hindi language channel called Zee TV. Another private channel operating out of India is called Asia Television, and uses the Russian Stationar satellites. It is carried in Europe on cable systems aimed at Indian expatriates and people of Indian descent. India has some 30,000 small cable systems distributing StarTV, Asia TV, CNN, and anything else they can pick up, as well as movies from VCRs plugged into the headends. They are unorganized and uncontrolled, but not illegal, because under the India Telegraph Act of 1885, it is not illegal to send signals by wire as long as wires do not cross public property such as roads or parks.[38]

Like the French, many developing country policy makers resent the advent of "Coca-Cola satellites" that bring Western commercial television, yet the impact may be as much in style as in substance. Dubbed and subtitled American series and films are popular and cheap to run. They augur a fundamental change that goes beyond the introduction of foreign content—introducing viewers to flashy graphics, fast pacing, and soap operas. These formulas are being emulated by broadcasters in Asia as well as Europe, with Asian versions of MTV, Malaysian game shows, and Thai soap operas.

The argument against imported signals is usually made in both cultural and economic terms. Imported signals may undermine indigenous broadcasting by siphoning off viewers. Foreign programs may also expose viewers to different cultural values and languages they may find more appealing than their own. While there are certainly examples of these phenomena, broadcasters might better take the arrival of foreign programming as a wake-up call to reexamine their own fare. In general, people prefer programs in their own language and about their own country and culture. In six Asian nations (Hong Kong, Indonesia, the Philippines, Singapore, South Korea, and Thailand), more than 90 percent of the top 20 programs were locally produced in 1992.[39]

The very limited choices available from their own broadcasters may make foreign programming more attractive. For example, Asians have an average of only 2.5 channels available, compared to 25 available to their European and North American counterparts.[40] Another indication of the unmet demand for video programming is the growth of VCRs in developing regions. In Asia, the number of VCRs increased more than tenfold from 1981 to 1991—from 4.9 to 51.4 million.[41]

Satellites also offer an opportunity for national and regional coverage, a greater choice of programs, program exchanges, and satellite news gathering. However, taking advantage of these opportunities will require entrepreneurial and innovative strategies that require a change in management style and incentives. For example, to win back viewers, India's Doordarshan (national television) is launching three new channels for sports, entertainment, and general and business news.

The result is likely to be a surge in the amount of transnational television, as well as growth in local cable systems and over-the-air controlled access systems such as MMDS. Some governments are launching pay or cable channels of their own, carrying a mix of popular foreign programming. This approach is designed to satisfy domestic demand while exerting some control over what viewers watch. Both Singapore and China are installing cable systems so that their viewers have access to more channels, but the channels must be approved by the government. Other countries such as Korea, Indonesia, and Sri Lanka are also licensing commercial broadcasters for the first time. We may also see the growth of productions designed for regional audiences through program exchanges via satellite such as Asiavision.

CHAPTER *16*

The International Policy Environment: The Players

"Standardization is no longer regarded as an exclusively technical task best left in the hands of technocrats—it is recognized more and more that standards are decisive tools in international marketing and competition for securing future markets."
—THEODORE IRMER, Director, ITU Standardization Bureau[1]

THE CHANGING INTERNATIONAL POLICY ENVIRONMENT

This chapter examines the major international organizations concerned with telecommunications policies and the major forces that are shaping their agendas. Several challenges are confronting international bodies concerned with telecommunications:

- **Technological change:** The rapidity of change and innovation in technology requires much more rapid resolution of issues such as standards setting and allocation of frequencies. Also, the convergence of technologies is blurring traditional regulatory distinctions such as telecommunications vs. information processing, fixed vs. broadcasting services, and fixed vs. mobile services.

- **Globalization:** It is increasingly difficult to draw a line between governments' national and international interests and policies.
- **Competition:** Whereas most national interests were once represented by a government-operated telecommunications administration, many countries now have private and/or competitive operators whose various interests need to be reflected in international negotiations.
- **Regionalization:** The rising importance of regional organizations is challenging the role of global organizations.
- **Services:** The proliferation and growing importance of telecommunications services have made them a highly visible component of international trade negotiations.

SETTING STANDARDS

The Importance of Standards

Changes in technologies and services have come about so rapidly that "in less than a decade, the proliferation of new and competing networks has turned the world of standards upside down."[2] Therefore, emerging global markets will need global standards. Similarly, the introduction of new wireless technologies from direct broadcasting and low earth-orbiting satellites to personal communications services and global positioning have created new demands for spectrum and challenges for frequency management.

Standards are set in many forums. The most important at the global level are the International Telecommunication Union (ITU) and the International Standards Organization (ISO). The Geneva-based ITU, established in 1865 and now a specialized agency of the United Nations, sets telecommunications standards, allocates radio frequencies, and provides technical assistance to its developing country members. The ISO, based in Geneva, is involved with telecommunications standardization, among many other issues. Regional telecommunications organizations and industry associations also play roles in managing frequencies and setting standards.

The number of people who rely on international telecommunications standards has grown dramatically as a result of technological innovation, competition, and globalization. International standardization is becoming increasingly important because only global solutions can satisfy the needs of geographically dispersed and vertically integrated industries. New mechanisms are needed to facilitate global collaboration on standardization questions at early stages of technological innovation.

Network providers rely on international standards to ensure interoperability and interconnectivity of systems around the world. Major equipment companies and network operators devote considerable time to standards making because of their commercial interests. Equipment suppliers face complexity, high production costs, and the need to amortize across large production runs, with minimum engineering to satisfy individual customer or country requirements. In general, ". . . standards will be beneficial for manufacturers because they can increase market predictability, reduce risk and effort in development, assist in training the technical and marketing staff, and broaden the market volume."[3]

Standards can influence the viability of an industry, for example, by enabling large and small players to build their products and services around known protocols, and by providing a degree of certainty about interoperability and availability of compatible equipment. Thus, standards can affect the viability of a new product or industry; for example, facsimile did not take off before standardization of Group 2/Group 3 in the ITU's CCITT (Consultative Committee on Telegraphy and Telephony—see below). Modems also employ ITU-developed protocols (such as V32 and V34) that facilitate global connectivity.

De Facto vs. De Jure Standards

De facto standards are typically developed by a single manufacturer or small group of suppliers, or through informal consensus among vendors or users that one solution or protocol is superior. The setting of open standards occurs where a wide range of interests, suppliers, network operators, and users is represented. In contrast, standards adopted by the ITU, the ISO, and various regional bodies could be called de jure standards. However, although the agreements have treaty status, they are actually voluntarily entered into and not legally enforceable.

Xerox's Ethernet, IBM's SNA, and the Hayes modem instruction set are a few examples of de facto standards. Microsoft's DOS followed by Windows became the de facto operating system standard for most of the world's personal computers. In North America, at least before the divestiture of AT&T, public network standards were essentially set by AT&T and its research arm, Bell Labs. Bell Labs and Bellcore (Bell Communications Research, the research arm of the Bell Operating Companies) are now major de facto standards setters for the domestic telecommunications industry. However, with the growth of global markets, U.S. suppliers must pay increasing attention to global standards, many of which are set by the ITU and the ISO.

Sometimes it appears that life would be simpler if there were some global standard rather than a proliferation of mutually exclusive systems. For example, there is no global television broadcasting standard; countries have adopted NTSC, which originated in the United States, several versions of PAL, which originated in Germany; or SECAM, a French transmission system. Global broadcasters must be able to convert from one system to another; the consumer electronics industry has also adapted by producing multisystem television sets and VCRs.

To avoid this scenario in the future, the ITU proposed setting standards for high-definition television (HDTV). However, after years of discussion, no consensus was reached. The Europeans and Japanese proposed different versions of analog systems, while the Americans proposed several digital solutions. Meanwhile, the Federal Communications Commission (FCC) pressured U.S. industry to adopt a single digital HDTV standard. If successful, this approach could not only spur the growth of digital television in the United States, but also create a de facto HDTV standard for the rest of the world.

The growth of digital mobile telephony would also seem to depend on global standards. Europe's GSM (Global System for Mobile Communications), using TDMA (time division multiple access), is being widely adopted in many countries as a de facto standard; however, TDMA is being challenged in the United States and in some Asian countries by a technique called CDMA (code division multiple access), which offers significantly more efficient spectrum use, increasing the call-carrying capacity of cellular networks. The lack of a single standard may make it difficult to implement "global roaming," or the ability of cellular phone customers to use their phones on any network. However, some vendors are now offering dual-mode handsets to overcome this problem.

Steps in Standards Setting

International and regional bodies typically follow four steps in standards making:

- establishing a framework or reference model before embarking on detailed specifications;
- agreeing on base standards for various parts of the model;
- specifying requirements for products to meet the standard;
- standardizing testing of the product for conformity to the standard.

While the standards concept appears to imply rigidity and mandatory compliance, the challenges outlined at the beginning of this chapter are forcing a rethinking of the role of standards and the standards-setting process. For example, in its report entitled "CCITT Interactions with Other Standards Organizations," the Strategic Planning Group of the U.S. National Committee of the CCITT proposed six principles for international standards development, stating that standards should be:

- global: to reflect the trend toward global markets;
- voluntary;
- flexible: mandatory standards can be used as nontariff barriers, to discourage R&D or vendors from designing special solutions for

particular customer needs. Mandatory standards create disincentives for efficient and creative product development and higher than necessary costs for manufacturers and therefore for consumers;

- open: anyone with direct and material interest should be allowed to participate; working methods should be transparent;
- efficient: there should be coordination, to avoid duplication, and save time and money;
- supportive of technical evolution: to meet market demand and be protective of intellectual property rights.[4]

THE INTERNATIONAL STANDARDS ORGANIZATION (ISO)

Traditionally, the ITU has been the forum for developing telecommunications standards, while the ISO has dealt with information systems. However, as the technologies converge, the work has begun to overlap. An example is the seven-layer Open Systems Interconnection (OSI) model, which is the foundation for many data communications protocols.

The ISO, established in 1947, is a worldwide federation of national standards bodies from more than 100 countries representing 95 percent of the world's industrial production. Some 80 percent of these bodies are full members with voting rights, while the remainder are corresponding members. The objective of the ISO is to promote standardization and related activities with a view to facilitating international exchange of goods and services, and to developing cooperation in intellectual, scientific, technological, and economic activity.[5] Its scope covers all fields except electrical and electronic engineering, which is the responsibility of the ISO's partner organization, the International Electrotechnical Commission (IEC). The IEC acts as a division of the ISO and carries out some work relevant to telecommunications standards.

The ISO develops and publishes international standards, its work is carried out by some 2,500 technical committees, subcommittees, and working groups in which more than 20,000 experts participate annually. The first step is to define the technical scope of the standard. In the second phase, countries negotiate detailed specifications. The third phase of formal approval requires approval of two-thirds of ISO members that have actively participated in the standards development process, and approval by 75 percent of all ISO members that vote.

Because committee work is administered by national standards bodies and consensus-oriented, the ISO's activities are decentralized as procedures require the broadest agreement possible before standards can be published. Compliance is voluntary. Incentives for adoption of standards are usually market driven. National regulators may also enforce the standards.

The OSI reference model was published in 1984. The OSI model established a seven-layer hierarchy of protocols and services; standards for each layer were worked out in a joint ISO/IEC Committee on Information Technology working closely with the CCITT. The speed of development resulted from the commitment of the industry plus flexibility in procedures. The OSI model required the establishment of a new discipline: functional standardization with selection of options from base standards for specific applications.[6]

OSI challenged the traditional assumptions that standardization lags behind market realization and that existing mechanisms are inflexible, slow, and concentrate too much on trying to harmonize minor differences in national documents. The requirement for vendor-independent computer-to-computer communication and interoperability placed market demand ahead of technical development.

THE INTERNATIONAL TELECOMMUNICATION UNION (ITU)

The ITU originated in 1865 as the International Telegraph Union; it became a specialized agency of the United Nations in 1947. Currently, ITU membership includes 185 nations grouped into five regions. The nations are typically represented by their telecommunications administrations. Members are eligible to participate and vote in the conferences and committees of the ITU. Private operating agencies, international organizations, and scientific and industrial groups also contribute to its work as nonvoting members. The ITU coordinates the establishment and operation of telecommunications networks through management of the radio frequency spectrum, standards setting, harmonization of national policies, and technical assistance to developing countries.

The ITU structure comprises:

- a Plenipotentiary Conference, which is the supreme authority of the Union
- the Administrative Council, which acts on behalf of the conference
- world conferences on international telecommunications
- a radio communication sector
- a telecommunication standardization sector
- a development sector
- a general secretariat (see Figure 16.1)

The Plenipotentiary Conference consists of all members of the ITU, and meets every four years. The Administrative Council consists of 43 ITU members elected by the plenipotentiary. Its role is to consider broad telecommunications policy issues in the interval between conferences, to ensure the efficient coordination of the work of the ITU, and to exercise financial control over ITU operations.

World international telecommunication conferences are held at the decision of the plenipotentiary, and are empowered to revise telecommunications regulations. Radio communication conferences are held every two years to review and/or revise the radio regulations. The radio communication assembly provides the technical basis for the work of these conferences. Telecommunication standardization conferences are held every four years to approve, reject, or revise draft recommendations proposed by study groups. Development conferences

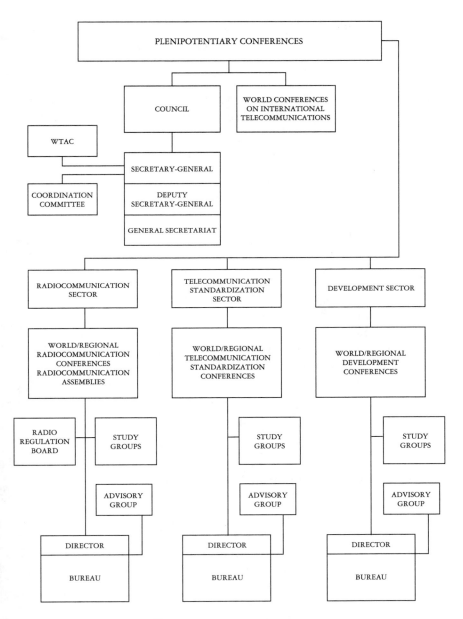

FIGURE 16.1 Structure of the ITU.

are held at regular intervals to promote cooperation in planning and implementation of services and facilities in developing countries.

Much of the work of the ITU is done in study groups composed of government officials and representatives of the telecommunications industry who review and analyze issues and make recommendations to be considered at the conferences. The bureaus coordinate the work of the study groups, communicate the results of the preparatory work to the members, organize the conferences, and provide ongoing technical expertise.

The Original Structure

The CCITT

Originally, two international consultative committees were established to bring technical expertise into intergovernmental meetings: the CCIT for telegraphy and CCIF for telephony (Fernsprechen in German). In 1956, the CCIT and CCIF were merged into the CCITT. The CCITT's mandate evolved to include standardization of telecommunications in the widest sense, from terminals and equipment to transmission, switching, and signalling to ISDN, as well as tariff methods and accounting principles. Its work included everything except radio communications, which was handled by the International Consultative Committee for Radiocommunication (CCIR).

The consultative committees' work is done by study groups whose members consist of administrations; recognized private operating agencies (RPOAs) such as AT&T, MCI, BT; scientific and industrial organizations (such as manufacturing companies); and other international organizations. In 1991, the CCITT's membership included 165 administrations, 65 RPOAs, 164 scientific/industrial organizations, and 37 international organizations.[7] There are also "invisible participants," such as governments and supranational organizations such as the European Union (EU). The CCITT included competitors, but developed consensus to benefit all: a compromise standard was generally considered better than no standard at all. However, the process was often very slow and cumbersome. There were also problems of how to translate political will and objectives into technical work.

Each study group develops standards which then have to be approved by the administrations. The standards are called "Recommendations," implying that they constitute voluntary nonbinding agreements. However, they are widely adopted because they ensure global interconnectivity and interoperability. At one time, the wait for approval by the Plenary Assemblies was up to four years; now it is done by correspondence. The last major ITU global standards conference took place in Melbourne in 1988. The WATTC (World Administrative Telegraphy and Telephony Conference) took into consideration trade in telecommunications services and reforms in ITU procedures discussed below.

Radio Regulations: The CCIR and IFRB

The 1947 Atlantic City Plenipotentiary Conference laid the foundation for an international system of regulations that is still in force today. It created a compulsory Table of Frequency Regulations governing the use of frequencies by bands, by services, by region (group of countries), or by country. It also created a Master International Frequency Register containing frequency assignments notified by administrations. The conference also drafted a set of administrative and regulatory procedures governing the use of frequencies and enabling members to use assignments and secure for them rights of recognition and protection.

The Atlantic City Conference also established the International Frequency Registration Board (IFRB), consisting of independent members acting as custodians of an international public trust for worldwide management and administration of the spectrum. Before 1947, the use of frequencies was communicated to the ITU, but there was no examination of usage to ensure freedom from harmful interference. The IFRB originally consisted of 11 members; it was reduced to five in 1965.[8] By the early 1990s, the five elected members from different regions, plus a staff of 120, processed 1,200 notices per week.

The work of the CCIR has been devoted to the compatible and efficient use of radio spectrum and the geostationary orbit (GSO), and to recommendations on radio system characteristics. When approved by the members at World or Regional Radio Conferences (WRCs or RRCs), the recommendations become part of the Radio Regulations. The objective of the Radio Regulations is: "an interference-free operation of the maximum number of radio stations in those parts of the radio frequency spectrum where harmful interference may occur."[9] As supplements to the International Telecommunication Constitution/ Convention, which is an international treaty entered into by sovereign states, the Radio Regulations have the force of an international treaty. They are complex and voluminous, amounting to over 1,500 pages. Revision of the radio regulations can only be done by a Radio Conference.

National assignments of frequencies are carried out in the framework of mutually accepted international regulations based on:

- frequency band allocations to services;
- protection of assignments recorded in the Master International Frequency Register or previous planning of portions of the spectrum;
- bilateral or multilateral agreements and coordination.

Recent Changes in the ITU

The Need for Reform

The rapid evolution of technologies and services coupled with trends toward globalization and restructuring of the telecommunications sector in many countries has put many new pressures on the ITU. Rapid technological change requires rapid standardization, but growing complexity retards the process: this could be called a "push-pull" effect. There are more standardization topics, and more time and effort is required to complete them. For example, the number of recommendations grew arithmetically, but the number of pages of CCITT books increased geometrically from less than 400 in 1968–1972 to 18,000 in 1985–1988.[10]

There are also more players; not only are the PTTs represented, but service providers, manufacturers, and other carriers and vendors. A national delegation of five people a decade ago might now have 60 participants. These factors posed a challenge for reforming, revamping, and reorganizing the CCITT. Meetings that had 50 to 80 delegates in 1980 had 500 to 600 a decade later.[11]

Reforms were begun at the 1988 WATTC in Melbourne, including changes in procedures that reduced approval time from several years to a few months, and cooperation with regional standards organizations (RSOs). The need for comprehensive reform resulted in the establishment by the Nice Plenipotentiary in 1989, of a High Level Committee (HLC) charged with recommending structural reforms in the ITU. The HLC chairman noted: "The ITU has an enviable track record. Unless, however, it puts itself quickly into fighting trim to meet the challenges of the rapidly changing telecommunications environment, its vital role in coordinating international telecommunications could become increasingly marginal."[12]

The New ITU Structure

The High Level Committee concluded that the ITU should not seek to broaden or change its overall mandate, but should play a stronger and more catalytic role in stimulating and coordinating cooperation among the increasing number of bodies concerned with telecommunications. It should also seek stronger relationships and roles for the industry.[13] Thus, the HLC recommended restructuring the ITU: a new organization would combine standardization functions of the CCIR and CCITT. A second would manage radio-related issues, including frequency management currently under the IFRB. A third would focus on telecommunications in developing countries.

An Additional Plenipotentiary was held in December 1992 to consider those HLC recommendations requiring amendment of the constitution and convention. As a result of the recommendations adopted, the ITU was reorganized into three vertical sectors: Standardization, Radiocommunication, and Development:

- **Telecommunication Standardization Sector:** This unit includes the work of the CCITT and some of the former work of the CCIR on interconnection of radio systems in public telecommunications networks.
- **Radiocommunication Sector:** The current IFRB and most activities of CCIR are merged into a Radiocommunication Sector. New radiocommunications activities will be directed by World Radiocommunication Conferences every two years that will combine responsibilities of the current WARCs and CCIR Plenary Assemblies. The permanent five-member IFRB board is replaced with a part-time, nine-member Regulations Board.
- **Development Sector:** Issues specifically related to developing countries are handled by the Bureau for Telecommunications Development (BDT) established in 1989.

Each sector is headed by an elected director, who also chairs an advisory body. The director is responsible not only for member administrations, but for industries and organizations participating in the work of the sector.

This restructuring is designed to make the ITU more responsive to the changing global telecommunications environment. For example, new, streamlined procedures will allow the committee to adopt technical recommendations before plenaries because, in order to make the standards process more flexible and market driven, the approval process is independent of the plenary cycle.

Spectrum allocation for promising new technologies also needs to be an ongoing activity. Frequency notices requiring possible new or different interpretation are handled by the part-time Radio Regulation Board, whereas all other cases are handled by the director. Radio conferences are held every two years; if an issue is not ripe for decision, it can be held over to the next conference instead of requiring last-minute decisions to avoid four- to five-year delays.

In order to give the private sector greater opportunity to join in the ITU priority setting and advisory process, special advisory groups may now be established, made up from representatives of the private sector, including telecommunications manufacturers, service providers, and users. New approaches to manage standards setting include initiatives spanning several study groups and electronic document transmission among members. Another recent ITU initiative was the creation of a Voluntary Group of Experts (VGE) to deal with simplification of the radio regulations, a need recognized by the 1991 Nice Plenipotentiary.

A coordinating committee chaired by the secretary general is intended to facilitate better horizontal integration of ITU policies and activities. The secretary general has set up a new strategic policy and planning unit, which is to prioritize standards development and technical assistance for developing countries, and technology transfer to develop national standards. The secretary general has also established a business advisory forum to consult with business leaders.

New Terminology

The terms CCITT and CCIR have been replaced by the terms ITU-Telecommunication Standardization Sector (ITU-T) and ITU-Radiocommunication Sector (ITU-R). The study groups are now called ITU-T and ITU-R study groups, and are designated by Arabic numbers instead of Roman numerals. Advisory groups known as the Telecommunication Sector Advisory Group (TSAG) and the Radiocommunication Sector Advisory Group (RSAG) oversee and provide guidance to the study groups. Intersector Coordination Groups coordinate work going on in both the ITU-T and the ITU-R, where both groups share equally in responsibility.[14]

The former World Administrative Radio Conferences (WARCs), which are to be held every two years, are now known as World Radiocommunication Conferences (WRCs). The CCITT Plenary has been replaced by the World Telecommunication Standardization Conference (WTSC).

The ITU's Role in Telecommunications Development

While in the early days of the ITU the membership was comprised primarily of industrialized countries, today the majority of the members are developing countries. These nations face problems not only of underinvestment in telecommunications, but also lack of management and technical skills for planning and operation of their networks (see chapters 8 and 9). The 1989 Plenipotentiary significantly elevated the status of and funding for ITU support for developing countries by establishing the Telecommunications Development Bureau (known as the BDT, for its French name). The BDT took over the functions of the former Technical Cooperation Department and the Centre for Telecommunications Development, a small unit that was established as a follow-up to the Maitland Commission, which advocated much greater investment in telecommunications in developing nations as well as training and technical assistance in its 1984 report *The Missing Link.*[15]

The BDT was set up to reflect the dual role of the ITU as a UN specialized agency and as the executing agency for projects financed by the United Nations Development Program (UNDP). A major role of the BDT is to provide guidance and advice to member states on telecommunications development policies and strategies through such activities as training, feasibility studies, world and regional development conferences, and dissemination of information. The BDT also executes telecommunications projects for the UNDP, which responds to requests from developing countries, such as planning of regional and national networks, preparation of master plans, and human resource development projects. The total volume of expenditure is approximately $30 million for some 200 projects per year.[16] However, it should be noted that telecommunications projects are only a small part of UNDP activity. For example, in 1990, UNDP programmable resources for all technical cooperation activities in all sectors totalled approximately $800 million. Implementation by the ITU of UNDP-funded or shared-cost projects amounted to $27.5 million, or about 3 percent of the total.[17]

The ITU is being called upon to play an expanded role in advising developing countries on telecommunications issues: "Recognizing that multilateral institutions provide only about 5 percent of the resources for telecommunications development, the ITU must play a more clearly-defined catalytic role by working co-operatively with international, regional and bilateral development and financial agencies, and by presenting to developing countries the range of policy and structural options that will generate greater resources for telecommunications development."[18] At the 1994 World Telecommunications Development Conference in Buenos Aires, developing countries requested ITU assistance not only in technical planning and evaluating technologies, but in management education, options for restructuring and financing for the telecommunications sector, and applications for socioeconomic development.

One aspect of the ITU's development activity will focus on the least developed countries (LDCs). In 1994, there were 59 LDCs throughout the world, including 33 in Africa south of the Sahara, 17 in Asia and the Pacific, three in the Arab States Region, and four in the Americas and two in Europe.[19] The ITU is planning to provide special assistance to the LDCs in reform and restructuring of the sector, improvement of maintenance, network development, and human resources development.[20]

REGIONAL TELECOMMUNICATIONS ORGANIZATIONS

In addition to the ITU, there are also numerous regional telecommunications organizations, whose membership is also typically national telecommunications administrations. These organizations may facilitate frequency coordination among their members, and may develop their own policies and standards.

Europe

European standards organizations are addressed in chapter 7 on the European Union. To summarize briefly, the key organizations are:

- **CEPT:** the Conference of European Posts and Telecommunications Administrations was founded in 1959 to provide a forum for Western Europe in the fields of posts and telecommunications. Its 26 members have grown to more than 30 with the addition of Central and Eastern European countries in 1990.[21] Its working group structure parallels that of the ITU.
- **ETSI:** the European Telecommunications Standards Institute was established in 1988 as an outcome of the 1987 European Commission Green Paper. There are now more than 260 members from the European Union and Central and Eastern Europe. The membership is composed of manufacturers, public network operators, administrations, users, research bodies, public service providers, and others.[22] ETSI works through Technical Committees, Subtechnical Committees, and Project Teams for rapid creation of standards which are highly product-specific.

The formation of ETSI and EU policies may challenge the ITU on two fronts. ETSI has proved to be a much more open and responsive mechanism for standards setting than the ITU. Its success contributed to the drive to restructure and streamline ITU activities. Also, the European Commission has stated that Community law will take precedence over the standards and recommendations of the ITU. Commission directives, in line with the Treaty of Rome, take precedence over international agreements.

Other Regions:

Other regional telecommunications organizations include:

Africa

- the Pan African Telecommunication Union (PATU);
- RASCOM: Regional African Satellite Communications, an organization planning to use satellites for communications within Africa.

The Americas

- CITEL (Inter-American Telecommunications Commission), a branch of the Organization of American States (OAS), which represents all countries in the Americas except Cuba.

Arab Region

- the Telecommunications Permanent Committee of the Arab League: composed of ministers of communications of the Arab-speaking world.

Asia-Pacific

- the Asia-Pacific Telecommunity (APT): based in Bangkok and composed of 28 member states, four associate members that are not autonomous states, and 32 affiliate members that are telecommunications service providers.

Commonwealth of Independent States (CIS)

- Regional Commonwealth for Communications (RCC): composed of the PTT ministries of the CIS; based in Moscow.

User Organizations

Telecommunications policy making has been dominated by telecommunications administrations and their national governments, within their countries and through their representatives in the ITU. However, these policies impact users of telecommunications services in terms of the pricing of services and equipment, standards users must adopt, and availability of bandwidth.

In order to influence policy making and standards setting, users have formed organizations or "user groups," typically composed of large corporate users of telecommunications represented by their telecommunications and information systems professionals. These user groups present the needs of their members to their national regulators and the major carriers and equipment suppliers. Although users do not directly participate in the ITU, they can advise their governments on the impact of ITU proposals and attempt to influence the positions taken by their government representatives at the ITU.

The major international and regional user organizations include:

- International Telecommunications Users Group (INTUG), based in the United Kingdom; active in policy forums of the ITU and ETSI;
- International Chamber of Commerce (ICC), representing large business users;
- International Communications Association (ICA), based in the United States, and with membership primarily from large U.S. users.

There are also national user organizations such as:

- the Canadian Business Telecommunications Association (CBTA)
- the Australian Telecommunications Users Group (ATUG)
- the French Association of Telecommunications Users (AFUT)

Equipment manufacturers and major users have also formed national and regional standards organizations, which have been very influential in developing what have become de facto standards for many new services, as noted earlier in this chapter. Among these major organizations are:

- American National Standards Institute (ANSI): is the coordinating committee for the United States' national system of standards and is comprised of about 900 companies. ANSI does not set standards, but coordinates development of voluntary national standards. It is the U.S. voting participant in the ISO and ITU-T;
- Institute of Electrical and Electronic Engineers (IEEE): a professional society in the United States that develops local area network standards. Its British counterpart is the Institution of Electrical Engineers (IEE);
- Corporation for Open Systems (COS): a nonprofit organization of computer and communications equipment vendors and users, established to accelerate introduction of products based on international standards, principally the ISO-OSI model;
- the European Computer Manufacturers Association (ECMA): represents equipment suppliers' interests in development of standards;
- the Telecommunications Technology Committee of Japan (TTC): established in 1985 in response to changes in Japanese legislation, including the Business Communications Law.

CHAPTER *17*

The International Policy Environment: Current Issues and Challenges

"Global Telecom Talks: A Trillion Dollar Deal."
—Institute for International Economics[1]

TRADE IN TELECOMMUNICATIONS SERVICES

Two Models for International Telecommunications

Today, telecommunications policy issues are the focus of international trade negotiations as well as deliberations by technical bodies. Telecommunications trade attracts large investment, and is the basis for an impressive percentage of the world economy. For example, global telecommunications revenues exceeded $513 billion in 1994, or more than 2 percent of the global GDP. International traffic exceeded 53 billion minutes in 1994, compared to 12 billion minutes a decade earlier, representing a compound annual growth rate of 15.2 percent.[2] Unmet demand for basic services in the developing world plus the proliferation of new and enhanced services worldwide is ex-

417

pected to generate investment of more than $567 billion between the years 1996 and 2000.[3]

The recent focus on trade in telecommunications results from evolving notions concerning international trade in services as contrasted with trade in goods. Until the mid-1980s, telecommunications and trade were considered to be relatively separate activities at the national and international levels. At the international level, two regimes coexisted: the trade regime, represented by the General Agreement on Tariffs and Trade (GATT), and the telecommunications regime, which was considered to be primarily technical, and largely the responsibility of the International Telecommunication Union (ITU). With the recognition of telecommunications as an important service, as well as a facilitator of other services, the issues of trade in telecommunications services were introduced in the General Agreement on Trade in Services (GATS) negotiations.[4]

The Role of the ITU

The ITU's 1988 WATTC introduced for the first time the concept of addressing trade concepts through bilateral agreements. As is typical of ITU conferences, the final outcome was a compromise. The United States wanted an overarching free trade regime in telecommunications; others wanted varying degrees of regulation and control. The final agreement was that there should be no international regulations enforceable on individual countries' internal communications regimes and that international standards would be recommended and should be applied whenever possible, but that individual countries could form "special arrangements" in interconnecting their networks.[5] For example, since 1988, there have been international value-added network (IVAN) agreements between the United States and Japan, the United States and the United Kingdom, Japan and the United Kingdom, the United States and Hong Kong, and the United States and the Republic of Korea.[6]

However, the ITU has been overshadowed by negotiations on international telecommunications in the trade arena. The trade model views telecommunications as an end product and vital input; it is seen by some as protectionist, with strategies of reciprocity and retaliation. In contrast, the ITU model emphasizes the need for interconnection and interoperability; it is flexible and minimalist. But there are problems with the ITU model. National policy makers in economics and trade cannot adequately participate because it is a specialized technical body; also regulators, who are supposed to defend the public interest, are not sufficiently separated from operators in many countries. In addition, the ITU has no system for reflecting users' opinions; the focus is on service providers, not customer needs. Thus, the ITU approach makes it hard to introduce structural as opposed to technical changes, although structural changes may be more beneficial for users if they result in lower tariffs and new service offerings.

A New Framework: The GATS

The traditional vehicle for international trade policies has been the General Agreement on Tariffs and Trade (the GATT). There are currently 123 members of the GATT; about one-third of them joined around or after the Uruguay Round of tariff negotiations in December 1993, indicating the desire of an increasing number of countries to enjoy the benefits of free trade. Significantly, to date, China has not become a member.

Signed in 1947, the GATT was the first international agreement to deal with trade issues on a comprehensive, multilateral basis; it dealt only with trade in goods, not services, but provided a model for countries to emulate in fashioning their own bilateral and multilateral agreements. The GATT embodied principles of most favored nation (MFN) status, national treatment, and transparency. It also contained a dispute settlement mechanism, although the process could be slow and cumbersome.[7]

The 1986 Punta del Este Declaration established the Uruguay Round of GATT negotiations, which included negotiations on trade in services to foster "expansion of (services) trade under conditions of transparency and progressive liberalization and as a means of promoting economic growth of all partners and the development of developing countries."[8] The declaration accepted a number of principles promoted by developing countries, in particular that they should have the same rights and obligations under a services agreement as all other national parties, but be allowed to determine coverage and liberalize their service sectors and subsectors at a gradual pace and according to domestic regulations they determine as necessary.[9]

Significant changes were made to the GATT as a result of the Uruguay Round of multilateral trade negotiations, one of which was inclusion of services, such as telecommunications, for the first time in the agreement. At first, the intention was to incorporate services into the existing GATT framework, but it soon became apparent that services would require a separate chapter in the GATT agreement. The result was the General Agreement on Trade in Services (GATS). Basically, the same provisions that apply for industrial goods also apply for services, such as MFN treatment, national treatment, rules relating to economic integration, prohibition of subsidies, transparency, recognition of standards, and the settlement of disputes. The Uruguay Round negotiations for the GATS were concluded in December 1993. The Final Act of the Uruguay Round went into effect in July 1995, with most terms to be phased in over the next decade, under the aegis of the new World Trade Organization (WTO).

Telecommunications was recognized by the GATT as serving a dual role: as a distinct sector of economic activities and as an underlying means of transport for other economic activities. However, the support provided by telecommunications services to the production and sale of other kinds of services led to a formal agreement by GATS participants to develop a separate Telecommunications Annex that would state explicitly the rights of telecommunications service users.[10] The major proponent of the Telecommunications Annex was the United States, whose goal was to open the world market to U.S. carriers and service providers, and, through open markets, to foster competition that would reduce prices of telecommunications services, thereby benefiting users.

The Telecommunications Annex is intended only for those countries that have decided to introduce competition into their telecommunications market. It seeks to extend GATS to measures affecting access to and use of public telecommunications transport networks and services (PTTNS) such as telephone, telegraph, telex, facsimile, modem, and data transmission. It does not apply to broadcast or cable transmission of radio or TV programming. According to the Annex, GATT members must ensure that relevant information and conditions affecting access to and use of PTTNS are publicly available, including:

- tariffs and other terms and conditions of service;
- specification of technical interfaces with these networks and services;
- information on bodies responsible for preparation and adoption of standards affecting access and use;
- conditions applying to attachment of terminal or other equipment;
- any notification, registration, or licensing requirements.

GATT members must also ensure that:

- there is no discrimination between suppliers;
- there are no restrictions on movement of information across borders, including communications between businesses and access to information contained in databases;
- security and confidentiality of messages are protected.[11]

To allow further liberalization of trade in telecommunications, the GATT members decided to set up a Negotiating Group on Basic Telecommunications (NGBT), which is designed to encourage more countries to introduce competition in basic telecommunications services, and persuade those countries that have already introduced competition to adopt nondiscriminatory policies in the future. The NGBT debated six main objectives in support of competition:[12]

- Regulatory reform: regulatory agencies must have autonomy, transparency, and the resources to foster competitive behavior;

- Interconnection: new operators must be allowed to connect with the existing PSTN under equitable conditions and at reasonable rates;
- Structural and accounting separation: companies active in different market segments should create separate subsidiaries to avoid cross-subsidization of competitive services;
- Number portability: customers need to be able to change carriers without changing telephone numbers;
- Pricing policy: countries need guidance in rate-setting policies that can be used in a competitive environment;
- Accounting rate reform: the role of the WTO in setting rates that more accurately reflect the costs to monopoly operators than the current accounting rate framework (see Accounting Rates, below).

The NGBT aimed to produce a final report by the end of April 1996, with commitments arising from the report to be added to the GATS. However, by the deadline, only 11 countries offered full commitments to mutual, unrestricted market access and open investment. In order to obtain commitments from a "critical mass" of countries, the talks were extended until February 15, 1997.

GATS and the Developing World

The Telecommunications Annex contains two special provisions for developing countries, aimed at assisting them in strengthening their own domestic telecommunications sectors. The first allows them to use protection when opening up their telecommunications markets:

> . . . a developing country may, consistent with its level of development, place reasonable conditions on access to and use of public telecommunications transport networks and services necessary to strengthen its domestic telecommunications infrastructure and service capacity, and to increase its participation in international trade in telecommunications services.[13]

The second provision encourages the transfer of information technology to developing countries and their involvement in telecommunications development programs at the international and regional levels.

The reduction and elimination of barriers to services and investment under the GATS present formidable challenges. For policy makers, the challenge will be to balance national and regional goals and traditions with the need for open markets and the need for ongoing coordination on technical issues such as standards and spectrum allocation. The debate is over whether the objective of the services agreement is the restructuring of the domestic and international telecommunications markets of the countries that sign the Final Acts, or whether the mandate is limited to promoting principles and rules which facilitate crossborder trade of services products of many sectors. The widely shared view is that telecommunications markets should not, as a consequence of the agreement, be forced to restructure.

ACCOUNTING RATES

One of the reasons that European countries came together in 1865 to form what became the ITU was to develop a means of dividing revenues of international telegraph traffic between originating and destination countries.[14] The mechanism for dividing revenues among international carriers, known as accounting rates, is modeled on the international settlements process between postal administrations. Today, the accounting rates framework is determined by Study Group III of the ITU Telecommunication Standardization Bureau (ITU-T, formerly the CCITT).

The accounting rate is the per minute revenue to be shared between correspondent carriers. The settlement rate is usually half the accounting rate. (If carriers have no correspondent agreement, traffic is sent via transit service, using an intermediary carrier's carrier.) The collection rate is the "retail" rate charged to the customer. Under the accounting rate system, an originating carrier pays the destination carrier a fee for completing a call. These fees, dating from the era of international monopoly carriers, are negotiated bilaterally, and typically bear no relationship to the rates charged to the customer or the cost of completing the call. Since each pair of carriers terminates calls for the other, the settlement charge is the net amount paid by one carrier to the other.

The accounting rate system was developed during the era of operator-assisted international monopoly service, and assumed relatively balanced flows of traffic between countries. However, traffic has grown dramatically with the introduction of direct dial, digital networks, calling cards, country direct services, and so on, so that "the luxury overseas call of the 1960s has become the routine global chat of the 1990s."[15] Prices have declined, but not as much as costs. For example, a three-minute daytime call from New York to London that cost $22.50 in 1975 cost the customer $2.20 in 1995, but the actual cost of providing the call has plummeted to 33 cents.[16]

Countries with lower "collection rates" became sizable net "exporters" of calls. International carriers in these competitive environments are victims of their own success; if they originate twice as many calls as they terminate, they must pay out twice as much as they receive. (The FCC imposed uniformity in accounting rate settlements for each route and each service so that foreign monopoly carriers could not secure revenue concessions by "whipsawing," that is playing one U.S. carrier off against another.)[17] Accounting rate schemes reward countries that keep international accounting rates high and/or generate less traffic, and are therefore a disincentive to price cutting and efficient operation.

High accounting rates, which are reflected in inflated international charges, can suppress demand; customers have a disincentive to originate calls, and try to find ways to communicate by originating traffic outside the high-cost country. However, this traffic actually contributes more to the monopoly carrier, because it increases its net settlement surplus. Countries with monopoly PTOs are able to extract huge subsidies from low rate countries that have competitive carriers; countries that generate more traffic than they receive end up with settlement deficits. Thus, the U.S. industry pays more than $3 billion per year in net settlement charges to foreign carriers.[18]

For developing countries, settlements are an important source of foreign exchange; in some developing countries, inbound settlements constitute the single largest source of foreign currency. Developing country PTOs typically claim that they use this surplus to invest in their domestic networks, although some actually funnel the foreign exchange directly to their central government. However, such rates in effect act as an export tax on the country as a whole, as its economy is subjected to higher international communication costs.

Some analysts have pointed out that the U.S. outpayment equalled 35 percent of the total U.S. budget for foreign aid in the early 1990s.[19] Yet, as the ITU notes, accounting rate subsidies are poorly targeted if they are perceived as helping the poorest countries to improve their infrastructure. For example, in 1992, only 3.9 percent of U.S. settlement payments went to Africa, and 3.3 percent went to Eastern Europe.[20]

An innovative industry known as "callback" has sprung up to bypass the charges of the high-priced carrier. Rather than dialing the receiving party directly, the customer in the high-cost country calls a number in a low-cost country (typically the United States). The callback company calls him back, and then allows him to dial the receiving party. The caller pays the lower rate plus a surcharge to the callback operator. Using intelligence in the network and in PBXs, this procedure is now usually automated. The customer calls the callback company, but hangs up before the call is answered, thus incurring no charges. Without answering, the callback company's equipment is able to recognize the originating number, which it then dials, setting up the circuit.

Countries with high accounting rates typically complain that callback deprives them of revenue and ties up their equipment. Many governments have declared that callback is illegal, although the Federal Communications Commission (FCC) in the United States (where most callback firms are based) has defended its legality, and Hong Kong, which also has declared callback legal, actually encourages its government departments to use callback services to save money.

Another industry that exploits the accounting rate discrepancies is audiotext services, for example, so-called "dial-a-porn" services. The service providers set up shop in countries that tolerate their business and have high accounting rates. They advertise that only direct-dial rates apply, but profit from agreements with the monopoly carriers to share revenues from settlement charges generated by the traffic.

The accounting rate regime remains largely a cartel arrangement among monopolies. However, several proposals have been made for accounting rate reform, ranging from gradual reduction in rates toward actual cost to facilities-based payments, sender-keeps-all approaches, volume-based payments, and call termination fees. Under facilities-based payments, the sender pays according to the use of the terminating operator's facilities. Sender-keeps-all eliminates settlements, allowing each originator to keep all the fees. Volume-based payments are set based on the volume of originated traffic: the higher the volume, the lower the rate. For call termination fees, each PTO would charge the same fee to terminate calls regardless of origin.[21]

While all of these approaches appear to be more equitable than the current system, they do not address the issue of alternatives to cross-subsidy from international revenues to invest in developing countries' infrastructures. One response is that the two issues of revenue and infrastructure expansion should not be linked; other sources of funding are needed, such as those discussed in previous chapters. Regardless of ideology, developing country planners must realize that such high profits will inevitably be squeezed out of their revenues as expanding international competition drives pricing toward cost. However, there may be other approaches that would allocate some international revenues to infrastructure development. For example, a surtax on international traffic or a small percentage of the revenues could be allocated to the international equivalent of a universal service fund, based on the U.S. domestic model. An earlier funding mechanism along these lines, called WorldTel, was proposed by the ITU's Maitland Commission in 1983. The WorldTel concept has resurfaced as a revolving loan fund that would be financed by investors rather than calling revenues.

CONTENT VS. CONDUIT

Telecommunications regulation has typically focused on technical and pricing issues, drawing a legal "bright line" between common carriers, which have no control over the content they transmit, and broadcasters, which are typically accountable for their content. However, convergence in technologies and services has led to new considerations of the content vs. conduit distinction.

Information in electronic form has become difficult to protect, as pirated software proliferates through copies on audiotapes, videocassettes, floppy disks, and CD-ROMs. U.S. government sources have estimated losses to the U.S. content industry of $12 to $15 billion per year, even before widespread Internet availability.[22] However, the Internet makes content even easier to disseminate, as anyone with access can download pirated content placed on the Net from anywhere in the world. The explosive growth of the Internet and its increasing internationalization have added new concerns and complexity to the content vs. conduit issue. In 1995, 70 percent of the Internet's host computers were in North America, 21 were percent in Europe, 7 percent in Asia, and the remainder in other parts of the world. Growth in Internet services and use is phenomenal, not only in industrialized countries, but also in the developing world. In 1994, the number of host computers grew 116 percent in Latin America and nearly 150 percent in Africa.[23] Twenty-two countries joined the Net in 1995.

Just as the Internet relies on a "network of networks," the international protection of intellectual property relies on a "system of systems."[24] International copyright protection is governed by the Berne Convention, administered by the United Nations' World Intellectual Property Organization (WIPO). All members of the new WTO are required to follow the substantive obligations of the Berne Convention, and its protections are bolstered by mechanisms for enforcement and dispute resolution resulting from the Uruguay Round of GATT negotiations. While the Berne convention does consider electronic storage and copying as protected reproduction of the original work, it has been criticized for having few substantive restrictions and weak compliance and dispute resolution mechanisms.[25] Also, content providers are offered limited protection from electronic piracy in the short run, as the WTO provisions allow a 12-year phase-in period for developing countries.[26]

The U.S. Information Industry Task Force, formed to coordinate public-private sector activity in development of a national information infrastructure (NII), has pointed out ambiguities in U.S. copyright law that also appear to weaken the protection that U.S. copyright holders could expect if they make their content available online, even assuming that they could trace electronic pirates and attempt to enforce the law. Operators of online services and bulletin boards continue to grapple with issues including responsibility for content and protection of intellectual property, as well as prevention of libel and indecency. Some courts believe that they do indeed have responsibility for content, although operators have argued either that they are like common carriers, and therefore not responsible, or like a library or newsstand that cannot be expected to monitor all the content it carries.[27]

Encryption may provide some protection from online piracy, as the tools to protect electronic commerce and private communications become more sophisticated, driven by the demand for electronic transactions and intracorporate communications on the Internet. However, encryption will not be a fail-safe solution, as clever hackers continue to demonstrate their ability to intercept supposedly secure communications and to breach electronic "firewalls."

From another perspective, electronic access to content from around the world is perceived as a threat by many developing country governments. Popular Western culture, particularly from the United States, now widely available on audio- and videocassettes and via satellite, threatens to undermine indigenous cultures, and to create demand for imported consumer goods. Although limited in its accessibility, the Internet is seen as the latest of the electronic invaders. Some governments such as China and Singapore have tried to keep this information from their citizens by blocking access to certain Internet sites, and prohibiting ownership of satellite receivers (TVROs).

An alternative to outright prohibition is to try to put new media in old policy boxes. For example, cable television may be considered like broadcasting, or online services like common carriers or perhaps libraries. Singapore has stretched this metaphor by considering the Internet as a broadcast medium for the purposes of regulating content, so that the standards that apply to broadcasters are to apply to Internet content accessible from Singapore.

Although the United States has generally dismissed other countries' concerns about perceived cultural imperialism, many Americans share concerns with people from other cultures about an electronic invasion of violence and pornography through television, "dial-a-porn," and Internet sites. The priority given to protection of individual rights in the United States has led to policies and technologies designed to give individuals more control over access to content for themselves and their families. These include computer programs that allow users to block access to Internet sites (sometimes called "network nannies"); "V chips" that can be installed in new television sets (as mandated in the U.S. Telecommunications Act of 1996) to enable parents to block access to violent or otherwise inappropriate television programs; and on-demand blocking of access to telephone 900 numbers (used to access pornographic and other fee-based audiotext services), as mandated by many U.S. public utilities commissions.

Of course, clever users are likely to be able to circumvent electronic censorship, whether mandated by their governments or built into their computers and television sets. A more promising strategy is for governments to encourage the use of these tools to strengthen their own national cultures and to disseminate their creative works. For examples, many Asian and Middle Eastern countries have developed thriving local music industries based on audiocassettes, and India has developed a huge film and video production industry. As Phillip Spector points out, this strategy may also benefit content providers in industrialized countries, because as other countries develop their domestic content industries, they have an incentive to protect intellectual property.[28]

CHALLENGES FACING THE ITU

Spectrum Scarcity: The Revival of Wireless

One of the responsibilities of the ITU is global management of the radio frequency spectrum. The need for a spectrum "traffic cop" was recognized early in this century when new radio stations chose frequencies that caused interference with those already on the air, and two-way radio systems in one country could interfere with similar systems across the border. Since radio waves do not respect national boundaries, international agreements are necessary to prevent chaos. Separate frequency bands are allocated for various services, and procedures for sharing are implemented when bands are allocated for more than one service. The actual assignment of frequencies within these bands is done by national governments.

The advent of satellite communications and the proliferation of wireless services have created new challenges for spectrum management. Wireless communications are booming. Once thought of as a means of broadcasting radio and television signals and reaching ships and isolated outposts beyond the range of wire and cable, wireless applications now include cellular telephone networks, personal communications networks using "microcells," data links to handheld devices, and wireless local area networks. Satellite systems also rely on radio frequencies to link earth stations with "transmitters in the sky."

While some consider the radio spectrum to be a limited resource, a better characterization is that its value is dependent on many factors that can involve expensive trade-offs. Some frequencies such as VHF and UHF are very attractive for point-to-point communications although they are also used for broadcasting. Higher frequencies offer much more bandwidth, but may be more expensive to exploit.

When geostationary satellites were first proposed, they were allocated frequencies in bands used by terrestrial microwave systems. As these became congested, higher frequencies were allocated for satellite communications. However, these frequencies, at 11 to 14 GHz, known as Ku band, are prone to signal attenuation in heavy rainfall, snow, or dust storms. As these frequencies became congested, a new allocation at 20 to 30 GHz was added, but these frequencies, known as Ka band, are even more prone to rain attenuation, requiring higher-powered and more costly transmitters to ensure reliability.

The commercial viability of new services such as personal communications services (PCS) and low earth-orbiting satellites (LEOs) depends on allocations of sufficient bandwidth, but at frequencies that can be readily exploited. These industries develop recommendations for the conferences and lobby to convince ITU members to support them. Building consensus is both arduous and time-consuming; most major decisions are made in late-night sessions at the end of the conference after weeks of lobbying and negotiation.

The ITU has never endorsed leasing or auctioning spectrum; however, current ITU policies have created the opportunity to create a market for international satellite slots. Some countries are realizing that they can file for domestic slots for "paper satellites" and then lease them to other satellite operators. The Kingdom of Tonga achieved notoriety when it filed for several slots over the Pacific Ocean, and then leased some to Rimsat, a U.S. based company that acquired Russian-built satellites and moved them to the Pacific.

Domestic strategies to reap financial returns from spectrum allocations are likely to proliferate. In the past, cash-strapped national governments tapped their telecommunications revenues for the national treasury. In an era of privatized operators, this source of funds has disappeared, but industry demand for the government-controlled spectrum may present new opportunities to help balance national accounts. The United States has auctioned spectrum for direct broadcast satellite frequencies and PCS microcellular networks. New Zealand has auctioned cellular mobile as well as broadcasting frequencies. Other countries are considering spectrum auctions and taxes.

If the ITU's member governments are seeking financial gain from national spectrum assignments, will the ITU itself eventually turn to spectrum fees to finance its operations? Although the members have incentives to seek alternatives to paying dues as a means to support the ITU, they are not likely to favor paying the ITU directly for spectrum usage. However, they could, of course, pass this cost on to the operators they license. Operators, in turn, would pass the cost on to end users in the form of higher prices. While this approach will not be popular with users, it may be a fairer way of paying for international telecommunications regulation than using general tax revenues. Developing countries are likely to resist such proposals, pointing out that neither they nor their users can afford these charges. One solution might be to include frequency set-asides or discounts for low-income countries, analogous to the special conditions for minority- and women-owned telecommunications businesses in the U.S. auctions.

Congestion in Orbit

With the proliferation of communications satellites for domestic and international communications, the orbit spectrum available for geostationary satellites is becoming increasingly congested. Satellite operators want to occupy orbital slots that provide good coverage for their satellite footprints and high "look angles"—the satellite should not appear close to the horizon, where buildings, hills, and trees could block transmission and reception. The most desirable orbital locations may be for domestic satellites serving adjacent countries such as the United States and Canada; for competing regional systems such as those by proliferating in Asia; and for international systems such as Intelsat and PanAmSat, which can occupy only certain locations to provide global coverage. Conflicts have already emerged over prime slots between domestic, regional, and international satellite operators in Asia.

Requests for orbital slots are filed with the ITU on a "first come, first served" basis. Thus, as noted above, there are incentives to file for "paper satellites" that may actually never be built if financing cannot be found, simply to reserve the orbit location. The "first come, first served" principle has long been criticized by developing countries, which fear that spectrum for new services and orbital locations may not be available when they are ready to use it.

The ITU is attempting to simplify the procedure and to develop more stringent criteria for slot allocation. Meanwhile, some operators are adopting innovative strategies to get prime orbital slots. Intelsat has entered into agreements with several national agencies to share satellites and orbital locations. For example, Intelsat will lease transponders on the Insat-2E satellite to be launched in late 1997. Intelsat will also lease transponders on China's Chinasat-5 satellite, and has an agreement to place a future Intelsat satellite in an orbital slot registered to China.

There are no effective international legal sanctions against operators that do not register with the ITU, but they risk being boycotted by potential customers who do not want to risk having their signals subject to interference from adjacent satellites. A case in point was the Apstar-1 satellite, designed to cover Asia, and launched without coordination with adjacent satellites. Several of Apstar's television customers took their business elsewhere. The orbital congestion issue was resolved when Apstar leased an orbital location registered to the Kingdom of Tonga.

To Plan or Not to Plan?

In the 1970s, the ITU acceded to demands of many members to change from "first come, first served" to a priori planning of orbit and spectrum for broadcasting satellites. European members advocated planning as a way to ensure that each country got access to broadcasting satellite service (BSS) channels, and to enable operators to proceed with designing BSS systems. American delegates argued that planning was not necessary because demand for broadcasting services was not clear, and the technology was changing rapidly. They convinced other countries in Region 2 (the Americas) not to plan the BSS spectrum until 1983 (when a much more limited and flexible plan for the Americas was adopted by countries in Region 2), while other countries developed regional plans.

The U.S. position appears to have been justified. Because the market for Direct Broadcasting Satellites in Europe was uncertain, the first DBS service was not offered until the mid-1980s. By that time, the technology had changed significantly, undermining many of the technical assumptions in the original plan. It also became apparent that the best way to enter the market was to offer services that could be received by several countries, rather than having separate channels for each country, as proposed under the original plan.

The planning debate raged again at the 1988 World Administrative Radio Conference (WARC), when many developing countries argued that the fixed satellite service should be planned so that they would not be left out when they were able to afford their own satellite systems. However, members eventually decided not to plan spectrum already in use, but to set aside some frequencies known as "expansion bands" that could be planned. To date, these frequencies have not been used.

While a priori planning has major disadvantages in the rapidly changing technological and business environment of international telecommunications, its advocates are attempting to address the very real problem of the "latecomer penalty." By the time many countries are able to take advantage of new technologies and services, the best frequencies may no longer be available. Although filing for paper satellites may be one way to reserve slots and obtain revenue from them, satellites that remain only unfulfilled dreams will not solve the problem of providing reliable and affordable communications in the developing world.

The Knowledge and Resources Gap

A more fundamental issue confronting the ITU, as well as other international organizations concerned with telecommunications such as the ISO and the WTO, is the gap between the wealthiest and poorest countries in technological expertise and resources to participate in policy-making activities of the organization. Much of the work of the ITU is done in study groups, which are dominated by representatives of industrialized countries and major vendors with a stake in the standards or frequency policies to be recommended. Developing countries lack the resources to participate—both in terms of funding to attend meetings and in personnel with the expertise or the time to devote to such matters.

This gap is magnified at the policy-making conferences. While ITU funds help developing country administrations to attend, their delegations of one or two people cannot hope to cover all of the topics on the agenda and the committees where the issues are thrashed out. And because they are swamped with responsibilities at home, they have little time to prepare for the conferences by submitting their own proposals or even reviewing other countries' proposals. In contrast, the United States, for example, sends large delegations of government and industry officials that are backstopped by analysts and computer programs in Washington. Some of these delegates will have spent much of the previous year on the road, visiting developing country officials whose support at the conference could be crucial in adopting their proposals.

Faced with proposals they have not had time or expertise to analyze fully, developing country delegates are either likely to acquiesce under the pressure of lobbying or attempt to block the proposals completely on the assumption that there must be detrimental consequences for them. Neither strategy may serve their interests very well. They may, however, turn to developing countries with relevant expertise. India played a leading role in championing the cause of developing countries at the ITU in the 1970s and 1980s. Other countries in developing regions may also have relevant expertise, such as China, Brazil, and South Africa. The ITU has attempted to reduce the knowledge gap by holding preparatory meetings to brief developing country delegates on the issues. The ITU's Bureau for Telecommunications Development (BDT) also offers training and technical assistance to update developing country administrations.

While many members complain about the cost in time and money of participating in the work of the ITU, the burden is heaviest on the least developed countries. One technological solution that may facilitate broader participation in study groups and conference preparatory activity would be use of the Internet for exchanges of drafts and interaction among committee members. Internet access is still quite limited in poorer countries, but most have at least capacity for electronic mail. Use of the Internet may not only broaden the participation of developing country members in the work of the ITU, but may also make them aware of the benefits of Internet access, for which there is growing demand in the developing world.

CHAPTER 18

Future Directions in International Telecommunications

"The death of distance as a determinant of the cost of communications will probably be the single most important economic force shaping society in the first half of the next century."

—The Economist [1]

THE CHALLENGE OF CHANGE

Distance began to die with the inventions of the telegraph and telephone, followed by Marconi's demonstration of wireless communications across the ocean. The 20th century has made these technological advances available to much of the world. Today, the telecommunications environment is changing dramatically, driven by technological innovation that results in new equipment and services, but also new entrants and alliances between companies with experience in a wide range of information industries from telecommunications to broadcasting to computer hardware and software to publishing. Three major trends driving these changes are:

- the rapid evolution of technologies and services;
- the restructuring of the telecommunications sector;
- the globalization of economies and of communications.

Together these developments are not only changing the world of telecommunications, but the ways people work, learn, and interact.

OVERCOMING DISTANCE

Communicating through Cyberspace: The Internet

Global voice communications is made possible by a network of telecommunications networks. While we may be far from a broadband network of networks with connections to everyone, we are witnessing the phenomenal growth of access to a data network of networks, the Internet. With the introduction of the World Wide Web, information on the Internet has expanded to include not only text, but graphics, audio, and video. The Internet will also become a more navigable resource as browsing software incorporates intelligent agents to roam the Internet looking for information on topics that users want to track. New programming languages, such as Sun's Java, which operate on any computer system, will make it possible for users to download specialized chunks of software to perform specific tasks. Perhaps, as some in the industry now predict, the network will become the computer, as users rely more on the network to upgrade the capabilities of their computers rather than on new hardware and new versions of bloated programs.

More than half the computers in U.S. offices are linked to local area networks (LANs). National and multinational corporations now routinely link their employees for data communications, including electronic mail, often using private or virtual public networks. Increasingly, businesses are also linking to the Internet to reach counterparts in other organizations, specialized databases, and potential customers. Each month, some 2,000 businesses join the more than 20,000 that have already set up "virtual shop" on the Internet.[2] While business use of the Internet is still in its infancy, it is bound to grow as large and small businesses figure out how to reach their customers electronically. Internet commerce also represents a major trend in the electronic delivery of goods and services. Small businesses without proprietary networks now have a low-cost means of reaching customers, while large businesses are finding that they now must also add Internet access so that their proprietary networks do not isolate them from other users.

As demand for these services grows, users will need access to more bandwidth to speed searches, download software, and transmit video and graphics. Some telecommunications companies, such as AT&T and MCI, are becoming Internet service providers, having decided that the Internet must be viewed as an opportunity rather than a threat. Others fear that the Internet will steal traffic, as users opt for flat rate voice and video transmissions rather than paying for time or bits. Traditional telephone companies will have to respond to the demand for new and cheaper services if they are to survive. Rather than local monopolies providing telephone services over copper wires, in many countries we may find cable television companies, electric utilities, and wireless operators competing with telephone companies to reach the end user, and offering a combination of voice, data, and video services.

Communication Anytime, Anywhere

In addition to the extension of "information superhighways" in the form of broadband networks, another major technological trend will be the increasing ubiquity of communications using wireless technologies that will initially provide access to squirts rather than floods of information. Personal communications networks using microcellular technology will allow people in urban areas not only to talk on pocket-sized telephones, but to transmit and receive data using wireless modems. In rural and developing areas, these services may be available from low earth-orbiting (LEO) satellite systems.

The death of distance could have profound implications for both individuals and organizations. The ability to work "anytime, anywhere" allows "road warriors" to work without offices on planes, in hotels, and at client sites, and enables information workers to telecommute from their homes rather than travel to work. This flexibility can be two-edged for individuals, who can work wherever they choose, but may never escape the "virtual workplace." Organizations may reduce their overhead costs and improve their productivity, but they must also learn how to manage their decentralized work force.

Internationally, the death of distance has profound implications for the globalization of industries and national economies. Rural regions in Europe and North America may lure businesses with their pleasant environments and lower labor costs; however, they are no longer competing only with cities in their own countries. Companies may hire information workers in developing countries where labor is far cheaper, not only for data entry and word processing, but for writing computer programs. Similarly, developing countries now find themselves competing in global markets, where quality and suitability of products may be as important as price.

TOWARD GII

National and Regional Initiatives

These technological and economic trends have led policy makers to call for the construction of "information highways" linking communities and nations. The phenomenal growth of the Internet as an information resource, communications tool, and electronic marketplace has focused attention on the need for national and global information infrastructure (NII and GII) to bring the Internet and other forms of electronic communications within reach of people around the world.

In the United States, the Clinton administration announced the National Information Infrastructure (NII) initiative in 1993, calling for joint industry and government efforts to create a seamless and interoperable national broadband infrastructure, an "information superhighway" to link all Americans. The NII is to be built by the private sector, with the government providing the policy and regulatory frameworks. To this end, the administration also established a federal

Information Infrastructure Task Force chaired by the Secretary of Commerce and a private sector Information Infrastructure Advisory Council to identify user needs and plan for the implementation of the information superhighway.

As part of the NII initiative, Vice President Al Gore challenged U.S. industry to connect all of the country's schools, libraries, hospitals, and clinics to the information highway by the year 2000. Federal and state governments are funding research and pilot projects to spur innovative applications. The High Performance Computing and Communications Program is supporting university research. NTIA initiated the Telecommunications and Information Infrastructure Assistance Program (TIIAP) to provide funds for telecommunications applications in distance education and health care delivery. The Rural Utilities Service has established Distance Learning and Medical Link Grant Programs to fund educational and health care projects providing advanced telecommunications technology and services for rural Americans.

Many states are also reexamining the information needs of their citizens and the status of their existing infrastructure. They are offering inducements to the carriers to accelerate the upgrading of their facilities and to provide Internet access to schools and other community locations. State governments are also providing seed money to communities and economic development agencies to help them plan and initiate projects using telecommunications to support development.

In 1993, the European Commission published a White Paper on "Growth, Competitiveness and Employment," which emphasized an urgent need to develop a Europe-wide information infrastructure to help restore economic growth and competitiveness, open up new markets, and create jobs.[3] Acting on the White Paper's recommendations, the European Commission called on a high-level group of information industry representatives to produce a report recommending practical measures for implementation. This report, entitled "Europe and the Global Information Society," recognized the significance of the emergence of an information society. The group urged the European Union (EU) to trust market forces and private sector initiatives, but noted that spending on education, health, and research may have to be retargeted toward new priorities, and a new form of public/private sector partnership would be needed to implement the group's recommended action plan. It recommended:

- accelerated liberalization of the telecommunications industry;
- identification of the degree of regulation required;
- interconnection of networks and interoperability of services to avoid fragmentation of information infrastructure;
- reduction in tariffs, to bring them in line with those of other advanced industrialized regions;
- review of the standardization process to increase its speed and responsiveness to the market.[4]

In response, the European Commission has set out a detailed work program in four key areas: regulation, applications, social and societal aspects, and the promotion of the information society. The EU has also approved $3.8 billion under its 1994–1998 Fourth Framework Programme for Research and Development to fund development of a new information infrastructure. This investment will support research and development in communications technologies and development of applications in distance education, health care, and other social services.

Canada also has plans to build a network of networks linking Canadian communities, businesses, government agencies, and institutions. The Canadian government sees the information highway as a catalyst to help Canadians share information, and to gain an edge in productivity and information industries in global markets. It has identified three key objectives for the Canadian information highway:

- to create jobs through innovation and investment;
- to reinforce Canadian sovereignty and identity;
- to ensure universal access at reasonable cost.[5]

These objectives emphasize cultural as well as economic priorities. Canada's Information Highway Advisory Council, composed of representatives of communications industries, business users, academics, performers, and consumers, takes a distinctively cultural perspective:

> The Information Highway, in our view, is not so much about information as it is about communication in both its narrowest and broadest senses. It is not a cold and barren highway with exits and entrances that carry traffic, but a series of culturally rich and dynamic intersecting communities. . . . Rather than a highway, it is a personalized village square where people eliminate the barriers of time and distance and interact in a kaleidoscope of different ways.[6]

In Japan, the NTT has announced its intent to wire every school, home, and office with fiber-optic cable by the year 2015. Japan's Ministry of Posts and Telecommunications (MPT) estimates the cost of building this network to be between $150 billion and $230 billion. The Japanese are also investing in projects and trials to ensure that users will be able to access a wide variety of services. For example, in 1994, the MPT launched a $50 million, three-year pilot project to assess the feasibility of integrating telecommunications and broadcast services such as video-on-demand, high definition television, videoconferencing, teleshopping, and telemedicine through fiber-to-the-home networks.

In 1995, the G-7[7] nations together embraced the goal of global information infrastructure, and initiated a series of demonstrations and pilot projects using high-capacity networks and switching. The APEC (Asia Pacific Economic Conference) nations are also sponsoring projects using telecommunications networks for distance education and training, health care delivery, and economic development. Yet there are still areas in the industrialized G-7 countries with very limited access to telecommunications. Also, APEC members include not only the industrialized economies of Japan, Singapore, the United States, and Canada, but also countries with much greater gaps in their infrastructure such as China, Thailand, and the Philippines.

GII: Promise or Hype?

Against this background, why all the hyperbole about electronic superhighways? Several themes recur in these information highway initiatives. There are twin assumptions that converging technologies will result in information services with both social and economic benefits, and that both public and private sectors must be engaged to ensure the installation of national broadband networks. These assumptions need to be carefully examined. Each new communications technology has been heralded as offering numerous benefits. Satellites and cable television were to provide the courses taught by the best instructors to students in schools, homes, and workplaces. Videoconferencing was to largely eliminate business travel. Telemedicine was to replace referral of patients to specialists. Computers were to replace traditional teaching with more personalized and interactive instruction.

To some extent, all of the prophesies have been fulfilled, yet the potential of the technologies is far from fully realized. In many cases, it took institutional incentives to change and innovate for these technologies to have much effect. In North America, the more remarkable change is in these incentives, rather than the technologies. As school districts face shrinking budgets and new curricular requirements, as spiraling health care budgets are targeted by governments and insurance companies, and as business realizes that people must "work smarter" to compete in a global economy, they find new and compelling reasons to turn to telecommunications and information technologies.

Thus, investment in technology alone will not likely result in major social benefits. Policy makers seem to be aware that public sector stimulus is needed to foster new educational and social service applications. There is widespread belief in the need to fund trials and demonstration projects, yet seed money for pilot projects may not ensure long-term implementation. Schools with ISDN access will benefit if the services they can access turn out to be cost-effective means of achieving their educational priorities. If the services are perceived as frills, or if there is no budget allocation to buy computers or pay monthly usage charges, connection to the information highway will mean little. Similarly, if insurers will not authorize payment for tele-consultations, or physicians are not authorized to practice beyond their borders, telemedical applications will remain limited. And if prices for connection and usage are beyond the reach of low-income and rural residents, small businesses, and nonprofit organizations, the much-heralded information society will be very narrowly based.

The U.S. communications industry has adopted the banner of the "information superhighway," with the assumption that there is an enormous new market in information services, including video-on-demand. Thus, telephone and cable companies plan interactive broad-band networks that would enable viewers to select digitized movies delivered from central servers. Direct Broadcasting Satellite systems offer similar capabilities, with less interactivity, but multiple chan-nels of compressed video to allow users to select and schedule pro-gramming. While these applications are viewed in the United States in business terms, in other countries, the concern is focused on cul-tural impact. Both Canada and the European Union stress the need to use these networks to strengthen their own cultures. Yet, the prolifer-ation of cable- and satellite-delivered channels in Canada and Western Europe tells a different story: the demand for content is so great that operators turn to inexpensive sources of content to fill them, and this content is overwhelmingly American.

Another recurring theme is the "we will be left behind" argument. In the late 1980s, U.S. telephone companies sought to convince American policy makers that the United States was at a disadvantage because its citizens did not have Minitels, small computer terminals provided to French households by France Telecom. But Americans had much of the functionality of the Minitel through widely available facilities including telephone access to audiotext services and growing access to personal computers equipped with modems. Today, Canada, the European Union, and Japan are all concerned that they will be left behind the United States if they do not implement their own infor-mation infrastructures. Notably, the report to the European Union states: "The first countries to enter the information era will be in a position to dictate the course of future developments to the late-comers."[8] But is this really so? It may be that their technology companies will have an advantage if they have a ready market for fast packet technologies such as Asynchronous Transfer Mode (ATM) switches, servers, set-top boxes, and multiplexers that can also be exported. But their major equipment vendors are already seeking international alliances to develop these products. The real advantage will be from the application of these technologies to access and share information that can contribute to the development of their own societies and the competitiveness of their economies.

GLOBAL NETWORKING: CHANGING THE GEOGRAPHY OF BUSINESS

Telecommunications networks now link manufacturers with assembly plants, designers with factories, software engineers with hardware vendors, suppliers with retailers, and retailers with customers. No longer is it necessary to have all the expertise in-house. Software engineers in Silicon Valley complain that they are laid off, while overseas contractors are hired to transmit programs from Russia and India. Freelance designers can now send clothing patterns directly to an automated garment factory. And customers can order anything from airline tickets to winter clothing online, and do their own banking and bill paying electronically.

These trends open opportunities for innovative entrepreneurs around the world. For consumers, they offer more choice and lower prices because there is no overhead of salesclerks and order-takers. These changes pose threats to traditional businesses as well as to employees. Increasingly, companies that want to compete on price will have to "work smarter" to reduce costs and respond to market changes, while others will have to rethink how to add value to attract customers. High levels of customer service and individualized attention are likely to become more important. As Wells Fargo found, a bank that offers human assistance 24 hours a day, in addition to online electronic banking, can attract new customers. And computer vendors that offer free and easy-to-reach customer support may be able to charge a premium, or at least not lose customers to commodity discounters.

Multinational companies are learning to use the Internet's new technologies to improve efficiency and reach their customers. Federal Express's 30,000 employees around the world are linked via the Internet to "intranet" sites within the company's Memphis headquarters; some 12,000 customers a day track their own packages using Federal Express's Internet Web site, rather than calling a human operator. Ford Motor Company engineers in Asia, Europe, and the United States worked together electronically to design the Taurus automobile. Eli Lilly, a pharmaceutical company, uses information compiled on its intranet sites to schedule clinical trials and submissions for approval of new drugs in countries around the world. Visa International provides an information service called Visa Vue for its 19,000 member banks on an internal Web site.[9] As electronic security improves in the form of "firewalls," which prevent unauthorized access to private networks, and encryption, which protects the privacy of personal and financial data, more companies will use the Internet to sell products and services as well as to link their employees.

The Internet opens a global market to the small business, and allows low-budget, nonprofit organizations to reach interested parties across the country or the world. While Reuters and Dow Jones are repackaging financial information for electronic subscribers, a start-up company in Silicon Valley called QuoteCom is selling financial information over the Internet for as little as $10 per month. The Future Fantasy Bookstore in Palo Alto, California, put its catalog on the Internet and suddenly became a global firm.[10] Small businesses with computer expertise can also set up shop as gateways to the Internet for their communities. In eastern Washington State, the Palouse Economic Development Council is establishing PalouseNet, a World Wide Web server and Internet aggregator for farmers and small businesses in rural Washington and Idaho.[11]

Telecommunications networks are creating a global information workforce, as employers seek the cheapest labor, ranging from clerical work such as data entry to software programming and research and development. In the short run, these clerical jobs offer attractive employment opportunities for developing countries, such as the Commonwealth Caribbean countries and the Philippines, with relatively high literacy rates and an English-speaking workforce. However, not only do these jobs have the same disadvantages as similar jobs in industrialized countries (low pay, boredom, little chance of advancement, stress from pressure to reach productivity targets or from computerized monitoring), but they may also be made obsolete by optical scanners and voice recognition technology that can transform hard copy and spoken words into digitized text.

Some countries and companies may together be able to use telecommunications to create a competitive advantage at the other end of the information work continuum. India has built software development parks equipped with satellite uplinks so that foreign high-tech companies can hire Indian engineers and programmers at a fraction of the cost of expanding its professional workforce in the United States, and India can retain professionals who might otherwise join its massive brain drain. Just as North American and European laborers complained in the 1980s about the growth of offshore manufacturing, highly skilled information workers in Silicon Valley now fear losing their jobs to lower-paid, offshore professionals.

PROMISES AND PARADOXES

New technologies and services are alluring, but they also present challenges and paradoxes for policy makers, the telecommunications industry, and users. Consider the following:

- **Technological Trojan Horses:** New technologies are changing faster than policy makers can respond. Callback services are introducing much lower international rates in countries that still have very high international tariffs designed to generate profits to be invested in domestic infrastructure. Satellite broadcasting has introduced foreign and commercialized programs in Western Europe and in much of Asia, forcing domestic broadcasters to innovate in order to hold on to their audiences. The Internet, seen by many policy makers as an important tool for their nation's industries to remain competitive, opens the door to unfiltered information that may be considered inappropriate or illegal in their countries.

- **Competition and Consolidation:** While telecommunications services are increasingly being liberalized to attract competitive providers, there is also a growing tendency to consolidate. The result may be only a few major players or consortia in the international environment, as well as a few providers in major domestic markets. These new oligopolies will be able to offer a greater range of services than their predecessors, and may make it easier for users looking for "one-stop shopping" to meet their telecommunications needs. The danger, however, is that they will form cartels that will prevent significant competition in price, service, or innovation.

- **Access and Control:** Some governments that see information technology as critical to their economic development strategy are at the same time concerned about the sociopolitical implications of access. One of the most ironic examples of the simultaneously held goals of modernization and control is Singapore, which is staking its economic future on becoming an "intelligent island." Singapore intends to extend fiber optics throughout the island to every business, home, and school by early in the next decade.[12] While catering to business needs, Singapore has retained tight control over individual access to information. It is illegal to install a satellite antenna, so Singaporeans cannot watch satellite-transmitted programs uplinked from Singapore's industrial parks to regional satellites. A fiber-optic network will enable the government to continue to control the population's electronic access to information.

Cable television networks offer the advantage of controlling access so that network operators can charge for reception. In Singapore and China, the perceived advantage of cable is that the information its citizens receive can be monitored. Another technology with paradoxical capabilities to expand and control choice is video compression. Using video compression, programmers can pack many television signals on a single satellite transponder, enabling television viewers to choose from hundreds of digitally compressed channels. Video compression also provides a cost-effective means for program distributors to pre-censor programs by editing different versions for different countries, then compressing and encrypting them for distribution on regional satellites. Viewers in each country will be able to see only the programs their governments have approved.

- **Tariff Reform:** In countries where competition has been introduced, prices have dropped dramatically. Users in countries that retain very high accounting rates are also likely to see prices fall as competition in the form of new carriers or callback services forces their national carriers to adjust their prices. Yet local calling may become more expensive, as carriers rebalance their tariffs to cover more of their costs from access to the local loop and local calling.

    The traditional pricing structure in telecommunications has been based on distance and time. Most carriers still rely on these parameters, even for satellite traffic, where the call travels more than 44,000 miles whether the locations on earth are 400 miles or 4,000 miles apart. But now that enormous growth in network capacity has drastically reduced the cost of carrying traffic, distance-sensitive pricing is becoming harder to justify. In contrast, the computer communications industry uses different models, charging by the volume of data transmitted, or setting fixed prices for unlimited access—the approach used by many Internet service providers.

- **Universal Service as a Moving Target:** New technologies and services are forcing policy makers to rethink their goals of universality. Provision of services for the community, rather than the household, which is the near-term goal for many developing countries, is also the goal adopted by most industrialized countries for providing access to enhanced services. In both industrialized and developing regions, universal service has become a moving target as policy makers must adjust their goals to make new services more accessible.

CLOSING THE GAP

The World Bank estimates that investment in telecommunications in the developing world must double to meet the growing demand for telecommunications services.[13] Despite accelerated investment in many developing regions during the past decade, the vast majority of people living in developing countries still lack access to basic telecommunications. Yet there is cause for optimism. New technologies offer the possibility of technological leapfrogging; for example, subscribers can be connected through wireless local loops or small satellite terminals rather than stringing wire and cable. Digital transmission and switching are increasing reliability and lowering cost, as well as making it possible for subscribers in developing countries to use electronic mail and voice mesaging, and to access the Internet.

The newly industrializing economies of Eastern Europe, Asia, and Latin America are starting to close the gap. Their growing economies are attractive to the telecommunications industry and to investors who are looking for new markets. Most of these countries are also taking steps to encourage investment by privatizing their operators, providing investment incentives, and introducing competition.

Information gaps show the least signs of shrinking in the poorest countries, two-thirds of which have less than one telephone line per 100 inhabitants. Telecommunications is not a panacea for countries with populations near the subsistence level, with urgent demands on foreign exchange for food, fuel, and medicine. Yet, as these countries develop market economies and seek to take maximum advantage of scarce expertise, they will need to invest in telecommunications. Of course, these regions are less attractive to investors than more prosperous economies; in general, they have also been the most reluctant to reduce their government's role as monopoly operator. Their networks are also the least efficient in terms of reliability and the number of lines per telecommunications employee. Restructuring their telecommunications sectors to improve productivity and encourage investment will be necessary if they are to begin to close the gap.

NEW GAPS?

As investment in telecommunications infrastructure increases, the gap between information haves and have-nots may become based on price and choice, rather than technology. Countries that continue to favor telecommunications monopolies, or seek to control access to information, may limit user access even where technology is available. For example, in most of Europe, access to the Internet is much more expensive than in North America. As one commentator states: "Digital Europe has many medieval features: road tolls and extortion-like taxes, witch hunts, an oppressed citizenry, and powers-that-be in feudal towers."[14] Access is much less affordable in many developing countries. Even professionals in many African countries cannot afford to use basic telecommunications services.

China may be the world's largest market for telecommunications, but the government is reluctant to allow access to information from abroad. Although credited with introducing market reforms in the Chinese economy, Deng Xiaoping voiced his ambivalence about opening China's doors to the world: "When the door opens, some flies are bound to come in."[15] Government attempts to control access include banning satellite antennas, blocking access to Internet sites, and impeding access to the Internet itself and to other means of electronic communication.

Clever users will inevitably find means to bypass these roadblocks, as shown by dissidents' use of facsimile and electronic mail during the Tiananmen Square uprising in China, the proliferation of satellite antennas in countries where they are officially banned, and the widespread availability of supposedly illegal callback services that undercut international tariffs. Yet these strategies are likely to be limited to an elite few with the technical expertise or political connections to end-run the regulations. Only when governments recognize that suffocation is worse than a few flies will the gaps really disappear.

TOWARD THE GLOBAL VILLAGE

New technologies have eliminated distance for the international finance industry, which trades not only around the world, but around the clock; for employees who collaborate on projects across time zones; for "footloose" businesses that operate from rural communities; and for students and researchers who can search libraries and databases beyond their borders. But in many parts of the world, to paraphrase Mark Twain, "the news of its death has been greatly exaggerated." [16] Some people may live hours or days from the nearest telephone. Others have modern facilities available, but cannot afford to use them. Still others may not know how to use these new tools to find the information they need, or how to reorganize their work to take advantage of the information available to them. These barriers must also fall if we are all to participate in the global village.

APPENDIX A:
SOURCES OF INFORMATION
ON THE INTERNET

Increasingly, timely information on international telecommunications is available on the Internet's World Wide Web. Below are Internet addresses that may be useful in tracking new developments in international communications. This list does not include telecommunications companies, most of which can easily be found simply by searching for their corporate name. Rather, it lists sites of research institutions, government agencies, international organizations, and development agencies that post information about international telecommunications.

Benton Foundation: universal service, consumer policy reports, and research, Telecommunications Act of 1996:

    http://www.cdinet.com/Benton

Canadian International Development Research Centre (IDRC):

    http://www.idrc.ca

Canadian Radio-Television and Telecommunications Commission (CRTC):

    http://www.crtc.gov.ca

CIRCIT (Centre for International Research on Communications and Information Technology):

    teloz.latrobe.edu.au/circit

Columbia University Institute for Tele-Information (CITI):

    http//:www.ctr.columbia.edu/vii

Electronic Frontier Foundation: public interest organization concerned with privacy, freedom of expression, and other computer communications issues:

    http://www.eff.org

European Telecommunications Standards Institute (ETSI):
http://www.etsi.fr

European Union documents:
www2.echo.lu/eudocs/en/eudocshome.html

Eutelsat (European Telecommunications Satellite Organization):
http://www.eutelsat.org

Inmarsat (International Maritime Satellite Organization):
http://www.inmarsat.org

Intelsat:
http://www.intelsat.int

International Telecommunication Union:
http://www.itu.ch/index.html

Leland Project (USAID project providing Internet access for Africa):
info.usaid.gov/regions/afr/leland

Massachusetts Institute of Technology Media Lab:
http://www.media.mit.edu

Ministry of Posts and Telecommunications of Japan:
http://www.mpt.go.jp

National Computer Board of Singapore: links to sources on information
infrastructure initiatives in many countries:
http://www.ncb.gov.sg

Pacific Telecommunications Council:
http//:www.ptc.org

U.S. National Telecommunications and Information Adminstration:
http://www.ntia.doc.gov

U.S. Federal Communications Commission:
http://www.fcc.gov

U.S. National Information Infrastructure Task Force:
http://www.iitf.doc.gov

University of Michigan Telecommunications Policy Resources: contains
more than 2,000 links to telecommunications policy sites:
http://www.spp.umich.edu/telecom/telecom-info.html

University of San Francisco Telecommunications Management and Policy Program:

    http://www.usfca.edu/mclaren/Telecom/Telecom.html

World Bank: InfoDev Project:

    http://www.worldbank.org/html/fpd/infodev/infodev.html

NOTES

Chapter 1: Introduction

1. William McGowan, "The Part as Prologue: The Impact of International Telecommunications," in *Telecom 91 Global Review,* ed. Hugh Chaloner (London: Kline Publishing, 1991), p. 56.

2. Ibid.

3. Quoted in McGowan, ibid.

4. Quoted in McGowan, ibid.

5. International Telecommunication Union (ITU), *World Telecommunication Development Report,* (1994): p. 13.

6. ITU, p. 14.

7. Derived from ITU, *World Telecommunication Development Report* (1995).

8. ITU (1994): p. 14.

9. Gerard Santucci, "Telecoms Equipment," *XIII Magazine* 8 (October 1992): p. 7.

10. Hans-Peter Gassman, "Information, Technology and Telecommunications Policies in OECD Countries," in *Eastern Europe: Information and Communication Technology Challenges,* ed. G. Russell Pipe (Burke, VA: Transnational Data Reporting Service, 1991), p. 52.

11. Santucci, p. 7.

12. ITU (1995).

13. ITU (1995).

14. McGowan, p. 58.

15. Geoffrey Vincent, "Upheaval in the Global Market: Competition and Technology in Telecoms," *Global Review* (1991): pp. 137–38.

16. Vincent, p. 138.

Chapter 2: New Technologies and Applications

1. Ithiel de Sola Pool, *Technologies without Boundaries* (Cambridge, MA: Harvard University Press, 1990).

2. David Pilling, "A Demand for Fast Workers with Built-in Memories," *Financial Times,* October 7, 1991, p. XX.

3. Derived from International Telecommunication Union (ITU), *World Telecommunication Development Report* (1995).

4. Derived from ITU, *World Telecommunication Development Report* (1994).

5. Roger Pye and Martin Heath, "Overcoming Barriers to Entry and Distortion to Competition," *Pan European Mobile Communications* (Autumn 1991): pp. 43–47.

6. Neil McCartney, "A Vehicle for Introduction of Competition," *Financial Times,* October 7, 1991, p. VIII.

7. Steven D. Dorfman, "Satellite Communications: The Global Network," in *Proceedings of the ITU Forum 91 Policy Symposium* (Geneva: ITU, October 1991), pp. 8–12.

8. Patrick Flanagan, "VSAT: A Market and Technology Overview," *Telecommunications* (March 1993): pp. 19–24.

9. Flanagan, p. 22.

10. ITU (1995).

11. Rhonda Crane, "Advanced Television: An American Challenge," *Boston Globe,* November 6, 1988, p. 46.

12. ITU (1994).

13. Ibid.

14. James O. Jackson, "It's a Wired, Wired World," *InDepth,* June 1995, pp. 14–15.

15. Alix Christie, Montieth M. Illingworth, and Larry Lange. "One World? Online Advocates Dream of a World Linked by the Information Superhighway," *Information Week,* October 2, 1995, pp. 52–64.

Chapter 3: The Major Players

1. Gerard Santucci, "Telecommunications Equipment," *XIII Magazine* 8 (October 1992): p. 7.

2. International Telecommunication Union (ITU), *World Telecommunication Development Report* (1994).

3. Stephen F. Stine,"Hutchison Telecom's Robert Siemens Seeks to Expand Cellular Operations Beyond Asia," *Wall Street Journal Europe,* July 23, 1991.

4. ITU (1994).

5. Santucci, p. 8.

6. Santucci, p. 8.

7. Datapro, "Telecommunications in North America." *Datapro Reports on International Telecommunications* (November 1991): pp. 101–108.

8. ITU, *World Telecommunication Development Report* (1995).

9. ITU (1995).

10. Catherine Arnst, "The Giants Aren't Sleeping," *BusinessWeek,* April 8, 1996, pp. 70–72.

11. ITU (1995).

12. Paul M. Eng, "Surfing's Biggest Splash?" *BusinessWeek,* March 11, 1996, p. 86.

13. Ibid.

14. Society for Worldwide Interbank Financial Telecommunication, *Annual Report 1994* (Belgium: SWIFT, 1995).

15. "Bank of America to Launch Pilot EDI Scheme," *Financial Times,* September 9, 1995.

16. T. D. Cureton, "Global Banking in the Year 2000" (paper presented at Networking in the Year 2000 Conference, Hong Kong, Sept. 28, 1995).

17. Armando T. Filho, "Banco Bradesco Implements Extensive VSAT Network," *Communications News* (March 1993): p. 11.

18. Stephen McClelland, "A Global Company, a Global Network," *Telecommunications* (March 1993): pp. 65–67.

19. Patrick Flanagan, "VSAT: A Market and Technology Overview," *Telecommunications* (March 1993): p. 23.

20. Arthur Markowitz, "K-Mart Power Merchandising." *Discount Store News* (January 21, 1994): pp. 54–55.

21. "Furniture Maker Uses EDI to Stop Paper Blizzard." *Communications News* (September 1992): pp. 26–27.

22. Susanna Schweizer, "Increasing Profitability through Distributed Network Teaming," *Telecommunications* (March 1993): pp. 52–57.

23. ITU (1995).

24. "Workers Keep Their Offices in a Briefcase," *Communications News* (March 1993): p. 12.

25. Heather E. Hudson, *Communication Satellites: Their Development and Impact* (New York: Free Press, 1990).

26. Charles Clements, "Healthnet." (Cambridge, MA: SatelLife, May 1991).

27. Tony Johnson, "Microsatellite that Turns Information into Medical Power," *The (Manchester) Guardian,* April 26, 1991.

28. Quoted in Louis Uchitelle, "We're Leaner, Meaner, and Going Nowhere Faster," *New York Times,* May 12, 1996, p. 4–4.

Chapter 4: Structural Models of the Telecommunications Industry

1. Quoted in Jocelyn Cote-O'Hara, "Risk and Opportunity: Navigating the Waves of Change," in *Proceedings of the Telecom 95 Strategies Summit* (Geneva: ITU, October 1995).

2. Douglas Goldschmidt, "On Curbing Predation in Newly Competitive Markets" (unpublished paper prepared for PanAmSat, Greenwich, CT, 1991).

3. *Telecommunications Act of 1996,* Public Law 104-104, February 8, 1996, Section 254 (b)(6).

4. Ibid.

5. The ITU's *World Telecommunication Development Report* includes data on the largest city and rest of the country. These data are themselves quite striking in revealing disparities in access, but it would be useful to remove cities completely to get a better estimate of rural communications, particularly in countries that have many large cities.

Chapter 5: Restructuring the Telecommunications Sector

1. Robert Lucky, "Technological Advances from 2001 Onwards," in *Proceedings of the ITU Forum 91 Regulatory Symposium* (Geneva: ITU, October 1991), p. 71.

2. *Hush-A-Phone v. U.S.,* 238 F.2d 266, 269.

3. *Carterfone v. AT&T,* 1968, 13 F.C.C.2d, *reconsideration denied.* 18 F.C.C.2d 871.

4. *In the Matter of Allocation of Frequencies in the Bands Above 890 Mc.* 27 F.C.C. 359 (1959).

5. See Heather E. Hudson, *Communication Satellites: Their Development and Impact* (New York: Free Press, 1990).

6. *Computer Inquiry II,* 77 F.C.C.2d 384 (1980).

7. *Computer Inquiry III, Phase I,* 104 F.C.C.2d 958 (1986).

8. *Global Digest* (January–February 1993): p. 19.

9. Bryan Carsberg, "The Role and Purpose of Regulation: Regulator," in *Proceedings of the ITU Forum 91 Regulatory Symposium,* (Geneva: ITU, October 1991): p. 23.

10. Datapro, "Telecommunications in Europe," *Datapro Reports on International Telecommunications* (June 1990): p. 104.

11. International Telecommunications Union (ITU), *World Telecommunication Development Report* (1995).

12. *Financial Times,* October 7, 1991, p. XXVI.

13. Quoted in Martin Cave, "The 1995 Review of Telecommunications Policy in the United Kingdom," in *Proceedings of 13th IDATE International Conference* (Montpelier: November 1991), p. 480.

14. Mark Newman, "Oftel: A Difficult Balancing Act," *Financial Times,* October 7, 1991, p. VI.

15. Cave, p. 480.

16. Ibid., p. 477.

17. William Wigglesworth, "The Impact of Privatisation and Competition on Telecommunications Industry Development: Experiences in the UK" (paper presented at the 1993 Pan-Asian Summit, Hong Kong, June 1993), p. 4.

18. Cave, p. 479.

19. Wigglesworth, p. 5.

20. Sei Kageyama, "Example from the Industrialized Countries," in *Proceedings of the ITU Forum 91 Economic Symposium* (Geneva: ITU, October 1991), p. 75.

21. Ibid., p. 76.

22. Ibid.

23. *APT Yearbook 1995* (Bangkok: Asia-Pacific Telecommunity, 1995).

24. Paul W. Sage, "Telecommunications Competition, Privatization, and Industry Development in Japan: Opportunities for the Asia Pacific Region" (paper presented at the 1993 Pan-Asian Summit, Hong Kong, June 1993).

25. Kazuo Inamori, "The Role and Purpose of Regulation: User," in *Proceedings of the ITU Forum 91 Regulatory Symposium* (Geneva: ITU, October 1991), p. 4.

26. Neil W. Davis, "Japan: International Services: Big Six Houses Branch Out," *Financial Times,* October 7, 1991, p. XXXI.

27. *APT Yearbook 1995,* p. 327.

28. Derived from ITU (1995).

29. Inamori, p. 8.

30. Haruo Yamaguchi, "The Social and Economic Role of Telecommunications in a Peaceful World: New Avenues for Cooperation," in *Proceedings of the ITU Forum 91 Regulatory Symposium* (Geneva: ITU, October 1991), p. 77.

31. Kageyama, p. 76.

32. *Financial Times,* October 7, 1991, p. XXXI.

33. Neil Weinburg, "Japan: Domestic Market: Goliath Gnawed by Rivals," *Financial Times,* October 7, 1991, p. XXXI.

34. M. Koshiyama, "Changing Faces of the Japanese Telecommunications Industry: Post-Privatization Perspectives," in *Proceedings of IDATE 13th International Conference* (Montpelier, November 1991), p. 361.

35. ITU (1995).

36. Inamori, p. 9.

37. Ibid., p. 11.

38. Ibid.

39. Tomeo Kambayashi, "The Provision of Universal Service under a Privatized System," in *Proceedings of the ITU Forum 91 Economic Symposium* (Geneva: ITU, October 1991), p. 113.

40. Ibid., p. 113.

41. Koshiyama, p. 365.

42. Koshiyama, p. 369.

43. For more information on Japan's satellite projects, see Heather E. Hudson, *Communication Satellites: Their Development and Impact* (New York: Free Press, 1990).

44. Ralph Negrine, ed., *Satellite Broadcasting: The Politics and Implications of the New Media* (London: Croom Helm, 1988), p. 265.

Chapter 6: The Telecommunications Sector in Industrialized Countries

1. Information Highway Advisory Council, *Canada's Information Highway: Providing New Dimensions for Learning, Creativity and Entrepreneurship* (Ottawa: Industry Canada, November 1994).

2. Jean-Pierre Chamoux, "The French Telematique Experience," in *Eastern Europe: Information and Communication Technology Challenges,* ed. G. Russell Pipe (Burke, VA: Transnational Data Reporting Service, 1991), p. 182.

3. Simon Nora and Alain Minc, *The Informaticization of Society* (Cambridge, MA: MIT Press, 1978).

4. Chamoux, p. 184.

5. *Financial Times,* October 7, 1991, p. XXVI.

6. "The Death of Distance," *The Economist,* September 30, 1995, p. 9.

7. Marc Fossier and Marie-Monique Steckel, *France Telecom: An Insider's Guide* (Chicago: Telephony Division of Intertec Publishing, 1991), p. 37.

8. Ibid., p. 7.

9. *The Economist,* July 31, 1993, p. 61.

10. Fossier and Steckel, p. 9.

11. Ibid., p. 29.

12. Ibid., p. 22.

13. Ibid., p. 23.

14. John Williamson, "France: Characteristics of Enlightened Paternalism," *Financial Times,* October 7, 1991, p. XXVI.

15. Gail Edmondson and Karen Lowry Miller, "Europe's Markets are Getting Rewired, Too," *BusinessWeek,* April 8, 1996, p. 87.

16. *The Economist,* p. 61.

17. Christian Schwarz-Schilling, "Germany: Structural Reform," in *Telecom 91 Global Review,* ed. Hugh Chaloner (London: Kline Publishing, 1991), p. 107.

18. David Goodhart, "Germany: Reforms Driven by Unity," *Financial Times,* October 7, 1991, p. XXVI.

19. Eberhard Witte, "Liberalization without Privatization," in *Proceedings of the ITU Forum 91 Regulatory Symposium* (Geneva: ITU, October 1991), p. 2.

20. Helmut Ricke, "Management of Change, in Particular in East-West Relations," in *Proceedings of the ITU Forum 91 Policy Symposium* (Geneva: ITU, October 1991), p. 133.

21. Ibid.

22. Edward Russell-Walling, ed., *Global Telecom's Business Yearbook 1995,* (London: Euromoney Publications, 1995), p 11.

23. Ibid., p. 45.

24. *OECD Outlook* (1995).

25. International Telecommunication Union (ITU), *World Telecommunication Development Report* (1995).

26. Russell-Walling, p. 35.

27. Ibid., p. 45.

28. Ibid., p. 57.

29. ITU (1995).

30. Russell-Walling, p. 59.

31. Datapro, "Telecommunications in North America," *Datapro Reports on International Telecommunications* (November 1991), p. 105.

32. CRTC World Wide Web page, www.crtc.gov.ca (May 1996).

33. APEC Telecommunications Working Group, "The State of Telecommunications Infrastructure and Regulatory Environment in APEC Countries" (Seoul: Ministry of Communications and Korea Information Society Development Institute, November 1991), p. 44.

34. Stentor Telecom Policy, *Global Digest* (January–February 1993): p. 18.

35. Bernard Simon, "Canada: Difficult to Hold the Line on Barriers," *Financial Times,* October 7, 1991 p. XXIV.

36. Meriel V. M. Bradford, "Telecommunications in Transition: A Canadian Perspective on Ownership and Other Key Issues," in *Proceedings of the Pacific Telecommunications Conference* (Honolulu: Pacific Telecommunications Council, January 1996).

37. Ibid.

38. Datapro, p. 102.

39. Datapro, "An Overview of World Telecommunications," *Datapro Reports on International Telecommunications* (1993).

40. Derived from APEC, p. 39.

41. APEC, p. 34.

42. ITU (1995).

43. APEC, p. 3.

44. Kevin Brown, "Australia: Compromise over Reforms," *Financial Times,* October 7, 1991, p. XXX.

45. APEC, p. 3.

46. Ibid., p. 188.

47. Ibid.

48. Ibid., p. 186.

49. Terry Hall, "New Zealand: From Monopoly to Competitor," *Financial Times,* October 7, 1991 p. XXX.

50. *The Economist,* July 31, 1993, p. 62.

51. Ibid.

52. Remarks of the President of DDI at the ITU Strategy Summit, Telecom 95 (Geneva, October 1995).

53. William Wigglesworth, "The Impact of Privatisation and Competition on Telecommunications Industry Development: Experiences in the U.K" (paper presented at the 1993 Pan-Asian Summit, Hong Kong, June 1993), p. 7.

54. *The Economist,* July 31, 1993, p. 62.

55. Ibid.

Chapter 7: The European Union:

1. Herbert Ungerer, "Telecom Risks, Telecom Futures," in *Proceedings of the Telecom 95 Strategies Summit* (Geneva: ITU, October 1995, Session 17:8), p. 1.

2. Commission of the European Communities, *From Single Market to European Union* (Brussels: European Commission, 1992), p. 28.

3. Commission of the European Communities, Directorate-General XIII, *Information and Communications Technologies in Europe* (Brussels: European Commission, 1991), p. 34.

4. John Williamson, "European Commission: Intervention is the Slogan," *Financial Times,* October 7, 1991, p. VI.

5. Datapro, "Telecommunications in Europe," *Datapro Reports on International Telecommunications* (June 1990): pp. 101-110.

6. Commission of the European Communities (1991), p. 31.

7. Organization for Economic Cooperation and Development, *Communications Outlook 1995* (Paris: OECD, 1995), pp. 105-110.

8. Commission of the European Communities (1991), p. 32.

9. Commission of the European Communities (1991), p. 35.

10. Herbert Ungerer, "Pushing Towards New World Communications," in *Telecom 91 Global Review,* ed. Hugh Chaloner (London: Kline Publishing, 1991), p. 251.

11. Michael Sheridan, "1992: The Wider Impact," in *Telecom 91 Global Review,* ed. Hugh Chaloner (London: Kline Publishing, 1991), p. 62.

12. Cor Berben, "International Harmonization of Standards," in *Proceedings of the ITU Forum 91 Regulatory Symposium* (Geneva: ITU, October 1991) p. 24.

13. Commission of the European Communities (1991), p. 37.

14. Piero Ravaioli, "View from the European Commission," in *Proceedings of the ITU Forum 91 Regulatory Symposium* (Geneva: ITU, October 1991), p. 91.

15. Williamson, p. VI.

16. Datapro (1990), p. 101.

17. Commission of the European Communities (1992), p. 33.

18. Williamson, p. VI.

19. Quoted in Sheridan, p. 61.

20. Ungerer (1995), pp. 2–4.

21. Ungerer (1995), p. 5.

22. Gail Edmondson and Karen Lowry Miller, "Europe's Markets are Getting Rewired Too," *BusinessWeek,* April 8, 1996, p. 87.

23. Commission of the European Communities (1991), p. 40.

24. *XIII Magazine* 4 (1992): p. 4.

25. Commission of the European Communities (1992), p. 19.

26. Commission of the European Communities (1991), p. 43.

27. Commission of the European Communities, Directorate-General XIII, "ORA 1992: Research and Technology Development on Telematic Systems for Rural Areas" (Brussels: European Commission, 1992).

28. STAR (Special Telecommunications Actions for the Regions) is a European Community program for the development of certain less-favored regions of the Community by improving access to advanced telecommunications services.

29. Commission of the European Communities (1991), p. 39.

30. Commission of the European Communities, "Growth, Competitiveness, Employment: The Challenges and Ways Forward into the 21st Century" (White Paper.) (Brussels: European Commission, 1993).

31. Bangemann Group, *Europe and the Global Information Society* (Brussels: European Commission, 1994).

32. Commission of the European Communities, *Europe's Way to the Information Society: An Action Plan* (Brussels: European Commission, 1994).

33. Colin Turner, "Trans-European Networks and the Common Information Area: The Development of a European Strategy," *Telecommunications Policy* 19, no. 6 (August 1995): pp. 501–508.

34. Quoted in Turner, p. 506.

35. Commission of the European Communities, "European Industrial Policy for the 1990s," *Bulletin of the European Communities,* Supplement 3/91 (Brussels: European Commission, 1991), p. 22.

36. Ibid.

37. Gerard Santucci, "Telecommunications Equipment," *XIII Magazine* 8 (October 1992): p. 9.

38. Michael Goddard, "WARC-92: A Golden Opportunity for Mobile Radio," *Pan-European Mobile Communications* 7 (Autumn 1991): p. 30.

39. Karl H. Rosenbrock and John Ketchell "International Standards: A New Environment for the 90s," in *Telecom 91 Global Review,* ed. Hugh Chaloner (London: Kline Publishing, 1991), p. 32.

40. Derived from Rosenbrock and Ketchell, p. 32.

41. Karl H. Rosenbrock, "An Overview of Major Standardization Issues Today," in *Proceedings of the ITU Forum 91 Policy Symposium* (October 1991), p. 26.

42. Rosenbrock, p. 27.

43. Ibid., p. 31.

44. Ibid., p. 25.

45. Ungerer (1995), p. 250.

46. Commission of the European Communities (1991), p. 46.

47. Ibid., p. 58.

48. See, for example, Parker, Hudson et al., *Rural America in the Information Age* (1989) and Parker and Hudson, *Electronic Byways* (1995).

Chapter 8: The Role of Telecommunications in Socioeconomic Development

1. International Commission for Worldwide Telecommunications Development (The Maitland Commission), *The Missing Link* (Geneva: International Telecommunication Union, 1985).

2. Ibid. (The author was a special advisor to the Maitland Commission and drafted sections of the report on the role of telecommunications in socioeconomic development.)

3. National Research Council Board on Science and Technology for International Development, *Science and Technology Information Services and Systems in Africa* (Washington, D.C.: National Academy Press, 1990).

4. MacBride Commission, *Many Voices, One World* (Paris: UNESCO, 1981).

5. See, for example, Robert Saunders, Jeremy Warford, and Bjorn Wellenius, *Telecommunications and Economic Development*, 2d ed. (Baltimore: Johns Hopkins University Press, 1994); Heather E. Hudson, *When Telephones Reach the Village: The Role of Telecommunications in Rural Development* (Norwood, NJ: Ablex, 1984); International Telecommunication Union, *Information, Telecommunications, and Development* (Geneva: ITU, 1986).

6. See, for example, Edwin B. Parker and Heather E. Hudson, *Electronic Byways: State Policies for Rural Development through Telecommunications*, 2d ed. (Washington, D.C.: Aspen Institute, 1995).

7. Andrew P. Hardy, "The Role of the Telephone in Economic Development," *Telecommunications Policy* (December 1980).

8. Heather E. Hudson, *Communication Satellites: Their Development and Impact* (New York: Free Press, 1990).

9. Hudson, 1984.

10. *SatelLife News* (May 1994).

11. Hudson, 1990.

12. Greg Halik, "Mexico Runs First Satellite-Delivered Digital Code," *Communications News* (March 1993), p. 15.

13. Saunders et al., 1994.

14. Heather E. Hudson, *Three Case Studies on the Benefits of Telecommunications in Socio-Economic Development* (Geneva: ITU, 1981).

15. Hudson, 1984.

16. William B Pierce and Nicolas Jequier, *Telecommunications for Development* (Geneva: ITU, 1983).

17. International Telecommunication Union (ITU), *Benefits of Telecommunications to the Transportation Sector of Developing Countries* (Geneva: ITU, 1988).

18. Pierce and Jequier, 1983.

19. Ibid.

20. John K. Mayo, Gary R. Heald, and Steven J. Klees, "Commercial Satellite Telecommunications and National Development: Lessons from Peru," *Telecommunications Policy* 16, no. 1 (January–February 1992): p. 78.

21. Saunders et al., 1994.

22. Hardy, 1980.

23. Hudson, 1990.

24. Chang-Bun Yoon, "Korea's Strategy for Developing Information/Communication Technology," in *Eastern Europe: Information and Communication Technology Challenges,* ed. G. Russell Pipe (Burke, VA: Transnational Data Reporting Service, 1991), pp. 191–232.

25. See Heather E. Hudson, *Communication Satellites: Their Development and Impact* (New York: Free Press, 1990).

26. See Parker, Hudson et al., *Rural America in the Information Age* (Lanham, MD: University Press of America, 1989).

27. Commission of the European Communities, Directorate-General XIII, *Information and Communications Technologies in Europe* (Brussels: European Commission, 1991).

28. Karen Tietjen, *AID Rural Satellite Program: An Overview* (Washington, D.C.: Academy for Educational Development, 1989).

29. Hudson, 1990.

30. International Development Research Centre, *Pan Asia Networking* (Ottawa: IDRC, 1995). (Also available on the Internet at http://www.idrc.org.sg)

31. Internet address at http://www.info.usaid.gov/regions/afr/leland

32. World Bank, "Information and Development Initiative" (Washington, D.C.: World Bank, 1995).

Chapter 9: Telecommunications Planning for Developing Regions

1. Al Gore, Address to the International Telecommunication Union (ITU) Plenipotentiary Conference (Kyoto, September 1994).

2. John Budden, "Meeting Telecommunications Needs for National and Regional Economic Development: Planning, Management and Funding Options" (Hong Kong, 1993 Pan-Asian Forum, June 1993), p. 3.

3. Ibid.

4. Richard C. Beaird, "Liberalization and Privatization: Prospects for the Future" (Hong Kong, 1993 Pan-Asian Forum, June 1993), p. 8.

5. Stanislawski, Stefan, "What is the Best Way to Organise the Telecommunications Sector? Results of Statistical Analysis and Other Studies" (1993 Pan-Asian Forum, Hong Kong, June 1993), pp. 7–8.

6. Ibid., p. 9.

7. Eliana A. Cardoso, "Privatization Fever in Latin America," *Challenge,* September–October 1991, p. 35.

8. Vineeta Shetty, "Latin America Comes to Market," *Communications International* (June 1996): p. 11.

9. Cardoso, p. 41.

10. Quoted in Beaird, p. 15.

11. Edwin B. Parker and Heather E. Hudson, *Electronic Byways: State Policies for Rural Development through Telecommunications,* 2d ed. (Washington, D.C.: Aspen Institute, 1995).

12. Cardoso, p. 36.

13. S. M. Lodgine, "The New Zealand Experience," *Telecommunications Policy* 16, no. 9 (December 1992): p. 776.

14. Beaird, p. 9.

15. Ibid., p. 13.

16. Ibid., p. 16.

17. Budden, p. 3.

18. William Cooperman, Lori Mukaida, and Donald M. Topping, "The Return of PEACESAT," in *Proceedings of the Pacific Telecommunications Conference* (Honolulu: Pacific Telecommunication Council, January 1991).

19. Arnold Djiwatampu, "Indonesia: The Search for Global Standards," in *Telecom 91 Global Review,* ed. Hugh Chaloner (London: Kline Publishing, 1991), pp. 80–83.

20. Joseph III Kraemer and Nicholas Williams, "Telecommunications Privatization: Managing the Change Process" (paper presented at the 1993 Pan-Asian Forum, Hong Kong, June 1993).

21. ITU, *World Telecommunication Development Report* (1994).

22. Vineeta Shetty, "Tigers to Elephants," *Communications International* (July 1996): p. 17.

23. Derived from ITU, *World Telecommunication Development Report* (1995).

24. Shetty (July 1996), p. 17.

25. Ibid.

26. Derived from ITU (1995).

27. Catherine Arnst, "The Last Frontier," *BusinessWeek,* September 18, 1995, pp. 58–66.

28. Shetty (July 1996): p. 15.

29. Catherine Arnst, "The Last Frontier," *BusinessWeek,* September 18, 1995, pp. 58–66.

30. ITU, *World Telecommunication Development Report* (1994).

Chapter 10: Telecommunications in Eastern Europe and Russia

1. Quoted by Kazimerz Cwiek in *Telecom 91 Global Review,* ed. Hugh Chaloner (London: Kline Publishing, 1991), p. 98.

2. Karlheinz Kaske, in *Telecom 91 Global Review,* ed. Hugh Chaloner (London: Kline Publishing, 1991), p. 88.

3. Ibid.

4. Timothy E. Nulty and Nikola Holcer, in *Telecom 91 Global Review,* ed. Hugh Chaloner (London: Kline Publishing, 1991), p. 46.

5. Blake S. Swensrud, ed., *Eastern European and Soviet Telecom Report* 2, no. 9 (September 1, 1991): p. 3.

6. Klaus W. Grewlich, "Eastern Europe—Cooperative Communications Strategies," in *Eastern Europe: Information and Communication Technology Challenges,* ed. G. Russell Pipe (Burke, VA: Transnational Data Reporting Service, 1991), p. 266.

7. International Telecommunication Union (ITU), *World Telecommunication Development Report* (1995).

8. Dermot Nolan, "Development of Satellite Services in Eastern Europe: Entering the Harsh Reality Phase," in *Telecom 91 Global Review,* ed. Hugh Chaloner (London: Kline Publishing, 1991), p. 198.

9. Kaske, p. 89.

10. Nolan, p. 198.

11. Kaske, p. 69.

12. Nulty and Holcer, p. 46.

13. Helmut Ricke, "Management of Change, in Particular in East-West Relations," in *Proceedings of the ITU Forum 91 Policy Symposium* (Geneva: ITU, October 1991), p. 133.

14. Kaske, p. 68.

15. Nulty and Holcer, p. 47.

16. Ibid., p. 49.

17. Ibid.

18. Thierry Baudon, "The European Bank for Reconstruction and Development: Operational Challenges and Priorities," in *Telecom 91 Global Review,* ed. Hugh Chaloner (London: Kline Publishing, 1991), p. 45.

19. Nigel Tutt, "PHARE." *XIII Magazine* 8 (October 1992): p. 23. (PHARE stands for Poland Hungary Aid for Restructuring the Economy.)

20. Nulty and Holcer, p. 50.

21. Mary Lu Carnevale, "Playing Politics," *Wall Street Journal,* October 4, 1991, p. R14.

22. Anthony Robinson, "Eastern Europe: Poised for a Quantum Leap," *Financial Times,* October 7, 1991, p. XXVIII.

23. Stentor Telecom Policy Inc., *Global Digest* (January–February 1993).

24. Grewlich, p. 269.

25. Wolfgang Kleinwachter and Angela Kolb. "German Unification and its Consequences for Telecommunications," in *Eastern Europe: Information and Communication Technology Challenges,* ed. G. Russell Pipe (Burke, VA: Transnational Data Reporting Service, 1991), p. 156.

26. Swensrud, p. 13.

27. Imre Bolcskei, "A Central European Approach," in *Proceedings of the ITU Forum 91 Regulatory Symposium* (Geneva: ITU, October 1991), pp. 9–13.

28. Paul Horvath, "Telecommunications in Hungary," in *Eastern Europe: Information and Communication Technology Challenges,* ed. G. Russell Pipe (Burke, VA: Transnational Data Reporting Service, 1991), p. 92.

29. Bolcskei, p.10.

30. Ibid., p. 9.

31. Swensrud, p. 13.

32. Horvath, p. 92.

33. Ibid.

34. Ibid., p. 95.

35. Ibid., p. 96.

36. Paul Horvath, "General Trends and Digitalization in Hungary" in *Proceedings of the ITU Forum 91 Technical Symposium* (Geneva: ITU, October 1991).

37. Bolcskei, p. 13.

38. Nulty and Holcer, p. 50.

39. Quoted by Cwiek in Chaloner, p. 98.

40. Jan Monkiewicz, "IT/Telecommunications in Poland," in *Eastern Europe: Information and Communication Technology Challenges,* ed. G. Russell Pipe (Burke, VA: Transnational Data Reporting Service, 1991), p. 119.

41. Swensrud, p. 14.

42. Cwiek, p. 100.

43. Ibid., p. 99.

44. John J. Keller, "Northern Wins Phone Order for $120 Million in Poland," *Wall Street Journal Europe,* July 25, 1991.

45. ITU (1995).

46. G. G. Kudriavtzev and L. E. Varakin, "Economic Aspects of Telephone Network Development: The USSR Plan," *Telecommunications Policy* (February 1990): p. 11.

47. Blake Swensrud and Gene Prilepski, "Russia Calls," *InDepth,* September 1995, p. 29.

48. Stentor Telecom Policy, p. 9.

49. Mary Lu Carnevale, "Soviets Select US West to do Phone Gateways," *Wall Street Journal Europe,* July 17, 1991.

50. Swensrud and Prilepski, p. 31.

51. Kathryn Graven, "GTE Forms Venture with Soviets to Provide Phone Service to Moscow," *Wall Street Journal Europe,* July 10, 1991.

52. Stentor Telecom Policy, p. 10.

53. Swensrud and Prilepski, p. 30.

54. Ibid.

55. Ibid., p. 31.

56. Ibid., p. 29.

57. Ken Schaffer, "A Television Window on the Soviet Union," in *Tracing New Orbits: Cooperation and Competition in Global Satellite Development,* ed. Donna Demac (New York: Columbia University Press, 1986), p. 306.

58. Joseph Pelton and John Howkins, *Satellites International* (New York: Stockton Press, 1988), p. 256.

59. Nulty and Holcer, p. 48.

60. Timothy E. Nulty, "Privatization as a Tool of Development: The Central and East European Situation," in *Proceedings of the ITU Forum 91 Economic Symposium,* (Geneva: ITU, October 1991), p. 106.

61. Karlheinz Kaske, "Universal and Advanced Services for Eastern Europe in a Competitive Environment," in *Proceedings of the ITU Forum 91 Policy Symposium* (Geneva: ITU, October 1991), p. 70.

Chapter 11: Asia

1. Personal interview by the author, 1986.

2. Derived from International Telecommunication Union (ITU), *World Telecommunication Development Report* (1995).

3. John Budden, "Meeting Telecommunications Needs for National and Regional Economic Development: Planning, Management and Funding Options" (1993 Pan-Asian Forum, Hong Kong, June 1993), p. 1.

4. Mark A. Hukill and Jussawalla, Meheroo, "Trends in Policies for Telecommunications Infrastructure Development and Investment in the ASEAN Countries" (Honolulu: Institute for Communication and Culture, East-West Center, 1991).

5. Budden, p. 1.

6. Ibid., p. 3.

7. *Wall Street Journal,* October 4, 1991, p. R7.

8. *Telecommunications Reports* 59, no. 28 (July 12, 1993): p. 19.

9. ITU (1995).

10. Ang-Zhao Di, "Development of Chinese Post and Telecommunications and its Effect on National Economy" (paper presented at the Conference on Telecommunications and the Integration of China, Hong Kong, June 17, 1993), p. 8.

11. Junjia Liu, "Basic Information on China's Telecommunications Infrastructure" (paper presented at the Conference on Telecommunications and the Integration of China, Hong Kong, June 17, 1993), p. 1.

12. Ibid., p. 3.

13. ITU (1995).

14. Luan Zhengxi, "China's Changing Telecommunications," in *Telecom 91 Global Review,* ed. Hugh Chaloner (London: Kline Publishing, 1991), pp. 122–124.

15. Di, p. 7.

16. *APT Yearbook 1995* (Bangkok: Asia-Pacific Telecommunity, 1995), p. 280.

17. Zhengxi, p. 123.

18. *APT Yearbook 1995,* p. 275.

19. Ibid., p. 291.

20. Datapro, "China: The Commercial and Regulatory Environment," *Datapro Reports on International Telecommunications* (March 1992): p. 153.

21. John Williamson, "Buoyant Regional Market," *Financial Times,* October 7, 1991, p. XXX.

22. ITU (1995).

23. Datapro, "Telecommunications in the Far East." *Datapro Reports on International Telecommunications* (January 1992), pp. 101–107.

24. *APT Yearbook 1995,* pp. 284–286.

25. Ibid., p. 286.

26. Ibid., p. 278.

27. Milton Mueller and Zixiang Tan, *China in the Information Age: Telecommunications and the Dilemma of Reform* (Washington, D.C.: CSIS Books), in press.

28. Ibid.

29. *APT Yearbook 1995,* p. 276.

30. Milton Mueller, "One Country, Two Systems: What Will 1997 Mean in Telecommunications?" *Telecommunications Policy* 18, no. 3 (April 1994): pp. 247–48.

31. Ibid.

32. *APT Yearbook 1995,* p. 277.

33. Lionel Olmer, "Financing The Global Information Infrastructure," in *Proceedings of the Pacific Telecommunications Conference* (Honolulu: Pacific Telecommunications Council, 1996), p. 987.

34. Mueller, 1994.

35. John Ure, ed., *Telecommunications in Asia: Policy, Planning and Development* (Hong Kong: Hong Kong University Press, 1995), p. 23.

36. *APT Yearbook 1995.*

37. Alan B. Kamman, "Hong Kong after 1997: The Effect on Telecommunications Services," in *Proceedings of the Pacific Telecommunications Conference* (Honolulu: Pacific Telecommunications Council, January 1996).

38. ITU, *World Telecommunication Development Report* (1994).

39. *APT Yearbook 1995.*

40. Ure, p. 28.

41. Quoted in Mueller, p. 244.

42. Wong Hung Khim, "Looking to the Future of Singapore," in *Telecom 91 Global Review,* ed. Hugh Chaloner (London: Kline Publishing, 1991), p. 76.

43. Ure, p. 30.

44. ITU (1995).

45. APEC Telecommunications Working Group, "The State of Telecommunications Infrastructure and Regulatory Environment in APEC Countries" (Seoul: Ministry of Communications and Korea Information Society Development Institute, November 1991), p. 203.

46. Mark Hukill, "The Privatisation of Singapore Telecom: Planning, Policy and Regulatory Issues" (paper presented at the 1993 Pan-Asian Summit, Hong Kong, June 1993), p. 1.

47. "The Region Advances," *Asian Communications,* September 1995, p. 87.

48. Ure, p. 31.

49. Hukill, p. 7.

50. Ibid., p. 6.

51. Hukill, pp. 7–8.

52. Ure, p. 31.

53. Hukill, pp. 8–9.

54. ITU (1995).

55. ITU (1995).

56. *APT Yearbook 1995,* p. 341.

57. Ibid., p. 340.

58. APEC, p. 153.

59. Ure, p. 36.

60. Ibid., p. 33.

61. Ibid. p. 34.

62. Chi-Young Kwak, "Telecommunications Liberalization, Industrial Development and Private Sector's Role: Experience from Korea" (paper presented at the 1993 Pan-Asian Summit, Hong Kong, June 1993), pp. 6–7.

63. Ibid., p. 35.

64. Source: Kwak, p. 3.

65. *APT Yearbook 1995,* p. 339.

66. Ure, p. 37.

67. Datapro (1992), p. 104.

68. Steven Y. Chen and Duei Tsai. "Network Construction Plan for NII in R.O.C.," in *Proceedings of the 1996 Pacific Telecommunications Conference,* (Honolulu: Pacific Telecommunications Council, 1996), pp. 662–668.

69. Ure, p. 38.

70. Bjorn Wellenius and Peter A. Stern, eds., *Implementing Reforms in the Telecommunications Sector: Lessons from Experience* (Washington, D.C.: World Bank, 1994), p. 661.

71. Ure, p. 40.

72. Ibid., p. 38.

73. Ibid., p. 39.

74. Mueller, 1994.

Chapter 12: Telecommunications in Asia's Emerging and Industrializing Economies

1. *Wall Street Journal,* October 4, 1991, p. R7.

2. Shahid Akhtar, A. Neelameghan, and Luc Laviolette, "Telecoms Policy Reaches Out to the Private Sector and Villages of India," *Intermedia* 24, no. 3 (June/July 1996): p. 30.

3. International Telecommunication Union (ITU), *World Telecommunication Development Report* (1995).

4. *APT Yearbook 1995* (Bangkok: Asia-Pacific Telecommunity, 1995).

5. Ibid., p. 302.

6. T. H. Chowdary, "Reforms in the Indian Telecom Sector," in *Proceedings of the 1996 Pacific Telecommunications Conference* (Honolulu: Pacific Telecommunications Council, 1996), p. 309.

7. ITU (1995).

8. As one Indian critic puts it: what the DOT is doing is "selling its privilege, derived from the Indian Telegraph Act of 1885, to make money without rendering any services." Chowdary, p. 308.

9. Nikhil Sinha, "The Political Economy of India's Telecommunications Reforms," in *Proceedings of the Pacific Telecommunications Conference* (Honolulu: Pacific Telecommunications Council, 1996), p. 921.

10. Akhtar et al., p. 32.

11. Chowdary, p. 308.

12. *APT Yearbook 1995*, p. 299.

13. Bjorn Wellenius and Peter A. Stern, eds., *Implementing Reforms in the Telecommunications Sector: Lessons from Experience* (Washington, D.C.: World Bank, 1994), p. 622.

14. Akhtar et al., p. 34.

15. Ibid., p. 32

16. Sinha, p. 924.

17. Chowdary, p. 312.

18. Sinha, p. 926.

19. Akhtar et al., p. 31.

20. Ravi Narain, "A Floorless Exchange," *Asian Communications* (June 1996): pp. 50–51.

21. ITU (1995).

22. Subhash Chandra, "The Roll-Out of Zee TV," *Intermedia* 24, no. 3 (June/July 1996): pp. 38–39.

23. Sevanti Ninan, "From Media Poor to Media Rich." *Intermedia,* 24, no. 3, (June/July 1996): pp. 35–37.

24. APEC Telecommunications Working Group, "The State of Telecommunications Infrastructure and Regulatory Environment in APEC Countries" (Seoul: Ministry of Communications and Korea Information Society Development Institute, November 1991), p. 103.

25. John Ure, ed., *Telecommunications in Asia: Policy Planning and Development* (Hong Kong: Hong Kong University Press, 1995), p. 52.

26. Ibid., p. 51.

27. Mirwan Suwarso, "New Public/Private Relationship in Indonesia's Telecommunications Sector," in Bjorn Wellenius and Peter A. Stern, eds., *Implementing Reforms in the Telecommunications Sector: Lessons from Experience* (Washington, D.C.: World Bank, 1994), p. 289.

28. *APT Yearbook 1995*, p. 311.

29. Thomas B. Subijanto, "Indonesia: Astra's Revenue Sharing Project" (paper presented at the 1993 Pan-Asian Summit, Hong Kong, June 1993), pp. 3–4.

30. Ibid., p. 7.

31. Ure, p. 55.

32. Ibid., p. 51.

33. Jonathan Parapak, "Bridging the Urban-Rural Divide in Telecommunication in Asia" (paper presented at the AMIC Conference on Communication, Technology, and Development: Alternatives for Asia, Kuala Lumpur, June 25–27), 1993, p. 7.

34. *APT Yearbook 1995,* p. 309.

35. APEC, p. 107.

36. ITU (1995).

37. *Satellite News* 16, no. 7 (July 5, 1993): p. 3.

38. Arnold Djiwatampu, "Indonesia: The Search for Global Standards," in *Telecom 91 Global Review,* ed. Hugh Chaloner (London: Kline Publishing, 1991), p. 82.

39. *APT Yearbook 1995.*

40. Syed Hussein Mohamed, "Corporatization and Partial Privatization of Telecommunications in Thailand," in *Implementing Reforms in the Telecommunications Sector: Lessons from Experience,* eds. Bjorn Wellenius and Peter A. Stern (Washington, D.C.: World Bank, 1994), pp. 269–270.

41. Ure, p. 57.

42. Wellenius and Stern, eds., p. 633.

43. *APT Yearbook 1995,* p. 352.

44. APEC, pp. 169–183.

45. A. A. Rubio, Timothy J. Logue, and Stuart MacPherson, "Developing a Communications Policy for the Philippines," in *Proceedings of the Pacific Telecommunications Conference* (Honolulu: Pacific Telecommunications Council, 1992).

46. *Wall Street Journal,* October 4, 1991, p. R8.

47. Teresa Dance, "Nurturing Interconnection," *Asian Telecommunications* (June 1996): p. 42.

48. Ure, p. 65.

49. Ibid., p. 60.

50. Ibid., p. 65.

51. Corazon P. B. Claudio, "The Philippines: From Natural Disaster to Rural Progress," in *Telecom 91 Global Review,* ed. Hugh Chaloner (London: Kline Publishing, 1991), pp. 70–75.

52. *APT Yearbook 1995,* p. 401.

53. Personal interviews, Manila, June 1996.

54. Rubio et al., 1992.

55. Teresa Dance, "The Philippines Comes of Age," *Asian Communications* (June 1996): p. 38.

56. *APT Yearbook 1995,* p. 404.

57. Ibid., p. 431.

58. Ure, p. 66.

59. Kosol Petchsuwan, "Telecommunications Development in Thailand," in *Implementing Reforms in the Telecommunications Sector: Lessons from Experience,* eds. Bjorn Wellenius and Peter A. Stern (Washington, D.C.: World Bank, 1994), pp. 259–264.

60. Vallobh Vimolvanich, "National Telecommunications Infrastructure Development Projects in Asia: Thailand" (paper presented at the 1993 Pan-Asian Summit, Hong Kong, June 1993), p. 11.

61. Ibid., pp. 5–6.

62. ITU (1995).

63. Ure, p. 71.

64. Ibid.

65. Ibid.

66. Ibid., pp. 71–72.

67. Ibid., p. 74.

68. *APT Yearbook 1995,* p. 440.

69. *Asian Telecommunications* (June 1996): p. 7.

Chapter 13: Telecommunications in Latin America and the Caribbean

1. Eliana A. Cardoso, "Privatization Fever in Latin America," *Challenge,* September–October 1991, p. 36.

2. Derived from International Telecommunication Union (ITU), *World Telecommunication Development Report* (1995).

3. Randy Zadra, "The Telecommunications Revolution in Latin America" (paper presented at INTERCOMM 93, Vancouver, Canada, February 24, 1993), p. 5.

4. ITU, "Acapulco Declaration" (Americas Regional Telecommunications Development Conference, Acapulco, May 1992).

5. Vineeta Shetty, "Latin America Comes to Market," *Communications International* (June 1996): p. 11.

6. Zadra, p. 8.

7. Stewart Wittering, "VSAT," *Communications International* (June 1996): p. 22.

8. Ibid., p. 21.

9. Judith O'Neill, quoted in Shetty, pp. 10–11.

10. Cardoso, p. 40.

11. Ibid., p. 37.

12. Ibid.

13. Datapro, "Argentina: The Commercial and Regulatory Environment," *Datapro Reports on International Communications* (February 1991): p. 301.

14. Ibid., p. 305.

15. Edward Russell-Walling, ed., *The Americas Yearbook* (London: Euromoney Publications, 1996), p. 7.

16. *Satellite News* 16, no. 27 (July 5, 1993): p. 6.

17. Russell-Walling, ed., p. 11.

18. Bjorn Wellenius and Peter A. Stern, eds., *Implementing Reforms in the Telecommunications Sector: Lessons from Experience* (Washington, D.C.: World Bank, 1994), p. 599.

19. Jonathan Wheatley, "Network in Line for a Shake-Up," *Financial Times,* September 19, 1996, p. XI.

20. Russell-Walling, ed., p. 11.

21. Ibid., p. 10.

22. Shetty, p. 16.

23. Wellenius and Stern, eds., p. 599.

24. Datapro, "Chile: Commercial and Regulatory Environment," *Datapro Reports on International Communications* (February 1991): p. 401.

25. Russell-Walling, ed., p. 14.

26. Wellenius and Stern, eds., p. 603.

27. Russell-Walling, ed., p. 15.

28. Wellenius and Stern, eds., p. 603.

29. Russell-Walling, ed., p. 15.

30. Ibid.

31. ITU, *World Telecommunication Development Report* (1995).

32. Datapro, "Mexico: The Commercial and Regulatory Environment," *Datapro Reports on International Communications* (February 1991): p. 251.

33. Carlos Mier y Teran, "A Latin Experience,s" in *Proceedings of the ITU Forum 91 Regulatory Symposium* (Geneva: ITU, October 1991), p. 33.

34. Ibid.

35. ITU (1995).

36. Mier y Teran, p. 31.

37. Russell-Walling, ed., p. 37.

38. Ibid.

39. Datapro, "Telecommunications in North America," *Datapro Reports on International Telecommunications* (November 1991): p. 105.

40. Ibid., p. 107.

41. Diane Solis, "The Long Wait," *Wall Street Journal,* October 4, 1991, p. R14.

42. Wellenius and Stern, eds., p. 673.

43. Zadra, p. 4.

44. Russell-Walling, ed., p. 55.

45. Wellenius and Stern, eds., p. 673.

46. Russell-Walling, p. 55.

47. Ibid., p. 9.

48. Ibid., p. 25.

49. Ibid., p. 19.

50. Wellenius and Stern, eds.

51. Ibid., p. 597.

52. Russell-Walling, ed., p. 21.

53. Wellenius and Stern, eds., p. 626.

54. Ibid., p. 652.

55. Annette Nabavi, quoted in Shetty, p. 16.

Chapter 14: International Satellite Communications

1. Steven D. Dorfman, "Satellite Communications: The Global Network," in *Proceedings of the ITU Forum 91 Policy Symposium* (Geneva: ITU, October 1991), p. 8.

2. *APT Yearbook 1995* (Bangkok: Asia-Pacific Telecommunity, 1995), p. 94.

3. Quoted in Jonathan F. Galloway, *The Politics and Technology of Satellite Communications* (Lexington, MA: D.C. Heath, 1972), p. 23.

4. 47 USC 701 et seq (The Communications Satellite Act) quoted in Jack Oslund, " 'Open Shores' to 'Open Skies': Sources and Directions of U.S. Satellite Policy," in *Economic and Policy Problems in Satellite Communications,* eds. Joseph Pelton and Marcellus Snow (New York: Praeger, 1977), p. 145.

5. *APT Yearbook 1995,* p. 97.

6. Intelsat, *Annual Report 1994* (Washington, D.C.: Intelsat, 1995).

7. Dean Burch, quoted in Joseph Pelton and John Howkins, *Satellites International* (New York: Stockton Press, 1988), p. 26.

8. Intelsat, *Annual Report 1994*.

9. Ibid., p. 11.

10. Intelsat, "Focus: Asia Pacific" (Washington, D.C.: Intelsat, 1993), pp. 8–14.

11. Intelsat, *Via Intelsat* 10, no. 3 (October 1995): p. 3.

12. Intelsat, *Annual Report 1994*.

13. David Tudge, "International Facilities: Cable and Satellites," in *Proceedings of the ITU Forum 91 Regulatory Symposium* (Geneva: ITU, October 1991), p. 195.

14. Quoted in Galloway, p. 123.

15. Pelton and Howkins, p. 129.

16. Ibid., p. 128.

17. Ibid., p. 129.

18. Ibid., p. 130.

19. Ibid., p. 29.

20. Ibid., p. 130.

21. Inmarsat, "Working Together: Sharing the Benefits of Global Mobile Satellite Communications" (London: Inmarsat, 1994).

22. Pelton and Howkins, p. 29.

23. Ibid., p. 133.

24. Ibid., p. 30.

25. Inmarsat, 1994.

26. Inmarsat, "Project 21: The Development of Personal Mobile Satellite Communications" (London: Inmarsat, 1994).

27. Rex Hollis, "Update on Mobile Satellite Services," in *Proceedings of the 1996 Pacific Telecommunications Conference* (Honolulu: Pacific Telecommunications Council, January 1996), pp. 56–57.

Chapter 15: International Networks and Competition

1. Douglas Goldschmidt, former telecommunications economics consultant for PanAmSat.

2. *APT Yearbook 1995* (Bangkok: Asia-Pacific Telecommunity, 1995), p. 237.

3. Stephen McClelland and Julian Bright, "AT&T: Towards Global Connectivity?" in *State of the Art Handbook* (London: Telecommunications International, October 1995), pp. S53–S54.

4. Bruce Elbert, *International Telecommunication Management* (Norwood, MA: Artech House, 1990), pp. 56, 59.

5. McClelland and Bright, p. S53.

6. Elbert, p. 56.

7. Ibid., p. 61.

8. Ibid., p. 63.

9. *APT Yearbook 1995,* p. 240.

10. John Warta, "Fiber Optic Submarine Cable Systems: An Opportunity for Emerging Markets to Access the Global Highways and Superhighways," in *Proceedings of the Pacific Telecommunications Conference* (Honolulu: Pacific Telecommunications Council, 1996), pp. 967–968.

11. Stephen McClelland and Julian Bright, "FLAG Sets New Standard," in *State of the Art Handbook* (London: Telecommunications International, October 1995), p. S54.

12. Elbert, p. 60.

13. Ibid., p. 64.

14. Warta, p. 968.

15. Elbert, p. 64.

16. Federal Communications Commission, "In the Matter of Establishment of Satellites Systems Providing International Communications" (CC Docket 84–1299), released September 3, 1985, p. 8.

17. Ibid.

18. Ibid., p. 7.

19. See Lee McKnight, "The Deregulation of International Satellite Communications: U.S. Policy and the Intelsat Response," *Space Communications and Broadcasting* 3 (1985): p. 57.

20. Douglas Goldschmidt, "Pan American Satellite and the Introduction of Specialized Communication Systems in Latin America," in *Proceedings of the Pacific Telecommunications Conference* (Honolulu: Pacific Telecommunications Council, 1988), p. 343.

21. Ibid., p. 345.

22. William Cooperman, "Communication Satellites: Narrow Beams to Broadband," *State of the Art Handbook* (London: Telecommunications International, 1995), p. S43.

23. Organization for Economic Co-operation and Development, *Communications Outlook 1995* (Paris: OECD, 1995), p. 135.

24. Durrell W. Hillis, "Mobile Satellite Personal Communications," in *Telecom 91 Global Review,* ed. Hugh Chaloner (London: Kline Publishing, 1991), pp. 230–232.

25. Heather E. Hudson, "Access to Information Resources: The Developmental Context of the Space WARC," *Telecommunications Policy* (March 1985).

26. Cynthia Boeke, "AsiaSat 1: Bringing a New Era of Communications to the Region," *Via Satellite* (March 1995): pp. 6–7.

27. Cynthia Boeke, "AsiaSat 2: Asia's Hot New Bird," *Via Satellite* (March 1995): pp. 10–11.

28. *APT Yearbook 1995,* pp. 229–236.

29. Cooperman, p. S43.

30. Quoted in Scott Chase, "Live via Satellite," *Via Satellite* (April 1988): p. 50.

31. Quoted in Kenneth Dyson and Peter Humphreys, eds., *The Politics of the Communications Revolution in Western Europe* (London: Frank Cass, 1986), p. 108.

32. For more detailed analysis of the origins of the European satellite industry, see Heather E. Hudson, *Communication Satellites: Their Development and Impact* (New York: Free Press, 1990).

33. William Mahoney, "With Success Comes Scrutiny," *Multichannel News International* (March 1996): p. 24.

34. International Telecommunication Union (ITU), *World Telecommunication Development Report* (1995).

35. Dyson and Humphreys, p. 110.

36. ITU (1995).

37. Georgette Wang, "Satellite Television and the Future of Broadcast Television in the Asia Pacific" (paper presented at the AMIC Conference on Communication, Technology and Development: Alternatives for Asia, Kuala Lumpur, June 25–27, 1993), pp. 7, 20.

38. Hugh Leonard, "Asian Broadcasting: The Changing Scene" (paper presented at the AMIC Conference on Communication, Technology, and Development: Alternatives for Asia. Kuala Lumpur, June 25–27, 1993), p. 3.

39. Wang, p. 19.

40. Ibid., p. 3.

41. Ibid., p. 5.

Chapter 16: The International Policy Environment

1. Theodore Irmer. "Challenge and Change for CCITT: Global Standardization of Telecoms in Transition," in *Telecom 91 Global Review,* ed. Hugh Chaloner (London: Kline Publishing, 1991), p. 16.

2. *Communications Week,* quoted by Bradley P. Holmes, "Standards and International Telecommunications Policy," in *Telecom 91 Global Review,* ed. Hugh Chaloner (London: Kline Publishing, 1991), p. 29.

3. Gerard Robin, "The Importance of Standards from a Manufacturer's Viewpoint," in *Proceedings of the ITU Forum 91 Regulatory Symposium* (Geneva: ITU, October 1991), p. 28.

4. The Strategic Planning Group of the U.S. National Committee of the CCITT, "CCITT Interactions with Other Standards Organizations" (Washington, D.C., 1993).

5. Lawrence D. Eicher, "The Convergence of Telecoms and IT: Standardizing in a Global Market," in *Telecom 91 Global Review,* ed. Hugh Chaloner (London: Kline Publishing, 1991), p. 22.

6. Ibid., pp. 22–25.

7. Irmer, p. 17.

8. M. Harbi, "International Regulation of Space Radio Services for the 90's," in *Telecom 91 Global Review,* ed. Hugh Chaloner (London: Kline Publishing, 1991), p. 34.

9. Gary C. Brooks, "International Regulation of Frequencies and Satellite Systems," in *Proceedings of the ITU Forum 91 Regulatory Symposium* (Geneva: ITU, October 1991), p. 30.

10. Irmer, p. 18.

11. Ibid.

12. Gabriel Warren, "The Mission of the ITU in the Fast Changing Environment," in *Proceedings of the ITU Forum 91 Policy Symposium* (Geneva: ITU, October 1991), p. 176.

13. Richard Kirby, "Radio Horizons," *Global Communications* (September–October 1991): pp. 21–26.

14. *Communications Standards Review,* 4, no. 3 (April 1993): pp. 1–3.

15. Independent Commission for Worldwide Telecommunications Development (The Maitland Commission), *The Missing Link* (Geneva: ITU, 1984).

16. International Telecommunication Union (ITU), "Telecommunications Development Worldwide: The Telecommunications Development Bureau at Your Service" (1991), p. 4.

17. Ibid., p. 6

18. Warren, p. 177.

19. Based on low-income countries (1993 per capita GDP of $695 or less). Derived from ITU, *World Telecommunication Development Report* (1995).

20. ITU, "Development Partners Agree on Telecommunications Development Plan for Asia and the Pacific," press release (May 17, 1993), pp. 5–6.

21. Michael Goddard, "WARC-92: A Golden Opportunity for Mobile Radio," *Pan-European Mobile Communications* 7 (Autumn 1991): p. 30.

22. Karl H. Rosenbrock and John Ketchell, "International Standards: A New Environment for the 90s," in *Telecom 91 Global Review,* ed. Hugh Chaloner (London: Kline Publishing, 1991), p. 32.

Chapter 17: The International Policy Environment

1. Ben Petrazzini, *Global Telecom Talks: A Trillion Dollar Deal* (Washington, D.C.: International Institute of Economics, 1996).

2. International Telecommunication Union (ITU), *World Telecommunication Development Report* (1995).

3. Ibid.

4. Jacques Arlandis, "Trading Telecommunications: Challenges to European Regulatory Policies," *Telecommunications Policy* 17, no. 3 (April 1993): pp. 171–185.

5. G. Russell Pipe, "GATT Uruguay Round Services Agreement—Impact on Telecommunications Development," in *Proceedings of the ITU Forum 91 Regulatory Symposium* (Geneva: ITU, October 1991), p. 73.

6. Tedson J. Meyers, "Settlement of Telecommunications Disputes Under the GATT-GNS Framework," in *Proceedings of the ITU Forum 91 Regulatory Symposium* (Geneva: ITU, October 1991), p. 178.

7. Pipe, p. 71.

8. Ibid.

9. Vary T. Coates, Todd M. LaPorte, and Mark G. Young. "Global Telecommunications and Export of Services: The Promise and the Risk," *Business Horizons* 36, no. 6 (November/December 1993): pp. 23–29.

10. General Agreement on Trade in Services, Annex on Telecommunications.

11. Quoted in Petrazzini, pp. 85–88.

12. Ibid., pp. 6–7.

13. Quoted in Petrazzini, p. 86.

14. ITU, *World Telecommunication Development Report* (1994).

15. Gregory Staple, quoted in James Savage, "The Future of the International Telecommunications Accounting Rate Scheme," in *Proceedings of the Pacific Telecommunications Conference* (Honolulu: Pacific Telecommunications Council, 1996), p. 773.

16. Savage, p. 775.

17. Robert M. Frieden, "Accounting Rates: The Business of International Telecommunications and the Incentive to Cheat," *Federal Communications Law Journal* 43, no. 2 (April 1991): p. 115.

18. Petrazzini, p. 65.

19. John Haring, Jeffrey Rohlfs, and Harry Shooshan III, "The U.S. Stake in Competitive Global Telecommunications Services: The Economic Case for Hard Bargaining" (Bethesda, MD: Strategic Policy Research, 1993), quoted in Savage, p. 775.

20. ITU (1994).

21. Ibid.

22. Phillip L. Spector, "The Internet Goes International: Intellectual Property Considerations for Industry," in *Proceedings of the Pacific Telecommunications Conference* (Honolulu: Pacific Telecommunications Council, 1996), p. 562.

23. Ibid., p. 561.

24. Ibid., p. 560.

25. Ibid., p. 564.

26. Ibid., p. 565.

27. Ibid., p. 563.

28. Ibid., p. 565.

Chapter 18: Future Directions in International Telecommunications

1. "The Death of Distance," *The Economist,* September 30, 1995.

2. John Browning, "Joys of the Express Lane," *Globe and Mail Report on Business Magazine.* January 1995, p. 103.

3. Commission of the European Communities, "Growth, Competitiveness, Employment: The Challenges and Ways Forward into the 21st Century" (White Paper) Brussels: European Commission, 1993).

4. Commission of the European Communities, *Europe and the Global Information Society* (Brussels: European Commission, 1994).

5. Industry Canada, *The Canadian Information Highway* (Ottawa: Industry Canada, April 1994).

6. Information Highway Advisory Council, *Canada's Information Highway: Providing New Dimensions for Learning, Creativity and Entrepreneurship* (Ottawa: Industry Canada, November 1994).

7. The G-7 is an association of seven major industrialized world powers: Canada, France, Germany, Japan, Russia, the United Kingdom, and the United States.

8. Commission of the European Communities (1994).

9. Amy Cortese, "Here Comes the Intranet," *BusinessWeek,* February 26, 1996, pp. 76–84.

10. Browning, p. 103.

11. Edwin B. Parker and Heather E. Hudson, *Electronic Byways: State Policies for Rural Development through Telecommunications,* 2d ed. (Washington, D.C.: Aspen Institute, 1995).

12. Steven Strasser, "An Island Wired for Jobs." *Newsweek,* July 19, 1993, p. 37.

13. Alan Cane, "Transforming the Way We Live and Work," in "International Telecommunications: Financial Times Survey," *Financial Times,* October 3, 1995, pp. 1–2.

14. Brent Gregston, "Power and Privilege," *Internet World* (November 1995): p. 96.

15. Steven Schwankert, "Dragons at the Gates," *Internet World* (November 1995): p. 112.

16. Mark Twain cabled to the Associated Press: "The reports of my death are greatly exaggerated."

GLOSSARY

ACP States: The countries in Africa, the Caribbean, and the Pacific that are associate members of the European Union under the Lomé Convention.

AMPS: Advanced Mobile Phone Service, an analog cellular standard developed by Bell Labs.

Analog: A signal that varies continuously (as contrasted with a digital signal).

ANSI: American National Standards Institute.

APEC: Asia Pacific Economic Conference, an association of countries in Asia and the Pacific Rim.

Aperture: The area of a parabolic antenna.

Arabsat: A regional satellite owned by a consortium of Arab countries.

ARPA: U.S. Department of Defense Advanced Research Projects Agency, developer of the Internet.

ASCII: American Standard Code for Information Interchange, the standard code of 8 bit characters used in data communications.

ASEAN: Association of Southeast Asian Nations.

ATM: Automatic teller machine for electronic banking; also Asynchronous Transfer Mode, a packet-switching technique that uses fixed length packets called cells, and effectively eliminates any delay in receiving packets, making it suitable for voice and video.

Audiographics: Graphic communications over narrowband channels.

Audiotext: Interactive audio services accessible via touchtone (DTMF) telephone.

Bandwidth: The capacity of a communications channel, expressed in hertz (cycles per second).

Baud: A unit of signaling speed equal to the number of discrete conditions or signal events per second.

Beam shaping: Configuring the antenna pattern of a satellite to cover a specific region with minimal spillover into adjacent areas.

Bellcore: Bell Communications Research, the research arm of the Bell Operating Companies.

Bit: Binary digit, i.e., either 0 or 1.

BOC: Bell Operating Company, one of the former AT&T subsidiaries that were divested by AT&T under the Modified Final Judgment in 1984.

BOO: Build-Operate-Own; a vehicle for attracting financing in developing regions that allows an investor to build, operate, and own telecommunications facilities on a joint venture basis with a national operator.

BOT: Build-Operate-Transfer; a vehicle for attracting financing in developing regions that allows an investor to build and operate a telecommunications network for a specified period, before transferring ownership to the national operator.

Broadband: High-speed, high-capacity transmission capacity with bandwidth sufficient for voice, data, and video.

Broadcasting: Transmission of signals for reception by the general public; also point-to-multipoint transmission (e.g., of data or video).

BSS: Broadcast Satellite Service.

BTO: Build-Transfer-Operate; a vehicle for attracting financing in developing regions that allows an investor to build a telecommunications network, transfer ownership to the national operator, and then operate the facilities through revenue-sharing arrangements.

Bypass: A service that avoids some or all of the public communications network.

C band: Portion of the electromagnetic spectrum in the 4-to-6 GHz range, used for satellite communications.

Cable television: A transmission system using broadband cable to deliver video signals directly to television sets, in contrast to over-the-air or direct satellite systems.

CAP: Competitive access provider; a carrier that provides telecommunications services, typically to large companies, in competition with the local telephone company.

Carrier's carrier: A telecommunications operator that carries traffic for other carriers and not for the public.

CCITT: International Consultative Committee for Telephony and Telegraphy, an international standards committee established by the International Telecommunication Union. Now replaced by the ITU Telecommunication Standardization Bureau (ITU-T).

CDMA: Code division multiple access, a transmission technique used in digital radio technology.

CEE: Central and Eastern Europe.

Cellular: A wireless communications system in which the coverage area is divided into small sections called cells, each with a transmitter.

CEPT: European Conference of Posts and Telecommunications Administrations.

Channel: A path for transmission of electromagnetic signals.

CIO: Chief information officer.

Coaxial cable: A cable constructed of two conductors separated by insulation, providing much greater bandwidth than copper wire.

COCOM: Coordinating Committee for Multilateral Export Controls.

Codec: Coder-decoder, a device that converts analog signals to digital signals, or vice versa.

Common carrier: An organization authorized to provide telecommunications services at established and stated prices.

Compression: The process of eliminating redundant characters from a data stream before it is stored or transmitted.

Comsat: Communications Satellite Corporation, the U.S. signatory of Intelsat.

CPE: Customer premises equipment.

CT2: A limited mobile communications service that allows users to transmit but not receive calls from portable telephones.

DBS: Direct broadcast satellite, designed to transmit to very small terminals for household television reception.

DCME: Digital circuit multiplying equipment, a form of digital compression used in submarine optical fiber networks.

DECT: Digital European Cordless Telecommunications, a standard for cordless telephones.

DGT: Direction Générale des Télécommunications.

Digital: A discrete or discontinuous signal that transmits audio, data, and video as bits (binary digits) of information.

Digital switching: A process in which connections are established by processing digital signals without converting them to analog signals.

DOC: Department of Communications (former Canadian and Australian ministries); also Department of Commerce (United States).

Downlink: An earth station used to receive signals from a satellite; the signal transmitted from the satellite to earth.

DS-3: A digital circuit with capacity of 45 Mbps.

DTH: Direct-to-home, referring to satellites designed to transmit video to very small satellite antennas.

DTMF: Dual-tone-multifrequency, a method of signaling by sending tones on the telephone line.

Earth station: The antenna and associated equipment used to transmit and/or receive signals via satellite. Also ground station.

EAS: Extended area service; a tariff that enables users to call an extended area for a flat monthly rate instead of paying a toll charge for each call.

EBRD: European Bank for Reconstruction and Development.

EC: European Community, now the European Union.

ECU: European Currency Units, a denomination used by the European Union.

EDI: Electronic data interchange; the computer-to-computer exchange of intercompany business documents such as orders, invoices, and customs forms in a standard format.

EDS: Electronic Data Services Corporation.

EIB: European Investment Bank.

Electronic mail (e-mail): The use of telecommunications and computers for sending textual messages.

Encryption: Transformation of data for transmission so that it cannot be understood if intercepted.

Equal access: The requirement that local telephone companies must provide all long distance companies access equal in type, quality, and price to that provided to the dominant carrier. In the United States, the ability to reach a preselected long distance carrier by dialing 1 plus the number (called "1+ dialing).

ESA: European Space Agency.

ETACS: Extended Total Access Communications System (see TACS), a cellular communications standard.

ETSI: European Telecommunications Standards Institute.

EU: European Union (formerly European Community).

Eutelsat: European Telecommunications Satellite Organization.

Facsimile: Transmission and reception of documents using telecommunications.

Fax: See **Facsimile.**

FCC: U.S. Federal Communications Commission.

FDMA: Frequency division multiple access; a multiplexing technique that divides the bandwidth of the transmission into partitions or slots.

Fiber optics: See **Optical Fiber.**

Footprint: The geographical coverage area of a satellite.

Frequency: The rate at which a current alternates, expressed in hertz cycles per second.

Frequency Spectrum: The range of frequencies useful for radio communications, from about 10 KHz to 3,000 GHz.

FSS: Fixed Satellite Service.

FTTH: Fiber to the home, high bandwidth optical fiber links to individual households.

G7: A grouping of the world's seven major industrialized nations: Canada, France, Germany, Japan, Russia, the United Kingdom, and the United States.

Gateway: The connection between networks using different protocols. Also the connection between a telecommunications carrier and an information provider.

GATS: General Agreement on Trade in Services.

GATT: General Agreement on Tariffs and Trade.

Gbps: Gigabits (billions of bits) per second.

GDP: Gross Domestic Product.

GEO: Geostationary satellite, a satellite with an equatorial orbit 22,300 miles (36,000 kilometers) above the earth. The satellite revolves around the earth in 24 hours, and thus appears stationary when viewed from earth.

GHz: Gigahertz; billions of cycles per second.

GII: Global information infrastructure.

GNP: Gross National Product.

GOES: Geostationary Orbital Environmental Satellites.

GPS: Global Positioning System.

GSM: Global System for Mobile Communications, a European digital cellular standard; originally Groupe Special Mobile.

GSO: Geostationary or geosynchronous orbit. See **GEO.**

GTE: General Telephone and Electronics, the largest "independent" (non-Bell) local telephone operator in the United States.

HDTV: High definition television; a television format with much higher resolution than the standard television picture.

HF: High frequency; the frequency band from 3 to 30 MHz (also known as shortwave).

IBS: Intelsat Business Service; an integral digital service designed to carry a full range of services, including voice, data, and video.

ICT: Information and communications technologies.

IFRB: International Frequency Registration Board of the ITU.

Inmarsat: International Maritime Satellite Organization.

Intelnet: An Intelsat service using VSATs for one-way or interactive low-volume data communications.

Intelsat: The International Telecommunications Satellite Organization, a consortium of national telecommunications organizations that provides global satellite services.

Internet: The worldwide network of interconnected computer networks.

Intersputnik: The international satellite system established by the U.S.S.R. and other Communist countries.

Intranet: Private network that uses the infrastructure and standards of the Internet and the World Wide Web, but is cordoned off from the public network by software programs known as "firewalls."

IRU: Indefeasible Right of Usage; e.g., right of an international carrier to use a submarine cable.

ISDN: Integrated Services Digital Network, an evolving set of standards for public digital telecommunications networks.

ISO: International Standards Organization, a global standards federation based in Geneva.

ITT: Information and telecommunications technologies.

ITU: International Telecommunication Union, the UN agency responsible for telecommunications.

ITU-R: ITU Radio Communications Bureau.

ITU-T: ITU Telecommunication Standards Bureau.

IXC: Interexchange carrier (United States).

Just-in-time: A production system in which parts are delivered from suppliers to the manufacturer as needed, rather than being stored on-site.

Ka band: Portion of the electromagnetic spectrum in the 20-to-30 GHz range, used for satellite communications.

Kbps: Kilobits; thousands of bits per second.

KHz: Kilohertz; thousands of cycles per second.

Ku band: Portion of the electromagnetic spectrum in the 12-to-14 GHz range, used for satellite communications.

LAN: Local area network; network linking computers at a single location.

LATA: Local Access and Transport Area; service area within which Bell Operating Companies carry traffic; established under the Modified Final Judgment.

LDC: Least developed country.

LEC: Local exchange carrier.

LEO: Low earth-orbiting satellite. In contrast with GEOs, LEOs appear to move across the sky, and must be tracked. A series of satellites is required to provide continuous coverage of a given geographical area.

Lifeline: Fund in the United States used to help low-income telephone subscribers maintain access to basic telephone service.

Local loop: The link, typically twisted copper wire pair, between the subscriber's premises and the local exchange.

MAN: Metropolitan area network; a network linking computers within a limited geographical area.

Mbps: megabits (millions of bits) per second.

MFJ: Modified Final Judgment, the legal decision that broke up AT&T.

MHz: Megahertz; millions of cycles per second.

Microwave: Radio transmissions in the 4-to-28 GHz range, requiring antennas located within line-of-site.

Modem: Modulator–demodulator; a device that enables digital data to be transmitted over analog telephone lines.

MMDS: Microwave multipoint distribution system.

MPT: Ministry of Posts and Telecommunications.

MSS: Mobile Satellite Service.

Multiplex: To combine two or more signals on a single channel for transmission.

Narrowband channel: A communications channel that transmits voice, facsimile, or data at relatively low data rates using limited bandwidth.

NCC: New common carrier (Japan).

NII: National information infrastructure.

Noise: An unwanted change that interferes with the transmission of the desired signal.

NTIA: National Telecommunications and Information Administration, in the U.S. Department of Commerce.

NTT: Nippon Telephone and Telegraph, Japan's major domestic carrier.

OECD: Organization for Economic Cooperation and Development.

OFTA: Office of the Telecommunications Authority, the regulator in Hong Kong.

Oftel: Office of Telecommunications, the British telecommunications regulatory authority.

ONA: Open Network Architecture, provisions established by the FCC, requiring competitive availability of, and access to, enhanced network services.

ONP: Open Network Provisioning, a policy established by the European Union similar to Open Network Architecture in the United States.

OPEC: Organization of Petroleum Exporting Countries.

Open Skies: A U.S. policy authorizing multiple domestic satellite systems.

Optical fiber: A communications medium made of hair-thin glass that conducts light waves.

OSI: Open Standards Interconnection, a reference model for data networking developed by the International Standards Organization.

Packet switching: A technique of switching digital signals by breaking the signal stream into packets and then reassembling the data in the correct sequence at the receiving end.

PATU: Pan African Telecommunications Union.

PBX: Private branch exchange; a private telephone switch located on the customer's premises, used for internal communications.

PCN; PCS: Personal communications network or service; a digital wireless cellular system for portable voice and data communications.

PCO: Public call office, a public or pay telephone.

Photovoltaic: Material that develops voltage and electrical current when light shines on it, used for solar power systems.

Point-to-multipoint service: Transmission of a signal from one originating point to many receiving points; broadcasting.

POP: Point of presence: the location within a LATA at which customers are connected to an interexchange carrier.

POTS: Plain old telephone service.

Price cap: The maximum price that telephone companies can charge for a designated group of services. Regulators may change the price cap over time, based on inflation and targets for improvement in productivity.

Propagation delay: The time necessary for a signal to travel from one point to another.

PSTN: Public Switched Telephone Network.

PTO: Public telecommunications operator, usually called a common carrier in North America.

PTT: Ministry of Posts, Telephone and Telegraph; a government-operated administration that provides telecommunications services.

PUC: Public utilities commission, state-level regulator in the United States.

Radio Paging: A service that broadcasts a special radio signal that activates a small portable receiver.

RARC: Regional Administrative Radio Conference of the ITU.

Rate of return: A method of regulation that defines the total revenue a telephone company requires to provide services, and a fair return on its investment.

RBOC: Regional Bell Operating Company; one of the seven companies formed by the divestiture of AT&T.

REA: Rural Electrification Administration, an agency in the U.S. Department of Agriculture, now called the Rural Utilities Service (RUS).

RUS: Rural Utilities Service, an agency in the U.S. Department of Agriculture.

S band: Frequencies in the 2.5 GHz range, used for community reception from satellites.

Satellite: A microwave relay system in the sky that receives a signal from an earth station, amplifies it, and retransmits it to one or more receiving stations.

SS7: Signaling System 7; a control system for the public telephone network used for routing calls on channels separate from those used for the calls themselves.

Shortwave: Frequencies in the 3-to-30 MHz range of the electromagnetic spectrum; high frequency.

SITA: Société Internationale des Télécommunications Aéronautiques, an international airline reservation network.

SMATV: Satellite master antenna television.

SNG: Satellite news gathering.

Solar cells: Devices that convert solar energy into electrical energy, for example, to power telecommunications equipment. See **Photovoltaics.**

Spillover: The signal from a satellite that can be received outside its intended service area.

SWIFT: Society for Worldwide Interbank Financial Telecommunications.

SPC: Stored program control; refers to analog switches that are controlled by computer software.

T-carriers: A family of high-speed digital transmission systems. A T1 carrier has a capacity of 1.544 Mbps; T3 carries 45 Mbps.

TACS: Total Access Communications System; an analog cellular standard developed for Europe by Motorola.

TCO: Telecommunications operator.

TCP/IP: Transmission Control Protocol/Internet Protocol; two of the standard protocols for data transmission used on the Internet.

TDMA: Time division multiple access; a multiplexing method that divides a circuit's capacity into time slots.

Teleconference: An audio and/or video interconnection that allows communications among three or more individuals at two or more locations.

Teledensity: Main telephone lines per 100 population.

Telematics: A term derived from the French "télématique" describing the study of new technologies resulting from the convergence of telecommunications and computers.

Telemedicine: Use of telecommunications for medical diagnosis, patient care, and health education.

Telepoint: British name for a limited mobile wireless service that allows customers to transmit but not receive calls, known in Europe as CT2.

Teleport: A communications facility or center that transmits and switches voice, data, and video communications

Telex: A public switched network connecting teletypewriters for transmission of text messages.

Touchtone: AT&T's term for dual-tone-multifrequency (DTMF) dialing.

Transponder: A microwave receiver, amplifier, and transmitter in a satellite that amplifies and changes the frequency of a signal from an earth station and retransmits it to earth.

TVRO: Television receive-only satellite earth station.

UHF: Ultra high frequency, the frequency band from 300 MHz to 3,000 MHz (3 GHz).

UNDP: United Nations Development Program.

UNESCO: United Nations Educational, Scientific and Technical Organization.

Universal service: The goal of providing telephone service to everyone, traditionally by providing basic telephone service to every household.

Uplink: An earth station used to transmit to a satellite; the transmission from earth to the satellite.

VAN: Value-added network; a data communications system in which special features such as protocol conversion or database access are added to the underlying transmission capabilities.

VAS: Value-added service; see **Value-added network.**

VCR: Videocassette recorder.

VHF: Very high frequency, the frequency band from 30 MHz to 300 MHz.

VOD: Video-on-demand; a service that enables viewers to order individual movies and television programs.

Voice mail: A voice messaging system in which spoken messages are recorded and stored in electronic "mailboxes" for later access.

VSAT: Very small aperture terminal, for satellite communications.

WAN: Wide area network; a computer network covering a large geographical area.

WARC: World Administrative Radio Conference of the ITU, now World Radio Conference (WRC).

WATTC: World Administrative Telephone and Telegraph Conference of the ITU, now replaced by the World Telecommunication Standardization Conference (WTSC).

Wireless: Communications via radio waves rather than wire or cable.

WLL: Wireless local loop, using radio technology rather than twisted copper pairs to reach subscribers' premises.

WRC: World Radio Conference (of the ITU).

WTO: World Trade Organization, responsible for the implementation of the General Agreement on Trade in Services (GATS).

WTSC: World Telecommunication Standardization Conference (of the ITU).

X.25: A packet-switching standard for data transmission over public works.

X.400: A standard for the transmission of electronic mail.

BIBLIOGRAPHY

Agata, Masahiko. "Financing Telecommunications Investment—Perspective of a Bilateral Official Financier," pp. 141–145. In *Proceedings of the ITU Forum 91 Economic Symposium.* Geneva: ITU, October 1991.

Akhtar, Shahid, A. Neelameghan, and Luc Laviolette. "Telecoms Policy Reaches Out to the Private Sector and Villages of India." *Intermedia* 24, no. 3 (June/July 1996): pp. 30–34.

Albernaz, Joao Carlos Fagundes. "Telecoms and Data Services: Brazilian Situation." In *Eastern Europe: Information and Communication Technology Challenges,* ed. G. Russell Pipe, pp. 275–81. Burke, VA: Transnational Data Reporting Service, 1991.

APEC Telecommunications Working Group. "The State of Telecommunications Infrastructure and Regulatory Environment in APEC Countries." Seoul: Ministry of Communications and Korea Information Society Development Institute, November 1991.

APT Yearbook 1995. Bangkok: Asia-Pacific Telecommunity, 1995.

Arcidiacono, Antonio. "Satellite Personal Business Services." In *Telecom 91 Global Review,* ed. Hugh Chaloner, pp. 215–218. London: Kline Publishing, 1991.

Arlandis, Jacques. "Convergence between Telecoms, Computing and Media: Visions for the Future," pp. 25–30. In *Proceedings of the ITU Forum 91 Economic Symposium.* Geneva: ITU, October 1991.

Arnst, Catherine. "The Giants Aren't Sleeping." *BusinessWeek,* April 8, 1996, pp. 70–72.

Arnst, Catherine. "The Last Frontier." *BusinessWeek,* September 18, 1995, pp. 58–66.

AT&T. *The World's Telephones.* Bedminster, NJ: AT&T, 1990.

Bace, Edward J. "Financing the Commercialisation and Privatisation of Telecommunications in Asia." 1993 Pan-Asian Forum, Hong Kong, June 1993.

Bajaj, K.K. "Information Systems and Technology in Local and Regional Development." Paper presented at the AMIC Conference on Communication, Technology, and Development: Alternatives for Asia, Kuala Lumpur, June 25–27, 1993.

Bangemann Group. *Europe and the Global Information Society.* Brussels: European Commission, 1994.

"Bank of America to Launch Pilot EDI Scheme." *Financial Times,* September 9, 1995.

Bartels, Tom. "Australia: Hubbing the Asia-Pacific." In *Telecom 91 Global Review,* ed. Hugh Chaloner, pp. 84–86. London: Kline Publishing, 1991.

Baudon, Thierry. In *Telecom 91 Global Review,* ed. Hugh Chaloner. London: Kline Publishing, 1991.

Beaird, Richard C. "Liberalization and Privatization: Prospects for the Future." Paper presented at the 1993 Pan-Asian Forum, Hong Kong, June 1993.

Beamon, Clarice. "Telecommunications: A Vital Link for Rural Business." *OPASTCO Roundtable* (Spring 1990).

Beikmann, Gary G. "ITV: The Coax Connection." *OPASTCO Roundtable* (Spring 1991): pp. 20–27.

Berben, Cor. "International Harmonization of Standards." In *Proceedings of the ITU Forum 91 Regulatory Symposium,* pp. 23–26. Geneva: ITU, October 1991.

Black, George, ed. "International Telecommunications: Financial Times Survey." *Financial Times,* October 3, 1995, p. 14.

Blair, Michael L. "VSAT Systems in Developing Countries." In *Proceedings of the Pacific Telecommunications Conference.* Honolulu: Pacific Telecommunications Council, February 1988.

Boeke, Cynthia. "AsiaSat 1: Bringing a New Era of Communications to the Region." *Via Satellite* (March 1995): pp. 6–7.

Boeke, Cynthia. "AsiaSat 2: Asia's Hot New Bird." *Via Satellite* (March 1995): pp. 10–11.

Bohlin, Ron, Allan Roth and David C. Wenner. "Do LECs Need Magic to Cut Costs?" *Telephony* (June 24, 1991): pp. 28–34.

Bolcskei, Imre. "A Central European Approach." In *Proceedings of the ITU Forum 91 Regulatory Symposium,* pp. 9–13. Geneva: ITU, October 1991.

Bournellis, Cynthia. "Internet '95." *Internet World* (November 1995): pp. 47–52.

Bradford, Meriel V. M. "Telecommunications in Transition: A Canadian Perspective on Ownership and Other Key Issues." In *Proceedings of the Pacific Telecommunications Conference.* Honolulu: Pacific Telecommunications Council, January 1996.

Brooks, Gary C. "International Regulation of Frequencies and Satellite Systems." In *Proceedings of the ITU Forum 91 Regulatory Symposium,* pp. 30–34. Geneva: ITU, October 1991.

Brown, Kevin. "Australia: Compromise over Reforms." *Financial Times,* October 7, 1991, p. XXX.

Browning, John. "Joys of the Express Lane." *Globe and Mail Report on Business Magazine,* January 1995, p. 103.

Budden, John. "Meeting Telecommunications Needs for National and Regional Economic Development: Planning, Management and Funding Options." 1993 Pan-Asian Forum, Hong Kong, June 1993.

Burns, Tom. "A Timely Change of the Company's Culture." *Financial Times,* October 7, 1991, p. XXVIII.

Burton, John. "Scandinavia: Mobile Sector is Nordic Flagship." *Financial Times.* October 7, 1991, p. XXVIII.

Butler, Steven. "Japan's Ministry: The Genie of Privatisation." *Financial Times,* October 7, 1991, p. VI.

Butler, Steven. "Japan: Mobile Phones: Market Blossoms in Japan." *Financial Times,* October 7, 1991, p. XXXI.

Cane, Alan. "Brussels Batters Down the Doors." In "International Telecommunications: Financial Times Survey." *Financial Times,* October 3, 1995, p. 2.

Cane, Alan. "Early Promise Surpassed." In "International Telecommunications: Financial Times Survey." *Financial Times,* October 3, 1995, p. 12.

Cane, Alan. "Era of 'The Information Society.' " In "International Telecommunications: Financial Times Survey." *Financial Times,* October 3, 1995, pp. 21–22.

Cane, Alan. "Transforming the Way We Live and Work." In "International Telecommunications: Financial Times Survey." *Financial Times,* October 3, 1995, pp. 1–2.

Cardoso, Eliana A. "Privatization Fever in Latin America." *Challenge* (September–October 1991): pp. 35–41.

Carnevale, Mary Lu. "Playing Politics." *Wall Street Journal,* October 4, 1991, p. R14.

Carnevale, Mary Lu. "Soviets Select US West to do Phone Gateways." *Wall Street Journal Europe,* July 17, 1991.

Carpentier, Michel. "Development of the Global Economy and the Role of Telecoms: Linking East and West, North and South." In *Proceedings of the ITU Forum 91 Economic Symposium.* Geneva: ITU, October 1991.

Carsberg, Bryan. "The Role and Purpose of Regulation: Regulator." In *Proceedings of the ITU Forum 91 Regulatory Symposium,* pp. 20–23. Geneva: ITU, October 1991.

Cave, Martin. "The 1995 Review of Telecommunications Policy in the United Kingdom." In *Proceedings of 13th IDATE International Conference,* pp. 473–481. Montpelier: November 1991.

Chaloner, Hugh, ed. *Telecom 91 Global Review.* London: Kline Publishing, 1991.

Chamoux, Jean-Pierre. "The French Telematique Experience." In *Eastern Europe: Information and Communication Technology Challenges,* ed. G. Russell Pipe, pp. 181–189. Burke, VA: Transnational Data Reporting Service, 1991.

Chandra, Subhash. "The Roll-Out of Zee TV." *Intermedia* 24, no. 3 (June/July 1996): pp. 38–41.

Chase, Scott. "Live via Satellite." *Via Satellite* (April 1988): p. 50.

Chen, Steven Y. and Duei Tsai. "Network Construction Plan for NII in R.O.C." In *Proceedings of the 1996 Pacific Telecommunications Conference,* pp. 662–668. Honolulu: Pacific Telecommunications Council, 1996.

Chowdary, T.H. "Reforms in the Indian Telecom Sector." In *Proceedings of the 1996 Pacific Telecommunications Conference,* pp. 308–312. Honolulu: Pacific Telecommunications Council, 1996.

Christie, Alix, Montieth M. Illingworth, and Larry Lange. "One World? Online Advocates Dream of a World Linked by the Information Super-highway." *Information Week,* October 2, 1995, pp. 52–64.

Claudio, Corazon P. B. "The Philippines: From Natural Disaster to Rural Progress." In *Telecom 91 Global Review,* ed. Hugh Chaloner, pp. 70–75. London: Kline Publishing, 1991.

Clements, Charles. "Healthnet." Cambridge, MA: SatelLife, May 1991.

Cleveland, Harlan. "The Twilight of Hierarchy: Speculations on the Global Information Society." *Public Administration Review* 45, no. 185 (1985).

Coates, Vary T., Todd M. LaPorte, and Mark G. Young. "Global Telecommunications and Export of Services: The Promise and the Risk." *Business Horizons* 36, no. 6 (November/December 1993): pp. 23–29.

Codding, George and Anthony Rutkowski. *The International Telecommunication Union in a Changing World.* Dedham, MA: Artech, 1982.

Commission of the European Communities. *Europe and the Global Information Society.* Brussels: European Commission, 1994.

Commission of the European Communities. "European Industrial Policy for the 1990s." *Bulletin of the European Communities,* Supplement 3/91, pp. 15–35. Brussels: European Commission, 1991.

Commission of the European Communities. *Europe's Way to the Information Society: An Action Plan.* Brussels: European Commission, 1994.

Commission of the European Communities. *From Single Market to European Union.* Brussels: European Commission, 1992.

Commission of the European Communities. "Growth, Competitiveness, Employment: The Challenges and Ways Forward into the 21st Century" (White Paper). Brussels: European Commission, 1993.

Commission of the European Communities, Directorate-General XIII. *Information and Communications Technologies in Europe.* Brussels: European Commission, 1991.

Commission of the European Communities, Directorate-General XIII. "ORA 1992: Research and Technology Development on Telematic Systems for Rural Areas." Brussels: European Commission, 1992.

Commission of the European Communities. *The Single Market in Action.* Brussels: European Commission, 1992.

Communications Standards Review 4, no. 3 (April 1993): p. 13.

Cooperman, William. "Communication Satellites: Narrow Beams to Broadband." *State of the Art Handbook,* pp. S43–S50. London: Telecommunications International, 1995.

Cooperman, William, Lori Mukaida, and Donald M. Topping. "The Return of PEACESAT." In *Proceedings of the Pacific Telecommunications Conference.* Honolulu: Pacific Telecommunications Council, January 1991.

Cortese, Amy. "Here Comes the Intranet." *BusinessWeek,* February 26, 1996, pp. 76–84.

Cote-O'Hara, Jocelyn. "Risk and Opportunity: Navigating the Waves of Change." In *Proceedings of the Telecom 95 Strategies Summit.* Geneva: ITU, October 1995.

Coulter, Kristin. "The Telco in Rural Development." *OPASTCO Roundtable* (Spring 1990).

Crane, Rhonda. "Advanced Television: An American Challenge." *Boston Globe,* November 6, 1988, p. 46.

Cureton, T. D. "Global Banking in the Year 2000." Paper presented at Networking in the Year 2000 Conference, Hong Kong, September 28, 1995.

Dadzie, K. K. S. In *Telecom 91 Global Review,* ed. Hugh Chaloner. London: Kline Publishing Ltd., 1991, pp. 40–42.

Dance, Teresa. "Nurturing Interconnection." *Asian Telecommunications* (June 1996): p. 42.

Dance, Teresa. "The Philippines Comes of Age." *Asian Communications* (June 1996): pp. 38–39.

Datapro. "Argentina: The Commercial and Regulatory Environment." *Datapro Reports on International Telecommunications* (February 1991).

Datapro. "An Overview of World Telecommunications." *Datapro Reports on International Telecommunications* (1993).

Datapro. "Chile: The Commercial and Regulatory Environment." *Datapro Reports on International Telecommunications* (February 1991).

Datapro. "China: The Commercial and Regulatory Environment." *Datapro Reports on International Telecommunications* (March 1992): p. 153.

Datapro. "Mexico: The Commercial and Regulatory Environment." *Datapro Reports on International Telecommunications* (February 1991).

Datapro. "Telecommunications in Europe." *Datapro Reports on International Telecommunications* (June 1990): pp. 101–110.

Datapro. "Telecommunications in North America." *Datapro Reports on International Telecommunications* (November 1991): pp. 101–108.

Datapro. "Telecommunications in the Far East." *Datapro Reports on International Telecommunications* (January 1992): pp. 101–107.

Davidson, William H., Anne C. Dibble, and Sandra H. Dom. "Telecommunications and Rural Economic Development." Redondo Beach, CA: MESA Inc., October 1990.

Davis, Neil W. "Japan: International Services: Big Six Houses Branch Out." *Financial Times,* October 7, 1991, p. XXXI.

"The Death of Distance." *The Economist,* September 30, 1995.

de la Cal, Martha. "Jose Ferreira de Areia of Telecom Portugal." In *Telecom 91 Global Review,* ed. Hugh Chaloner, pp. 102–104. London: Kline Publishing, 1991.

Demac, Donna, ed. *Tracing New Orbits: Cooperation and Competition in Global Satellite Development.* New York: Columbia University Press, 1986.

DeMaeyer, Bruce. "Status of the Cellular Industry in North America." In *Proceedings of the ITU Forum 91 Technical Symposium.* Geneva: October 1991.

Di, Ang-Zhao. "Development of Chinese Post and Telecommunications and its Effect on National Economy." Paper presented at the Conference on Telecommunications and the Integration of China, Hong Kong, June 17, 1993.

Dickson, Martin. "Federal Communications Commission: Overhaul of US System Looms." *Financial Times,* October 7, 1991, p. VI.

Dillman, Don A. and Donald M. Beck. "Information Technologies and Rural Development in the 1990s." *Journal of State Government* 61, no. 1 (January/February 1988).

Dixon, Hugo and Greg Staple, "New Perspective on Patterns of Power." *Financial Times,* October 7, 1991, p. II.

Dixon, Hugo. "The Sleeping Giants Awaken." *Financial Times,* October 7, 1991, p. I.

Djiwatampu, Arnold. "Indonesia: The Search for Global Standards." In *Telecom 91 Global Review,* ed. Hugh Chaloner, pp. 80–83. London: Kline Publishing, 1991.

Dorfman, Steven D. "Satellite Communications: The Global Network." In *Proceedings of the ITU Forum 91 Policy Symposium,* pp. 8–12. Geneva: ITU, October 1991.

Dorros, Irwin. "Two Major Forces Driving the Evolution of Public Switched Networks." In *Proceedings of the ITU Forum 91 Technical Symposium.* Geneva: ITU, October 1991.

Dyson, Kenneth and Peter Humphreys, eds. *The Politics of the Communications Revolution in Western Europe.* London: Frank Cass, 1986.

Edmondson, Gail and Karen Lowry Miller. "Europe's Markets are Getting Rewired, Too." *BusinessWeek,* April 8, 1996, p. 87.

Eicher, Lawrence D. "The Convergence of Telecoms and IT: Standardizing in a Global Market." In *Telecom 91 Global Review,* ed. Hugh Chaloner, pp. 22–25. London: Kline Publishing, 1991.

Elbert, Bruce. *International Telecommunication Management.* Norwood, MA: Artech House, 1990.

Eng, Paul M. "Surfing's Biggest Splash?" *BusinessWeek,* March 11, 1996, p. 86.

Englebardt, Robert. "The Pacific Region's Unique Requirements for Telecoms Development." In *Telecom 91 Global Review,* ed. Hugh Chaloner, pp. 66–69. London: Kline Publishing, 1991.

Federal Communications Commission. "In the Matter of Establishment of Satellites Systems Providing International Communications." (CC Docket 84–1299) Released September 3, 1985.

Filho, Armando T. "Banco Bradesco Implements Extensive VSAT Network." *Communications News* (March 1993): p. 11.

Flanagan, Patrick. "VSAT: A Market and Technology Overview." *Telecommunications* (March 1993): pp. 19–24.

Folkestad, Christian. "Norwegian Telecom: Facing the 90's." In *Telecom 91 Global Review,* ed. Hugh Chaloner, pp. 114–116. London: Kline Publishing, 1991.

Fossier, Marc and Marie-Monique Steckel. *France Telecom: An Insider's Guide.* Chicago: Telephony Division of Intertec Publishing, 1991.

Frieden, Robert M. "Accounting Rates: The Business of International Telecommunications and the Incentive to Cheat." *Federal Communications Law Journal* 43, no. 2 (April 1991): pp. 115–139.

"Furniture Maker Uses EDI to Stop Paper Blizzard." *Communication News* (September 1992): pp. 26–27.

Gallagher, Lynne and Dale Hatfield. "Distance Learning: Opportunities in Telecommunications Policy and Technology." Washington, D.C.: Annenberg Washington Program of Northwestern University, May 1989.

Gallottini, Giovanna T. "Infrastructure: The Rural Difference." *Telecommunications Engineering and Management* 95, no. 1 (January 1, 1991): pp. 48–50.

Galloway, Jonathan F. *The Politics and Technology of Satellite Communications.* Lexington, MA: D.C. Heath, 1972.

Gassmann, Hans-Peter. "Information Technology and Telecommunications Policies in OECD Countries." In *Eastern Europe: Information and Communication Technology Challenges,* ed. G. Russell Pipe, pp. 43–59. Burke, VA: Transnational Data Reporting Service, 1991.

Gawith, Philip. "South Africa: Socio-Economic Division." *Financial Times,* October 7, 1991, p. XXXII.

General Agreement on Trade in Services. Annex on Telecommunications.

Gilhooly, Denis. "Science Outruns Watchdogs." *Financial Times,* October 7, 1991, p. VI.

Glowinski, Albert. "Market Perspectives of Mobile Communications in the 90s." In *Proceedings of the ITU Forum 91 Technical Symposium.* Geneva: ITU, October 1991.

Goddard, Michael "WARC-92: A Golden Opportunity for Mobile Radio." *Pan-European Mobile Communications* 7 (Autumn 1991): pp. 29–31.

Goldschmidt, Douglas. "On Curbing Predation in Newly Competitive Markets." Unpublished paper prepared for PanAmSat. Greenwich, CT, 1991.

Goldschmidt, Douglas. "Pan American Satellite and the Introduction of Specialized Communication Systems in Latin America." In *Proceedings of the Pacific Telecommunications Council,* pp. 343–349. Honolulu: Pacific Telecommunications Council, January 1988.

Goodhart, David. "Germany: Reforms Driven by Unity." *Financial Times,* October 7, 1991, p. XXVI.

Gore, Albert, Jr. "Infrastructure for the Global Village: Computers, Networks and Public Policy." *Scientific American* 265, no. 3 (September 1991): p. 150.

Graven, Kathryn. "GTE Forms Venture with Soviets to Provide Phone Service to Moscow." *Wall Street Journal Europe,* July 10, 1991.

Gregston, Brent. "Power and Privilege." *Internet World* (November 1995): pp. 96–101.

Grewlich, Klaus W. "Eastern Europe—Cooperative Communications Strategies." In *Eastern Europe: Information and Communication Technology Challenges,* ed. G. Russell Pipe, pp. 265–273. Burke, VA: Transnational Data Reporting Service, 1991.

Halik, Greg. "Mexico Runs First Satellite-Delivered Digital Code." *Communications News* (March 1993): p. 15.

Hall, Terry. "New Zealand: From Monopoly to Competitor." *Financial Times,* October 7, 1991 p. XXX.

Halprin, Albert. "A Modest Proposal for International Telecommunications Regulation." In *Proceedings of the ITU Forum 91 Regulatory Symposium,* pp. 68–70. Geneva: ITU, October 1991.

Hamilton, Joan O'C. "The New Workplace." *BusinessWeek,* April 29, 1996, pp. 106–117.

Harbi, M. "International Regulation of Space Radio Services for the 90's." In *Telecom 91 Global Review,* ed. Hugh Chaloner, pp. 34–39. London: Kline Publishing, 1991.

Hardy, Andrew P. "The Role of the Telephone in Economic Development." *Telecommunications Policy* (December 1980).

Hellstrom, Kurt. "The Future of Mobile Telephony." In *Proceedings of the ITU Forum 91 Technical Symposium.* Geneva: ITU, October 1991.

Herabat, Jumpone. "Telecommunications Development and Privatisation in Thailand: Future Opportunities." 1993 Pan-Asian Summit, Hong Kong, June 1993.

Hillis, Durrell W. "Mobile Satellite Personal Communications." In *Telecom 91 Global Review,* ed. Hugh Chaloner, pp. 230–232. London: Kline Publishing, 1991.

Hills, Jill. "The Telecommunications Rich and Poor." *Third World Quarterly* 12 no. 2 (1990): pp. 71–90.

Hollis, Rex. "Update on Mobile Satellite Services." In *Proceedings of the 1996 Pacific Telecommunications Conference,* pp. 56–57. Honolulu: Pacific Telecommunications Council, January 1996.

Holman, Michael. "Wary Welcome for Telecom Breakthroughs." In "International Telecommunications: Financial Times Survey." *Financial Times,* October 3, 1995, pp. 21–22.

Holmes, Bradley P. "Standards and International Telecommunications Policy." In *Telecom 91 Global Review,* ed. Hugh Chaloner, pp. 26–29. London: Kline Publishing, 1991.

Hornik, Robert C. *Development Communications: Information, Agriculture and Nutrition in the Third World.* White Plains, NY: Longman, 1988.

Hornik, Robert C. "Communication as Complement in Development." *Journal of Communication* 30, no. 2 (1980).

Horten, Monica. "Value-Added Services: Surprising Collaborations Emerge." *Financial Times,* October 17, 1994, p. 515.

Horvath, Paul. "General Trends and Digitalization in Hungary." In *Proceedings of the ITU Forum 91 Technical Symposium.* Geneva: ITU, October 1991.

Horvath, Paul. "Telecommunications in Hungary." In *Eastern Europe: Information and Communication Technology Challenges,* ed. G. Russell Pipe. Burke, VA: Transnational Data Reporting Service, 1991.

Housego, David. "India: A Tangled Network." *Financial Times,* October 7, 1991, p. XXXII.

Hudson, Heather E. "Access to Information Resources: The Developmental Context of the Space WARC." *Telecommunications Policy* (March 1985).

Hudson, Heather E. *Communication Satellites: Their Development and Impact.* New York: Free Press, 1990.

Hudson, Heather E. "The Convergence of Computers and Telecommunications: Opportunities and Challenges for Development." In *Proceedings of Americas Telecom.* Rio de Janeiro: ITU, May 1988.

Hudson, Heather E. *Three Case Studies on the Benefits of Telecommunications in Socio-Economic Development.* Geneva: ITU, 1981.

Hudson, Heather E. *When Telephones Reach the Village: The Role of Telecommunications in Rural Development.* Norwood, NJ: Ablex, 1984.

Hudson, Heather E., ed. "Innovative Strategies for Telecommunications Development." Special issue of *Telematics and Informatics* 4, no. 2 (Spring 1987).

Hudson, Heather E., Andrew P. Hardy, and Edwin B. Parker. "Impact of Telephone and Satellite Earth Station Installations on GDP." *Telecommunications Policy* (December 1982).

Hudson, Heather E. and Lynn York. "Generating Foreign Exchange in Developing Countries: The Potential of Telecommunications Investments." *Telecommunications Policy* (September 1988).

Hukill, Mark. "The Privatisation of Singapore Telecom: Planning, Policy and Regulatory Issues." Paper presented at the 1993 Pan-Asian Summit, Hong Kong, June 1993.

Hukill, Mark A. and Meheroo Jussawalla. "Telecommunications Policies and Markets in the ASEAN Countries." *Columbia Journal of World Business* 24, no. 1 (Spring 1989).

Hukill, Mark A. and Jussawalla, Meheroo. "Trends in Policies for Telecommunications Infrastructure Development and Investment in the ASEAN Countries." Honolulu: Institute for Communication and Culture, East-West Center, 1991.

Igarashi, Mitsuo. "The Development of Telecommunications and Institution of New International Frameworks." In *Proceedings of the ITU Forum 91 Regulatory Symposium,* pp. 89–91. Geneva: ITU, October 1991.

"In the Matter of Allocation of Frequencies in the Bands Above 890 Mc." 27 F.C.C. 359 (1959).

Inamori, Kazuo. "The Role and Purpose of Regulation: User." In *Proceedings of the ITU Forum 91 Regulatory Symposium,* pp. 1–19. Geneva: ITU, October 1991.

Industry Canada. *The Canadian Information Highway.* Ottawa: Industry Canada, April 1994.

Information Highway Advisory Council. *Canada's Information Highway: Providing New Dimensions for Learning, Creativity and Entrepreneurship.* Ottawa: Industry Canada, November 1994.

Information Infrastructure Task Force. *The National Information Infrastructure: Agenda for Action.* Washington, D.C.: U.S. Department of Commerce, September 1993.

Inmarsat. "Project 21: The Development of Personal Mobile Satellite Communications." London: Inmarsat, 1994.

Inmarsat. "Working Together: Sharing the Benefits of Global Mobile Satellite Communications." London: Inmarsat, 1994.

Intelsat. *Annual Report 1994.* Washington, D.C.: Intelsat, 1995.

Intelsat. "Focus: Asia Pacific." Washington, D.C.: Intelsat, 1993.

Intelsat. *Via Intelsat* 10, no. 3 (October 1995).

International Commission for Worldwide Telecommunications Development (The Maitland Commission). *The Missing Link.* Geneva: ITU, 1985.

International Development Research Centre. *Pan Asia Networking.* Ottawa: IDRC, 1995. (Also available on the Internet at http://www. idrc.org.sg.)

International Development Research Centre. *Sharing Knowledge for Development: IDRC's Information Strategy for Africa.* Ottawa: IDRC, 1989.

International Telecommunication Union. "Acapulco Declaration." Americas Regional Telecommunications Development Conference, Acapulco, May 1992.

International Telecommunication Union. *Benefits of Telecommunications to the Transportation Sector of Developing Countries.* Geneva: ITU, 1988.

International Telecommunication Union. "Development Partners Agree on Telecommunications Development Plan for Asia and the Pacific." Press Release. Geneva: ITU, May 17, 1993.

International Telecommunication Union. *Information, Telecommunications, and Development.* Geneva: ITU, 1986.

International Telecommunication Union. "Telecommunications Development Worldwide: The Telecommunications Development Bureau at Your Service." Geneva: ITU, 1991.

International Telecommunication Union. "Tomorrow's ITU: The Challenges of Change." Report of the High Level Commission. Geneva: ITU, 1992.

International Telecommunication Union. *World Telecommunication Development Report.* Geneva: ITU, 1994.

International Telecommunication Union. *World Telecommunication Development Report.* Geneva: ITU, 1995.

International Trade Administration. "U.S. Industrial Outlook 1991—Telecommunications Services." Washington, D.C.: U.S. Department of Commerce, 1991.

Irmer, Theodore. "Challenge and Change for CCITT: Global Standardization of Telecoms in Transition." In *Telecom 91 Global Review,* ed. Hugh Chaloner, pp. 16–20. London: Kline Publishing, 1991.

Ito, Youichi. "Information Technologies and Telecommunications in the Process of Global Change." In *Eastern Europe: Information and Communica-*

tion Technology Challenges, ed. G. Russell Pipe, pp. 25–42. Burke, VA: Transnational Data Reporting Service, 1991.

Jackson, James O. "It's a Wired, Wired World." *InDepth* (June 1995): pp.14–15.

Johnson, Tony. "Microsatellite that Turns Information into Medical Power." *The (Manchester) Guardian,* April 26, 1991.

Johnston, William R. "Global Workforce 2000: The New World Labor Market," *Harvard Business Review* (March–April 1991): pp. 115–127.

Kageyama, Sei. "Example from the Industrialized Countries." In *Proceedings of the ITU Forum 91 Economic Symposium,* pp. 75–77. Geneva: ITU, October 1991.

Kambayashi, Tomeo. "The Provision of Universal Service under a Privatized System." In *Proceedings of the ITU Forum 91 Economic Symposium,* pp. 113–115. Geneva: ITU, October 1991.

Kamman, Alan B. "Hong Kong after 1997: The Effect on Telecommunications Services." In *Proceedings of the 1996 Pacific Telecommunications Conference.* Honolulu: Pacific Telecommunications Council, January 1996.

Kamman, Alan B. "Use of Global Networking for Worldwide Corporations and Institutions." In *Proceedings of the ITU Forum 91 Policy Symposium,* pp. 153–157. Geneva: ITU, October 1991.

Kaske, Karlheinz. "Universal and Advanced Services for Eastern Europe in a Competitive Environment." In *Proceedings of the ITU Forum 91 Policy Symposium,* pp. 67–70. Geneva: ITU, October 1991.

Kaske, Karlheinz. "Eastern Europe: Opportunities for Development." In *Telecom 91 Global Review,* ed. Hugh Chaloner. London: Kline Publishing, 1991.

Keller, John J. "Northern Wins Phone Order for $120 Million in Poland." *Wall Street Journal Europe,* July 25, 1991.

Keller, John J. "Spanning the Globe." *Wall Street Journal,* October 4, 1991, p. R1.

Keller, John J. "Worldwide Warriors." *Wall Street Journal,* October 4, 1991 p. R4.

Khim, Wong Hung. "Looking to the Future of Singapore." In *Telecom 91 Global Review,* ed. Hugh Chaloner, pp. 76–78. London: Kline Publishing, 1991.

Kirby, Michael. "The Ten Lessons of Budapest—Crossing the Bridge of Informatics." In *Eastern Europe: Information and Communication Technology Challenges,* ed. G. Russell Pipe, pp. 345–358. Burke, VA: Transnational Data Reporting Service, 1991.

Kirby, Richard. "Radio Horizons." *Global Communications* (September–October 91): pp. 21–26.

Kirvan, Paul F. "Teleconferencing: Bringing the World Together . . . Finally." In *Telecom 91 Global Review,* ed. Hugh Chaloner, pp. 220–224. London: Kline Publishing, 1991.

Kleinwachter, Wolfgang and Angela Kolb. "German Unification and its Consequences for Telecommunications." In *Eastern Europe: Information and Communication Technology Challenges,* ed. G. Russell Pipe, pp. 153–162. Burke, VA: Transnational Data Reporting Service, 1991.

Koshiyama, M. "Changing Faces of the Japanese Telecommunications Industry: Post-Privatization Perspectives." In *Proceedings of IDATE 13th International Conference,* pp. 361–371. Montpelier, November 1991.

Kraemer, Joseph III and Williams, Nicholas. "Telecommunications Privatization: Managing the Change Process." Paper presented at the 1993 Pan-Asian Forum, Hong Kong, June 1993.

Kudriavtzev, G. G. and L. E. Varakin. "Economic Aspects of Telephone Network Development: The USSR Plan." *Telecommunications Policy* (February 1990): pp. 7–14.

Kunihiro. "Communication System Supplier's Views on the Evolution of Networks in the 1990s." Panel Session on Evolution of the Networks. In *Proceedings of the ITU Forum 91 Policy Symposium.* Geneva: ITU, October 1991.

Kwak, Chi-Young. "Telecommunications Liberalization, Industrial Development and Private Sector's Role: Experience from Korea." Paper presented at the 1993 Pan-Asian Summit, Hong Kong, June 1993.

Laenser, Mohand. "Dealing with Global Operators: The Joint Ventures and the New Forms of Collaboration." In *Proceedings of the ITU Forum 91 Policy Symposium.* Geneva: ITU, October 1991.

Lane, David. "Italy: The Big Talk Show Takes Off." *Financial Times,* October 7, 1991, p. XXVIII.

Lawton, Raymond A. "Telecommunications Modernization: Issues and Approaches for Regulators." Columbus, OH: National Regulatory Research Institute, 1990.

Lehner, J. Christopher. "Rural Development at a Crossroads: The Emergence of a National Consensus." *Rural Telecommunications* (Fall 1989).

Lehner, J. Christopher. "Toward Rural Revival: The Telco-Community Partnership." *Rural Telecommunications* (Summer 1990).

Lehner, J. Christopher and Ingrid K. Young. "Conspicuous Personalities: Ideas from Rural Telephony." *Rural Telecommunications* (Winter 1989).

Leonard, Hugh. "Asian Broadcasting: The Changing Scene." Paper presented at the AMIC Conference on Communication, Technology, and Development: Alternatives for Asia, Kuala Lumpur, June 25–27, 1993.

Lerner, Norman. "Necessary Environmental Changes for Organization Efficiency: Government Telecommunications Privatization and Liberalization: Policy Issues in Developing Countries." In *Proceedings of the ITU Forum 91 Economic Symposium,* pp. 107–111. Geneva: ITU, October 1991.

Levine, Jonathan B. and Mark Landler. "Cable has a New Frontier: The Old World." *BusinessWeek,* June 28, 1993, p. 74.

Lewin, David. "Numbering International Telecommunications Services in the 21st Century." In *Proceedings of the ITU Forum 91 Economic Symposium,* pp. 42–44. Geneva: ITU, October 1991.

Lindow, Herbert. "U.S. Exports to Venezuela Soar." *Business America,* March 23, 1992.

Liu, Junjia, "Basic Information on China's Telecommunications Infrastructure." Paper presented at the Conference on Telecommunications and the Integration of China, Hong Kong, June 17, 1993.

Lloyd, Ann. "The Rural (Radio) Connection." *Rural Telecommunications* (Fall 1988).

Lodgine, S. M. "The New Zealand Experience." *Telecommunications Policy* 16, no. 9 (December 1992): pp. 768–776.

Lomax, David. "Entering the Information Age: Implications for Developing Countries." *Proceedings of IDATE 13th International Conference,* pp. 159–166. Montpelier, November 1991.

Long, Colin. "Agreements Between Operators for Interconnection, System Roaming and International Service." In *Proceedings of the ITU Forum 91 Regulatory Symposium,* pp. 66–68. Geneva: ITU, October 1991.

Lucky, Robert. "Technological Advances from 2001 Onwards." In *Proceedings of the ITU Forum 91 Regulatory Symposium,* pp. 71–73. Geneva: ITU, October 1991.

Lynch, Karen. "Wave of Privatisations Sweeps the Globe." *Financial Times,* October 7, 1991, p. III.

MacBride Commission. *Many Voices, One World.* Paris: UNESCO, 1981.

Mahoney, William. "With Success Comes Scrutiny." *Multichannel News International* (March 1996): p. 24.

Mansell, Robin and Michael Jenkins. "Electronic Data Networks: EDI and Beyond." In *Proceedings of IDATE 13th International Conference,* pp. 173–183. Montpelier, November 1991.

Mansell, Robin and Dimitri Ypsilanti. "New Network-Based Services: Challenges for Policy." In *Proceedings of the ITU Forum 91 Economic Symposium,* pp. 47–51. Geneva: ITU, October 1991.

Markowitz, Arthur. "K-Mart Power Merchandising." *Discount Store News* (January 21, 1994): pp. 54–55.

Mayo, John K., Gary R. Heald, and Steven J. Klees. "Commercial Satellite Telecommunications and National Development: Lessons from Peru." *Telecommunications Policy* 16, no. 1 (January–February 1992): pp. 67–79.

Mayo, John K., G. R. Heald, S. J. Klees, and M. Cruz de Yanes. *Peru Rural Communication Services Project Final Evaluation Report.* Washington, D.C.: Academy for Educational Development, 1987.

McCartney, Neil. "Beeper Image Fades." *Financial Times,* October 7, 1991, p. VIII.

McCartney, Neil. "Expansion Rate Slows Down." *Financial Times,* October 7, 1991, p. VIII.

McCartney, Neil. "A Vehicle for Introduction of Competition." *Financial Times,* October 7, 1991, p. VIII.

McClelland, Stephen. "A Global Company, a Global Network." *Telecommunications* (March 1993): pp. 65–67.

McClelland, Stephen and Julian Bright. "AT&T: Towards Global Connectivity?" *State of the Art Handbook,* pp. S53–S54. London: Telecommunications International, October 1995.

McClelland, Stephen and Julian Bright. "FLAG Sets New Standard." *State of the Art Handbook,* p. S54. London: Telecommunications International, October 1995.

McGowan, William. "The Past as Prologue: The Impact of International Telecommunications." In *Telecom 91 Global Review.* London: Kline Publishing, ed. Hugh Chaloner, pp. 56–58. 1991.

McKnight, Lee. "The Deregulation of International Satellite Communications: U.S. Policy and the Intelsat Response." *Space Communications and Broadcasting* 3, (1985).

McLuhan, Marshall. *Understanding Media: The Extensions of Man.* New York: New American Library, 1964.

McWilliams, Gary. "Small Fry Go Online." *BusinessWeek,* November 20, 1995, pp. 158–164.

Melody, William. "Privatization and Developing Countries." In *Telecommunications Politics: Ownership and Control of the Information Highway in Developing Countries,* eds. Bella Mody, Johannes M. Bauer, and Joseph D. Straubhaar. Mahwah, NJ: Lawrence Erlbaum Associates, 1995.

Meyers, Tedson J. "Settlement of Telecommunications Disputes Under the GATT-GNS Framework." In *Proceedings of the ITU Forum 91 Regulatory Symposium,* pp. 177–182. Geneva: ITU, October 1991.

Michna, Ales. "Views from Central and Eastern Europe." In *Proceedings of the ITU Forum 91 Economic Symposium,* pp. 89–90. Geneva: ITU, October 1991.

Mier y Teran, Carlos. "A Latin Experience." In *Proceedings of the ITU Forum 91 Regulatory Symposium,* pp. 31–35. Geneva: ITU, October 1991.

Minaguchi, Koichi. "Evolution of Telecom Industry to Information Technology Industry, New Avenues for Developments: A Vision for the 21st Century." In *Proceedings of the ITU Forum 91 Economic Symposium,* pp. 9–13. Geneva: ITU, October 1991.

Modoux, Alain. "Remarks to APEC Conference." Paper presented at the AMIC Conference on Communication, Technology and Development: Alternatives for Asia, Kuala Lumpur, June 25–27, 1993.

Mody, Bella, Johannes M. Bauer, and Joseph D. Straubhaar, eds. *Telecommunications Politics: Ownership and Control of the Information Highway in Developing Countries.* Mahwah, NJ: Lawrence Erlbaum Associates, 1995.

Mohamed, Syed Hussein. "Corporatization and Partial Privatization of Telecommunications in Thailand." In *Implementing Reforms in the Telecommunications Sector: Lessons from Experience,* eds. Bjorn Wellenius and Peter A. Stern, pp. 267–270. Washington, D.C.: World Bank, 1994.

Monkiewicz, Jan. "IT/Telecommunications in Poland. In *Eastern Europe: Information and Communication Technology Challenges,* ed. G. Russell Pipe, pp. 119–133. Burke, VA: Transnational Data Reporting Service, 1991.

Montes de Souza, Roberto B. "Views from Latin America." In *Proceedings of the ITU Forum 91 Economic Symposium,* pp. 91–92. Geneva: ITU, October 1991.

Mueller, Milton. "One Country, Two Systems: What Will 1997 Mean in Telecommunications?" *Telecommunications Policy* 18, no. 3 (April 1994): pp. 243–253.

Mueller, Milton and Zixiang Tan. *China in the Information Age: Telecommunications and the Dilemma of Reform.* Washington, D.C.: CSIS Books, in press.

Murari, P. "Communication and Development: South Asia." Paper presented at the AMIC Conference on Communication, Technology, and Development: Alternatives for Asia, Kuala Lumpur, June 25–27, 1993.

Mutambirwa, Raymond. "Views from Africa." In *Proceedings of the ITU Forum 91 Economic Symposium,* pp. 85–88. Geneva: ITU, October 1991.

Narain, Ravi. "A Floorless Exchange." *Asian Communications* (June 1996, pp. 50–51).

Nasuda, Yoshio. "The Next Century Smart Technologies; Redefining the Work Environment and the Quality of Life." In *Proceedings of the ITU Forum 91 Policy Symposium,* pp. 81–85. Geneva: ITU, October 1991.

National Research Council, Board on Science and Technology for International Development. *Science and Technology Information Services and Systems in Africa.* Washington, D.C.: National Academy Press, 1990.

National Telecommunications and Information Administration. *The NTIA Infrastructure Report: Telecommunications in the Age of Information.* Washington, D.C.: U.S. Department of Commerce, October 1991.

National Telecommunications and Information Administration. *Telecom 2000: Charting the Course for a New Century.* NTIA Special Publication 88–21. Washington, D.C.: Government Printing Office, October 1988.

National Telecommunications and Information Administration. *U.S. Spectrum Management Policy: Agenda for the Future.* NTIA Special Publication 91–23. Washington, D.C.: February 1991.

Negrine, Ralph, ed. *Satellite Broadcasting: The Politics and Implications of the New Media.* London: Croom Helm, 1988.

Neumann, Karl-Heinz. "Domestic Issues of Frequency Regulations in Europe." In *Proceedings of the ITU Forum 91 Regulatory Symposium,* pp. 38–41. Geneva: ITU, October 1991.

Newman, Mark "Oftel: A Difficult Balancing Act." *Financial Times,* October 7, 1991, p. VI.

Ninan, Sevanti. "From Media Poor to Media Rich." *Intermedia* 24, no. 3 (June/July 1996): pp. 35–37.

Nolan, Dermot. "Development of Satellite Services in Eastern Europe: Entering the Harsh Reality Phase." In *Telecom 91 Global Review,* ed. Hugh Chaloner, pp. 198–204. London: Kline Publishing, 1991.

Nora, Simon and Alain Minc. *The Informaticization of Society.* Cambridge, MA: MIT Press, 1978.

Nordman, Kurt. "The Impact of Competition on the Evolution of Net-works." In *Proceedings of the ITU Forum 91 Technical Symposium.* Geneva: ITU, October 1991.

Nulty, Timothy E. "Privatization as a Tool of Development: The Central and East European Situation." In *Proceedings of the ITU Forum 91 Economic Symposium,* pp. 97–106. Geneva: ITU, October 1991.

Nulty, Timothy E. and Holcer, Nikola. In *Telecom 91 Global Review,* ed. Hugh Chaloner, pp. 46–50. London: Kline Publishing, 1991.

Obuchowski, Janice. "Spectrum Management Reform in the U.S." In *Proceedings of the ITU Forum 91 Regulatory Symposium,* pp. 35–37. Geneva: ITU, October 1991.

Ohara, Hiroshi. "New Opportunities from the Information Explosion—An Experience of a New International Common Carrier in Japan." In *Proceedings of the ITU Forum 91 Economic Symposium,* pp. 53–57. Geneva: ITU, October 1991.

Olmer, Lionel. "Financing The Global Information Infrastructure." In *Proceedings of the 1996 Pacific Telecommunications Conference,* pp. 984–990. Honolulu: Pacific Telecommunications Council, January 1996,

Oniki, Hajime. "Construction of Telecommunications Infrastructure in Japan: 1950–1980." In *Eastern Europe: Information and Communication Technology Challenges,* ed. G. Russell Pipe, pp. 239–257. Burke, VA: Transnational Data Reporting Service, 1991.

Organization for Economic Co-operation and Development. *Communications Outlook 1995.* Paris: OECD, 1995.

Oslund, Jack. " 'Open Shores' to 'Open Skies': Sources and Directions of U.S. Satellite Policy." In *Economic and Policy Problems in Satellite Communications,* eds. Joseph Pelton and Marcellus Snow. New York: Praeger, 1977.

Oxman, Stephen A. In *Proceedings of the ITU Forum 91 Regulatory Symposium,* pp. 62–67. Geneva: ITU, October 1991.

Parapak, Jonathan. "Bridging the Urban-Rural Divide in Telecommunication in Asia." Paper presented at the AMIC Conference on Communication, Technology, and Development: Alternatives for Asia, Kuala Lumpur, June 25–27, 1993.

Parker, Edwin B. "MicroEarth Station Satellite Networks and Economic Development." *Telematics and Informatics* 4, no. 2 (1987).

Parker, Edwin B. and Heather E. Hudson. *Electronic Byways: State Policies for Rural Development through Telecommunications,* 2d ed. Washington, D.C.: Aspen Institute, 1995.

Parker, Edwin B., Heather E. Hudson, Don A. Dillman, Andrew D. Roscoe. *Rural America in the Information Age: Telecommunications Policy for Rural Development.* Lanham, MD: University Press of America, 1989.

Pelton, Joseph and John Howkins. *Satellites International.* New York: Stockton Press, 1988.

Pelton, Joseph N. "Editorial: WARC 1992." *Space Communications* 10, no. 1 (1992): p. 1.

Petchsuwan, Kosol. "Telecommunications Development in Thailand." In *Implementing Reforms in the Telecommunications Sector: Lessons from Experience,* eds. Bjorn Wellenius and Peter A. Stern, Washington, D.C.: World Bank, 1994.

Petrazzini, Ben. *Global Telecom Talks: A Trillion Dollar Deal.* Washington, D.C.: International Institute of Economics, 1996.

Petrazzini, Ben. "Telecommunications Policy in India: The Political Underpinnings of Reform." *Telecommunications Policy* (January 1996): pp. 39–51.

Phillips, Jonathan. "Regulation and Competition in the New Environment." In *Proceedings of the ITU Forum 91 Regulatory Symposium,* pp. 18–20. Geneva: ITU, October 1991.

Pierce, William B. and Nicolas Jequier. *Telecommunications for Development.* Geneva: ITU, 1983.

Pilling, David. "A Demand for Fast Workers with Built-in Memories: *Financial Times,* October 7, 1991, p. XX.

Pipe, G. Russell, ed. *Eastern Europe: Information and Communication Technology Challenges.* Burke, VA: Transnational Data Reporting Service, 1991.

Pipe, G. Russell. "GATT Uruguay Round Services Agreement— Impact on Telecommunications Development." In *Proceedings of the*

ITU Forum 91 Regulatory Symposium, pp. 71–73. Geneva: ITU, October 1991.

Pool, Ithiel de Sola. *Technologies without Boundaries.* Cambridge, MA: Harvard University Press, 1990.

Price, Christopher. "Face up to the Future of Conversations." *Financial Times,* October 7 1991, p. XV.

Pulver, Glen C. "The Changing Economic Scene in Rural America." *Journal of State Government* 61, no. 1 (January/February 1988).

Purton, Peter. "Still Waiting for Take-Off," *Financial Times,* October 7, 1991, p. XV.

Pye, Roger and Martin Heath, "Overcoming Barriers to Entry and Distortion to Competition." *Pan European Mobile Communications* (Autumn 1991): pp. 43–47.

Rajab, Tarik. "Saudi Arabia: Progress Against the Odds." In *Telecom 91 Global Review,* ed. Hugh Chaloner, pp. 118–120. London: Kline Publishing, 1991.

Ravaioli, Piero. "View from the European Commission." In *Proceedings of the ITU Forum 91 Regulatory Symposium,* pp. 91–94. Geneva: ITU, October 1991.

Raymond, J. Leopold. "Iridium on the Move: When Telephones Reach the Smallest Village." *Global Communications* (September/October 1991): pp. 28–35.

Rechendorff, Torben. "Denmark: The Liberalization of Telecommunications." In *Telecom 91 Global Review,* ed. Hugh Chaloner, pp. 110–112. London: Kline Publishing, 1991.

"The Region Advances." *Asian Communications* (September 1995).

Reimers, Alexis and Jorge Weitzner. "The Mexican Experience: MSS in Developing Countries." In *Telecom 91 Global Review,* ed. Hugh Chaloner, pp. 127–130. London: Kline Publishing, 1991.

Ricke, Helmut. "Management of Change, in Particular in East-West Relations." In *Proceedings of the ITU Forum 91 Policy Symposium,* pp. 132–134. Geneva: ITU, October 1991.

Robin, Gerard. "The Importance of Standards from a Manufacturer's Viewpoint." In *Proceedings of the ITU Forum 91 Regulatory Symposium,* pp. 27–29. Geneva: ITU, October 1991.

Robinson, Anthony. "Eastern Europe: Poised for a Quantum Leap." *Financial Times,* October 7, 1991, p. XXVIII.

Rosenberg, David. "Israel: Reforms are Piecemeal." *Financial Times,* October 7, 1991, pp. XXXI–II.

Rosenbrock, Karl H. "An Overview of Major Standardization Issues Today." In *Proceedings of the ITU Forum 91 Policy Symposium,* pp. 25–31. Geneva: ITU, October 1991.

Rosenbrock, Karl H. and John Ketchell. "International Standards: A New Environment for the 90s." In *Telecom 91 Global Review,* ed. Hugh Chaloner, pp. 30–33. London: Kline Publishing, 1991.

Roussel, Anne-Marie. "VSATs: No Nation is an Island." *Satellite Communications* (April 1993).

Rubio, A. A., Timothy J. Logue, and Stuart MacPherson. "Developing a Communications Satellite Policy for the Philippines." In *Proceedings of the Pacific Telecommunications Conference.* Honolulu: Pacific Telecommunications Council, 1992.

Russell-Walling, Edward, ed. *The Americas Yearbook.* London: Euromoney Publications, 1996.

Russell-Walling, Edward, ed. *Global Telecom's Business Yearbook 1995.* London: Euromoney Publications Yearbook, 1995.

Sage, Paul W. "Telecommunications Competition, Privatization, and Industry Development in Japan: Opportunities for the Asia Pacific Region." Paper presented at the 1993 Pan-Asian Summit, Hong Kong, June 1993.

Santucci, Gerard. "Telecommunications Equipment." *XIII Magazine* no. 8 (October 1992): pp. 7–10.

SatelLife News, 5th Issue, November 1993.

SatelLife News, 6th Issue, February 1994.

SatelLife News, 7th Issue, May 1994.

Satellite News, 16, no. 7 (July 5, 1993).

Saunders, Robert, Jeremy Warford, and Bjorn Wellenius. *Telecommunications and Economic Development,* 2d ed. Baltimore: Johns Hopkins University Press, 1994.

Savage, James. "The Future of the International Telecommunications Accounting Rate Scheme." In *Proceedings of the Pacific Telecommunications Conference,* pp. 772–780. Honolulu: Pacific Telecommunications Council, 1996.

Schaffer, Ken. "A Television Window on the Soviet Union." In *Tracing New Orbits: Cooperation and Competition in Global Satellite Development,* ed. Donna Demac. New York: Columbia University Press, 1986.

Schmandt, Jurgen, Frederick Williams, and Robert H. Wilson. *Telecommunications Policy and Economic Development: The New State Role.* Austin: University of Texas at Austin, 1988.

Schwankert, Steven. "Dragons at the Gates." *Internet World* (November 1995): pp. 109–112.

Schwarz-Schilling, Christian. "Germany: Structural Reform." In *Telecom 91 Global Review,* ed. Hugh Chaloner, pp. 106–109. London: Kline Publishing, 1991.

Schweizer, Susanna. "Increasing Profitability through Distributed Network Teaming." *Telecommunications* (March 1993): pp. 52–57.

Seciniaz, Laurent. "The U.S. Mobile Telecommunications Market." In *Telecom 91 Global Review,* ed. Hugh Chaloner, pp. 242–244. London: Kline Publishing, 1991.

Sekizawa, Tadashi. "ISDN: Development Towards the Twenty First Century." In *Telecom 91 Global Review,* ed. Hugh Chaloner, pp. 148–152. London: Kline Publishing, 1991.

Sheridan, Michael. "1992: The Wider Impact." In *Telecom 91 Global Review,* ed. Hugh Chaloner, pp. 60–62. London: Kline Publishing, 1991.

Shetty, Vineeta. "Latin America Comes to Market." *Communications International* (June 1996): pp. 8–22.

Shetty, Vineeta. "Tigers to Elephants." *Communications International* (July 1996): pp. 14–17.

Shimisaki, Nobuhiko. "An Overall View of Information Technology (IT) Industry Complex—Focussing on the C&C Aspect of Communication Industry." In *Proceedings of IDATE 13th International Conference*, pp. 67–80. Montpelier, November 1991.

Simon, Bernard. "Canada: Difficult to Hold the Line on Barriers." *Financial Times*, October 7, 1991 p. XXIV.

Sinha, Nikhil. "The Political Economy of India's Telecommunications Reforms." In *Proceedings of the 1996 Pacific Telecommunications Conference*, pp. 921–927. Honolulu: Pacific Telecommunications Council, January 1996.

Smith, Peter. "The Voice of the Small Customer and Residential User." In *Proceedings of the ITU Forum 91 Policy Symposium*, pp. 108–111. Geneva: ITU, October 1991.

Smyth, Maurice. "Asian Broadcasting: The Changing Scene." Paper presented at the AMIC Conference on Communication, Technology, and Development: Alternatives for Asia, Kuala Lumpur, June 25–27, 1993.

Society for Worldwide Interbank Financial Telecommunication. *Annual Report 1994*. La Hulpe: SWIFT, Belgium, 1995.

Solis, Diane "The Long Wait." *Wall Street Journal*, October 4, 1991, p. R14.

Solomon, Jonathan. "Conditions for Market Entry in the New Environment: Views of a Facility Provider." In *Proceedings of the ITU Forum 91 Policy Symposium*, pp. 37–57. Geneva: ITU, October 1991.

Spector, Phillip L. "The Internet Goes International: Intellectual Property Considerations for Industry." In *Proceedings of the 1996 Pacific Telecommunications Conference*, p. 560. Honolulu: Pacific Telecommunications Council, January 1996.

Stanislawski, Stefan. "What is the Best Way to Organise the Telecommunications Sector? Results of Statistical Analysis and Other Studies." Paper presented at the 1993 Pan-Asian Forum, Hong Kong, June 1993.

Staple, Gregory. "The Global Telecommunications Traffic Report, 1991." London: International Institute of Communications, 1991.

Steiner, Robert. "The Right Numbers." *Wall Street Journal,* October 4, 1991 p. R6.

Stentor Telecom Policy Inc. *Global Digest* (January–February 1993).

Stevenson, J.R.A. "Liberalization and Privatization in New Zealand." In *Proceedings of the ITU Forum 91 Regulatory Symposium,* pp. 26–30. Geneva: ITU, October 1991.

Stine, Stephen F. "Hutchison Telecom's Robert Siemens Seeks to Expand Cellular Operations Beyond Asia." *Wall Street Journal Europe,* July 23, 1991.

Stine, Stephen F. "Catching Up." *Wall Street Journal,* October 4, 1991, pp. R7–8.

Stokes, Bruce. "Beaming Jobs Overseas." *National Journal* (July 27, 1985): p. 1727.

Stover, William J. *Information Technology in the Third World: Can I.T. Lead to Humane International Development?* Boulder, CO: Westview Press, 1984.

Strasser, Steven. "An Island Wired for Jobs." *Newsweek,* July 19, 1993, p. 37.

The Strategic Planning Group of the U.S. National Committee of the CCITT. "CCITT Interactions with Other Standards Organizations." Washington, D.C., 1993.

Subijanto, Thomas B. "Indonesia: Astra's Revenue Sharing Project." Paper presented at the 1993 Pan-Asian Summit, Hong Kong, June 1993.

Suwarso, Mirwan. "New Public/Private Relationship in Indonesia's Telecommunications Sector." In *Implementing Reforms in the Telecommunications Sector: Lessons from Experience,* eds. Bjorn Wellenius and Peter A. Stern. Washington, D.C.: World Bank, 1994.

Swensrud, Blake S., ed. *Eastern European and Soviet Telecom Report.* Vol. 2, no. 9, September 1, 1991.

Swensrud, Blake and Gene Prilepeki. "Russia Calls." *In Depth* (September 1995).

Tanyi-Tang, Enoh. "Regulation in Developing Countries." In *Proceedings of the ITU Forum 91 Regulatory Symposium,* pp. 5–8. Geneva: ITU, October 1991.

Telecommunications Act of 1996. United States Congress. Public Law 104–104, February 8, 1996.

Telecommunications Reports 59, no. 18, May 3, 1993.

Thornhill, John. "Anxiety over Pattern of Developments." "International Telecommunications: Financial Times Survey." *Financial Times,* October 3, 1995, p. 4.

Tietjen, Karen. *AID Rural Satellite Program: An Overview.* Washington, D.C.: Academy for Educational Development, 1989.

Tudge, David. "International Facilities: Cable and Satellites." In *Proceedings of the ITU Forum 91 Regulatory Symposium,* pp. 195–196. Geneva: ITU, October 1991.

Turner, Colin. "Trans-European Networks and the Common Information Area: The Development of a European Strategy." *Telecommunications Policy* 19, no. 6 (August 1995): pp. 501–508.

Tutt, Nigel. "PHARE." *XIII Magazine,* no. 8 (October 1992): p. 23.

Uchitelle, Louis. "We're Leaner, Meaner, and Going Nowhere Faster." *New York Times,* May 12, 1996, pp. 4–1, 4–4.

UNESCO. *World Communication Report.* Paris: UNESCO, 1989.

Ungerer, Herbert. "Pushing Towards New World Communications." In *Telecom 91 Global Review,* ed. Hugh Chaloner, pp. 250–251. London: Kline Publishing, 1991.

Ungerer, Herbert. "Telecom Risks, Telecom Futures." In *Proceedings of the Telecom 95 Strategies Summit.* Geneva: ITU, October 1995.

Ure, John, ed. *Telecommunications in Asia: Policy Planning and Development.* Hong Kong: Hong Kong University Press, 1995.

U.S. Congress, Office of Technology Assessment. *Rural America at the Crossroads: Networking for the Future,* OTA-TCT-471. Washington, D.C.: U.S. Government Printing Office, April 1991.

U.S. Congress. House. *Bringing the Information Age to Rural America. Hearings before the Government Information, Justice, and Agriculture Subcommittee of the Committee on Government Operations.* Washington, D.C.: U.S. Government Printing Office, 1991.

U.S. Congress, Office of Technology Assessment. *Linking for Learning: A New Course for Education,* OTA-SET-430. Washington, D.C.: U.S. Government Printing Office, November 1989.

U.S. Department of Commerce. *The Global Information Infrastructure: Agenda for Cooperation.* Washington, D.C.: Department of Commerce, 1995.

U.S. Department of Commerce. *The National Information Infrastructure: Agenda for Action.* Washington, D.C.: Department of Commerce, 1993.

Vimolvanich, Vallobh. "National Telecommunications Infrastructure Development Projects in Asia: Thailand." Paper presented at the 1993 Pan-Asian Summit, Hong Kong, June 1993.

Vincent, Geoffrey. "Upheaval in the Global Market: Competition and Technology in Telecommunications." In *Telecom 91 Global Review,* ed. Hugh Chaloner, pp. 134–138. London: Kline Publishing, 1991.

Wang, Garget. "Satellite Television and the Future of Broadcast Television in the Asia Pacific." Paper presented at the AMIC Conference on Communication, Technology and Development: Alternatives for Asia, Kuala Lumpur, June 25–27, 1993.

Warren, Gabriel. "The Mission of the ITU in the Fast Changing Environment." In *Proceedings of the ITU Forum 91 Policy Symposium,* pp. 176–178. Geneva: ITU, October 1991.

Warta, John. "Fiber Optic Submarine Cable Systems: An Opportunity for Emerging Markets to Access the Global Highways and Superhighways." In *Proceedings of the Pacific Telecommunications Conference,* pp. 966–971. Honolulu: Pacific Telecommunications Council, 1996.

Weinburg, Neil. "Japan: Domestic Market: Goliath Gnawed by Rivals." *Financial Times,* October 7, 1991, p. XXXI.

Wellenius, Bjorn and Peter A. Stern, eds. *Implementing Reforms in the Telecommunications Sector: Lessons from Experience.* Washington, D.C.: World Bank, 1994.

Wellenius, Bjorn, Peter A. Stern, Timothy E. Nulty, and Richard J. Stern, eds. *Restructuring and Managing the Telecommunications Sector.* Washington, D.C.: World Bank, 1989.

Wigglesworth, William. "The Impact of Privatisation and Competition on Telecommunications Industry Development: Experiences in the UK." Paper presented at the 1993 Pan-Asian Summit, Hong Kong, June 1993.

Williams, Nicholas. "Economics of Local Loop Competition by Radio Access." In *Telecom 91 Global Review,* ed. Hugh Chaloner, pp. 234–237. London: Kline Publishing, 1991.

Williamson, John. "Buoyant Regional Market." *Financial Times,* October 7, 1991, p. XXX.

Williamson, John. "European Commission: Intervention is the Slogan." *Financial Times,* October 7, 1991, p. VI.

Williamson, John. "France: Characteristics of Enlightened Paternalism." *Financial Times,* October 7, 1991, p. XXVI.

Williamson, John. "Western Europe: Reality of Free for All a Long Way Off." *Financial Times,* October 7, 1991, p. XXV.

Witte, Eberhard. "Liberalization without Privatization." In *Proceedings of the ITU Forum 91 Regulatory Symposium,* pp. 1–3. Geneva: ITU, October 1991.

Wittering, Stewart. "VSAT." *Communications International* (June 1996): pp. 20–22.

"Workers Keep Their Offices in a Briefcase." *Communications News* (March 1993): p. 12.

World Bank. "Information and Development Initiative." Washington, D.C.: World Bank, 1995.

Wresch, William. "New Lifelines." *Internet World* (November 1995): pp. 102–106.

Wright, Karen. "The Road to the Global Village." *Scientific American* 262, no. 3 (March 1990): pp. 83–94.

Yamaguchi, Haruo. "The Social and Economic Role of Telecommunications in a Peaceful World: New Avenues for Cooperation." In *Proceedings of the ITU Forum 91 Regulatory Symposium,* pp. 75–79. Geneva: ITU, October 1991.

Yang, Dori J. and Joan Warner. "Hear the Muzak, Buy the Ketchup." *BusinessWeek,* June 28, 1993, pp. 70–71.

Yoon, Chang-Bun. "Korean Experience in Developing the Information/Communication Industry." In *Proceedings of the ITU Forum 91 Economic Symposium,* pp. 79–83. Geneva: ITU, October 1991.

Yoon, Chang-Bun. "Korea's Strategy for Developing Information/Communication Technology." In *Eastern Europe: Information and Communication Technology Challenges,* ed. G. Russell Pipe, pp. 191–232. Burke, VA: Transnational Data Reporting Service, 1991.

Zadra, Randy. "The Telecommunications Revolution in Latin America." Paper presented at INTERCOMM 93, Vancouver, Canada, February 24, 1993.

Zakaria, M. Nor. "Satellites and Pan Asian Broadcasting Networks: Malaysian Experience." Paper presented at the AMIC Conference on Communication, Technology, and Development: Alternatives for Asia, Kuala Lumpur, June 25–27, 1993.

Zhengxi, Luan. "China's Changing Telecommunications." In *Telecom 91 Global Review,* ed. Hugh Chaloner, pp. 122–124. London: Kline Publishing, 1991.

INDEX

banking industry, 50–52
Bashtelcom (Russia), 252
Basic Law, status of Hong Kong after
    mid-1997, 273
BBC (British Broadcasting
    Corporation)
    transmitted over Soviet television,
        234
    World Service Television, 363
Beaird, Richard, 217
bearer services
    See data transport services
Beijing Cable TV, 265
Belgium, telecommunications sector,
    128
Belize, telecommunications
    infrastructure, 326, 351
Bell Atlantic
    consortia, 332
    foreign investment, 69
    investments in Chile, 340
    joint ventures, 243–244, 297,
        313
    mobile communications, 40
    partnership agreement with Czech
        PTT, 41
Bell Canada, 132, 136, 344, 348
Bell Manufacturing of Belgium, 305
Bell Operating Companies (BOCs),
    40–42, 95, 212
BellSouth
    cellular services, 144–145
    consortium, 350
    investments in Chile, 340
    investments in Mexico, 344
    joint ventures, 294
    Moviecom consortium, 333
    operations, 41
Benin, regional telecommunications
    technology, 228
Bereq (Israel), 36
Berlin Wall, international television
    coverage, 363
Berne convention, 427
BETRS (Basic Exchange Telephone
    Radio Service), 29
Bhutan, challenges, 226
"Big LEO" systems, 99, 373, 385

See also Low Earth Orbiting (LEO)
    satellites
bilateral aid, 225
bilateral trade
    agreements, 418, 419
    U.S. and Canada, 137
Binariang Sdn Bhd (Malaysia), 310
Birla (India), 293
Board of Communications and
    Telecommunications Control
    Bureau (Philippines), 312
Board of Governors, role in Intelsat,
    359–360
Bolivia
    domestic satellite systems, 328
    GDP, 347
    telecommunications infrastructure,
        326
Bombay and Delhi Telephone
    Administration, 294–295
Bond, Alan, 331
Botswana, telecommunications
    infrastructure, 230–231
Botswana Telecommunications
    Corporation (BTC), 231
Brasilsat 1 domestic satellite, 337
Brasilsat 2 domestic satellite, 337
Brasilsat system, 339
Brazil
    banking industry, 51–52
    cellular and satellite systems,
        336–338
    Columbus II system, 377
    domestic outside financing, 224
    domestic satellite systems, 328
    economic development, 334–335
    privatization, 337
    subscriber investment through share
        capital, 212
    TeleBahia virtual telephone service,
        184
    telecommunications infrastructure,
        335–336
    telecommunications potential,
        338–339
    telecommunications trade, 327–328
    telephone lines, 181
    VSAT technology, 28, 213, 329, 337